Deadlock or Decision

Books by Fred R. Harris

Deadlock or Decision: The U.S. Senate
and the Rise of National Politics

Los Obstáculos para el Desarrollo
Editor

America's Government
(with Gary Wasserman)

Quiet Riots:
Race and Poverty in the United States
Editor (with Roger W. Wilkins)

Understanding American Government
(with Randy Roberts and Margaret S. Elliston)

Readings on the Body Politic
Editor

Estados Unidos y
Su Relación Bilateral con México
(with David Cooper)

America's Democracy:
The Ideal and the Reality

America's Legislative Processes:
Congress and the States
(with Paul L. Hain)

Potomac Fever

The New Populism

Social Science and Public Policy
Editor

Now Is the Time

Alarms and Hopes

Deadlock or Decision

The U.S. Senate and the Rise of National Politics

FRED R. HARRIS

A Twentieth Century Fund Book

New York   *Oxford*
OXFORD UNIVERSITY PRESS
1993

Oxford University Press

Oxford   New York   Toronto
Delhi   Bombay   Calcutta   Madras   Karachi
Kuala Lumpur   Singapore   Hong Kong   Tokyo
Nairobi   Dar es Salaam   Cape Town
Melbourne   Auckland

and associated companies in
Berlin   Ibadan

Copyright © 1993 by The Twentieth Century Fund

Published by Oxford University Press, Inc.,
200 Madison Avenue, New York, New York 10016

Oxford is a registered trademark of Oxford University Press

Library of Congress Cataloging-in-Publication Data
Harris, Fred R., 1930–
Deadlock or decision : the U.S. Senate and the rise
of national politics / Fred R. Harris.
p. cm.   "A Twentieth Century Fund book."
ISBN 0-19-508025-4.
ISBN 0-19-508026-2 (pbk.)
1. United States. Congress. Senate—History.
2. United States—Politics and government—1945–1989.
I. Title
JK1161.H26   1993   328.73'071—dc20
92-43279

1 3 5 7 9 8 6 4 2

Printed in the United States of America
on acid-free paper

To my wife, Margaret S. Elliston

and to Kathryn Harris Tijerina,
Byron Harris, and Laura Harris

and to Amanda Elliston
and Amos Elliston

and to my grandson,
Samuel Fred Goodhope,
who could be a senator

The Twentieth Century Fund is a research foundation undertaking timely analyses of economic, political, and social issues. Not-for-profit and nonpartisan, the Fund was founded in 1919 and endowed by Edward A. Filene.

Acknowledgments

This book would not have been possible, this project could not have been undertaken, except for the support of the Twentieth Century Fund. I am grateful to the Twentieth Century Fund, and in particular to Richard Leone, president; John Samples, vice president, program; Beverly Goldberg, vice president, director of publications; and others (including, earlier, Richard Sinopoli, former program officer, and the late director Murray Rossant), for their support, encouragement, and assistance in the conception, research, and writing of this book. My editor at Oxford University Press, David Roll, has been especially helpful.

I am grateful, too, for the initial suggestion by Samuel Patterson, Ohio State University, that I undertake a study of the Senate of the United States and for his encouragement, as well as that of Paul L. Hain, my former departmental chair at the University of New Mexico, now dean of humanities and social sciences at Corpus Christi University.

Outstanding congressional scholars Barbara Sinclair and Steven S. Smith were very helpful in talking with me about the project and in sharing some of their research, later to be published in two excellent books, Sinclair's *The Transformation of the U.S. Senate* and Smith's *Call to Order: Floor Politics in the House and Senate,* which I have referred to repeatedly in the preparation and writing of this book. I have also profited a great deal from having been associated in seminars or on professional panels with other experts on the Senate, including Ross Baker, Roger Davidson, Richard Fenno, Jr., John Hibbing, Gary Jacobson, Donald R. Matthews, Nelson Polsby, and Randall B. Ripley. My work was improved by access to them and to their unpublished papers and commentary, as well as to their published research and writing.

Lastly, I wish to express my gratitude to the fifty members of the Senate who made time for interviews and gave thoughtful, serious responses to my questions about the present-day Senate, as well as to the senatorial staff members, former senators, and others who granted me interviews or completed questionnaires, and to the extremely helpful personnel of the Congressional Research Service of the Library of Congress, who furnished me an office and more assistance than a scholar, or even a former senator, ought to ask for.

Albuquerque, New Mexico   F. R. H.
January 1993

Contents

Foreword

Much of the press and commentary about American elected officials seems premised on the notion that those who hold office are not a consequence of the public they represent. Indeed, a visitor from Mars might recommend direct election as the easy antidote to our present discontent. It is arguably true that the central problem of American democracy lies in the inattention of the American people. That said, it is also true that a large segment of the voting public feels dissatisfied and even angry about the sort of elected leadership we have in Washington.

Serving in the Senate, often described as membership in the "world's most exclusive club," also is generally considered to be the best job in politics. Senators stand only once every six years; they enjoy ample staff support; and some of their number traditionally have been able to rise to national prominence. In recent years, however, like other elements of American politics, the U.S. Senate has lost a fair share of its luster. Indeed, the constant pressure to raise millions of dollars for future re-election campaigns, coupled with routine negative press coverage, has made the job, for many, considerably less attractive than in the past. In 1992 several Senators chose not to seek re-election. Significantly, many others indicated that the chance for a graceful exit would not be unwelcome.

The causes of this "decline" are complex, but one can identify specific recent events that clearly have contributed to the problem. In the glare of television cameras and lights, the Senate has been found wanting. The Clarence Thomas hearings, featuring the all-male, all-white judiciary committee, revealed an institution that somehow had failed to move into the modern age, and sounded an alarm that seemed to motivate Americans. The calls for term limits, for campaign finance reform, and for an end to the careers of incumbents were prompted, of course, by more than this one incident. The accumulation of troubles such as the S&L scandals, revelations of individual senatorial excesses, and the national government's apparent inability to control the deficit all contributed to the notion that the "club" needed reform and new blood.

As a result of the 1992 election, the Senate, in fact, has acquired some enthusiastic and determined new members; the recently all-white male body is no longer quite so white or quite so male. These changes may suggest that an era of Senate resurgence is at hand. The Senate, after all, has had its share of shining moments in American history, most recently during the debate over the Gulf War and, somewhat earlier, in the late 1950s, a period marked by efficiency and responsibility.

Fred R. Harris, the author of this book, was elected to the Senate in 1964 and remained a member until 1972. Now a professor of political science at the University of New Mexico, he has been writing about the American legislative experience extensively since the latter part of his senatorial career. At once an admirer of the institution and a clear-sighted observer of its strengths and weaknesses, he seemed the ideal candidate to explore the changes that have occurred in this august body as special interests and nationalization have come to dominate it.

His analysis of the problems facing the Senate in the 1990s joins a distinguished list of Twentieth Century Fund works examining the way America is governed, including books ranging from Barry Blechman's analysis of relations between Congress and the president, *The Politics of National Security,* to Dennis Ippolito's *Congressional Spending* and papers and task forces focusing on the way we finance the elections that determine who will serve in Congress.

Senator Harris's portrait is of a body that seems to have become a collection of individuals with fragmented power, marked by obstructionism and legislative-executive conflict. And yet it is the home of many good and honorable citizens who have chosen it as the best place from which to serve their nation, a fact that may help us find ways to once again make the Senate the "greatest deliberative body" in the world. We thank him for his efforts, which have resulted in this remarkable study of an important subject.

RICHARD C. LEONE, President
The Twentieth Century Fund

Deadlock or Decision

Prologue

On the evening of October 15, 1991, an obviously pleased but properly somber Vice President Dan Quayle, presiding in the glare of the bright lights now required to televise U.S. Senate sessions, banged down the ivory gavel and intoned to a packed chamber: "On this vote, the yeas are 52, the nays 48. The nomination of Clarence Thomas, of Georgia, to be an Associate Justice of the Supreme Court is hereby confirmed."[1]

Republican leader Robert Dole of Kansas moved to reconsider the vote, Majority Leader George Mitchell of Maine moved to lay that motion on the table, and Vice President Quayle announced from the chair that, without objection, the motion to table had been agreed to—thus locking down, in the customary way, the finality of the Senate's decision. It was over.

Clarence Thomas would, indeed, take his seat as only the second African-American to serve on the Supreme Court, replacing the first African-American, Thurgood Marshall, who had retired. Thomas's 107-day confirmation ordeal had been especially mean and bitter at the last, and it was a battered nominee who, outside his Washington home that night, called for a "time of healing" in the country.[2]

The confirmation fight had been won in no small measure because Thomas and his backers were successful in deflecting the spotlight from the nominee to the Senate itself—and the Senate had suffered in the glare. From all sides came assertions that something was dreadfully wrong with the Senate's confirmation "process," and nationwide efforts to limit the terms of members of Congress gained new adherents.

3

It was clear that nobody wound up very happy about the way the Thomas confirmation was handled. Thomas's opponents were frustrated by their inability to pin down the conservative nominee on his past or present views on certain issues, including abortion and affirmative action.[3] Feminist leaders and others were upset when they learned that the all-male Senate Judiciary Committee had at first failed to take seriously a claim of sexual harassment leveled against Thomas by his former assistant at the Equal Employment Opportunity Commission, law professor Anita Hill. Thomas's Senate supporters cried foul when the private affidavit Hill gave to the Judiciary Committee's staff was leaked to the press just before the originally scheduled Senate vote, and they demanded that the leaker be searched out and severely punished.

Huge television audiences were drawn magnetically to the three days and nights of reopened Judiciary Committee hearings. Called circuslike by some, the sessions featured sensational testimony by Thomas, Hill, and corroborating witnesses on both sides and numerous repetitions of salacious words not normally heard on the public airwaves.[4] Committee Democrats adopted a neutral, fact-finding posture and often asked long and muddled questions of witnesses, while Thomas's White House "handlers" and Committee Republicans went on the offensive, launching a nasty attack on Hill and the Senate itself; Thomas ultimately called the hearings a "high-tech lynching for an uppity black."[5]

This conscious Republican strategy worked; public opinion switched in Thomas's favor.[6] The Senate wound up split largely along party lines, but there were sufficient defections among Senate Democrats to tip the vote in favor of confirmation—by the narrowest margin for a successful Supreme Court nominee since 1888.[7]

In the aftermath, observers wrote about the "mess of the Senate"[8] and "how to restore [its] good name."[9] But, in reality, it was not the confirmation process—the president nominating, the Senate approving or rejecting—that had caused the trouble. Senate President Pro Tempore Robert C. Byrd of West Virginia noted that "the 'process' is a constitutional process, and it's done us well over the centuries."[10] The problem was that, through the years, the process had inevitably become politicized.

Especially since the 1950s and the era of the liberal Warren Court, the public had become increasingly aware of the policymaking role of the Court. Presidents increasingly chose Court members because of their views and judicial philosophy, and in turn the Senate increasingly approved or rejected them on the same basis. As Democratic leader George Mitchell of Maine put it, "It is illogical and untenable to suggest that the President has the right to select someone because of that person's views and then say the Senate has no right to reject that person because of those very same views."[11] After the 1950s, too, came the trend toward divided government—a president and congressional majority from different parties—as well as, more recently, increasing Senate partisanship. Conflict over Court appointments grew. Charges involving a nominee's character often made the conflict nastier.

Public disapproval of the Senate because of the Thomas confirmation fight was intensified by contemporaneous revelations that some members of the House of Representatives had been allowed to write bad checks at the House bank and had failed to pay their bills at the House restaurant. Never mind that these malefactors were few and were House members, not senators.

Increasing numbers of Americans called for limits on how long senators and House members could serve.[12] The truth is that states probably have no power to apply such limits to federal officials. And it would be bad policy to do so, anyway. National legislators with limited experience would likely be less able to balance the power of the executive branch and its bureaucracy as the framers intended. Too, what would prevent voters from electing new congressional members just as "bad" as the members the limited-term proposal would throw out? Nothing, of course. What *is* needed is a major reform of the campaign-financing system that gives incumbent members of Congress such a great advantage in getting reelected and thus reduces the wholesome, restraining effect on them that stronger challenges would represent.

Politics was very much involved in the limited-term campaign—the Republican party pushing for limits, the Democrats generally opposed. Politics was also very much at the heart of the Clarence Thomas nomination fight. But that has always been true.[13] As a matter of fact, it was politics that produced the Senate itself.

"The United States Senate came into existence as a result of a political deal," political scientist Donald Matthews has written.[14] The deal was one made at the Constitutional Convention between the small states, which "demanded equal representation in the Congress as the price of union," and the larger states, which argued that "the states should be represented according to population." The deal was also one between the nationalists and the confederationists. The large states and the nationalists got the House of Representatives; the small states and the confederationists got the Senate.

From this prosaic beginning the Senate evolved into an institution that its members, and others, would come to regard as the world's greatest deliberative body and its most exclusive club.

In the modern era, the 1950s were a time of particular greatness in the Senate, according to its most devoted chroniclers. Matthews found the Senate of this period a bounded, inward-looking organization with forceful leaders and potent folkways, or norms of behavior, which facilitated action and decision. Journalist William S. White wrote, in 1956,

> For a long time, I have felt that the one touch of authentic genius in the American political system, apart, of course, from the incomparable majesty and decency and felicity of the Constitution itself, is the Senate of the

United States. It is a place upon whose vitality and honor will at length rest the whole issue of the kind of society that we are to maintain.[15]

Even in the 1950s, however, the Senate was not without its detractors. Matthews made that point at the time:

Americans have never been entirely happy about the Senate's unusual power. Its almost incredible system of representation, the filibuster, the chamber's fondness for seniority, its sometimes controversial role in making presidential appointments and ratifying treaties, its ability to launch and sustain investigations with brutal consequences, its tolerance of demagoguery—all these things have concerned thoughtful Americans for a long time.[16]

And they still do, more so now, perhaps, than ever, because after the 1950s the Senate began to change and, in the process, became less able to act responsibly. Why did the Senate change? How did it change? What are the consequences of those changes? What can be done to make the Senate more efficient? Those are the questions that are addressed in this book.

The Senate is, at the very least, a co-equal chamber in our bicameral national legislature, but it is quite different from the House. Chapter 1 traces the historical development of the Senate and its powerful role in our system, paying particular attention to the Senate of a remembered better time, the 1950s.

The Senate's external environment was transformed before the Senate itself was. Chapter 2 deals with first causes: a virtual nationalization of American society and of Senate campaigns. The Senate became much less insular, more accessible, more responsive.

There were other great changes, exogenous to the Senate. Chapter 3 details the national advocacy explosion that occurred, an immense growth in the numbers and influence of national interest groups and a nationalization of issues. The financing of Senate campaigns was nationalized, too. These developments both pressed and permitted senators to become national spokespersons, while at the same time causing them to become much more susceptible to interest-group leverage and increasing their individual sense of political and electoral vulnerability.

Chapter 4 shows the internal consequences, on senators, of all these developments, how senators became more outward-looking and more national in their viewpoints and more individualistic in their behavior—all as they also began to pay more attention to their constituents and feel more electorally vulnerable. Old norms were weakened or metamorphosed into others. Senators became less willing to conform, to follow.

National and individualistic senators produced, as Chapter 5 demonstrates, a nationalized Senate with internally fragmented power, a place where many more points of view were fully represented but where decisive action was more easily, and more often, prevented or delayed.

Chapter 6 discusses how political parties in the Senate have been nationalized, each becoming stronger, more internally homogeneous, and more unlike the other—as a result of changes in the electorate, primarily, but also because of the increased importance of Senate party leaders and campaign committees. The Senate has become a more partisan place. With divided government, greater partisanship has frequently contributed to deadlock, but it can also be made to serve as a bridge over separation.

As the Senate has become national and individualistic, the making of national budget policy has grown increasingly frustrating and difficult, as shown in Chapter 7. Improvements can and should be made to assist the Senate in reducing the inordinate time required in budget policymaking and to help it make decisions and set priorities in a more rational way.

The Senate's special role in the making of national security and foreign policy is discussed in Chapter 8. It is a role that a more assertive, more partisan Senate has often played out in confrontation with the chief executive, but a role that can be molded toward greater cooperation.

The Senate of the United States is still the world's greatest deliberative body, but it has become less responsible as it has grown more responsive, less efficient as it has grown more representative. It has frequently been a participant in deadlock and division. The Epilogue is a summarization of all this, together with a restatement of recommendations presented throughout the book for making the Senate more effective—for allowing it to serve the nation both as a great deliberative body and as a more decisive one, as well.

I

THE SENATE
IN THE
AMERICAN SYSTEM

1

The Evolution
of the United States Senate

The day was March 4, 1789; the place was a newly redecorated second-floor chamber in New York City's handsomely refurbished former city hall; the occasion was the first meeting of the new Senate of the United States.

Before dissolving itself forever, the old Congress of the Confederation had set the date of March 4 for the first meeting of the new Congress. After long, wrangling debate, however, it had been unable to decide on the place, the location for the capital of the new republic. In frustration, the old Congress had ended by making a kind of nondecision, simply saying that the new Congress would meet where the old Congress had met.

This was good news for New York merchants. They were only too happy to supply the funds to dress up their old city hall. A room on the first floor was redone for the House of Representatives. Appropriately, an upper room was prepared for the "upper" body, with two interior stairways, one private, one public, to connect it with the House. The spacious Senate chamber featured great marble fireplaces, sumptuous carpeting, and rich, red damask drapes for the tall windows and as a canopy over the presiding officer's elevated chair.

No spectators were present when the clerk began the roll call on that first Senate day. In fact, by design, there were no galleries for the public, like those that had been provided in the House. The Senate would conduct its business behind closed doors, in secret sessions, throughout its first six years.

When the first roll call was completed, the clerk reported the obvious: only eight senators were present. Without a quorum no business could be transacted. It was not an auspicious beginning.[1]

A little over a month was to pass before the Senate could begin work. On April 6, Senator Richard Henry Lee of Virginia showed up and answered to his name. He was the twelfth senator to do so, and a quorum was finally declared present.*

Early-day critics of the Senate claimed that the cavalier attitude of senators toward punctual attendance to their duties was clear evidence of "too much of the remains of Monarchical Government, where those promoted to office consider themselves as clothed with Majesterial [*sic*] Dignity instead of confidential servants of the People."[2] Many observers expressed fears that the new Senate would turn out to be an aristocratic House of Lords—and not without reason. Earlier, at the Constitutional Convention, John Dickinson of Delaware had, indeed, declared that the Senate should bear "as strong a likeness to the British House of Lords as possible," and Alexander Hamilton had proposed that senators should be chosen for life.[3] Too, the very actions of the first senators seemed to some contemporary observers like reaches for aristocratic status.

While the new House of Representatives went to work at once on the financial and other important problems facing the nation, the Senate devoted its first three weeks to matters of ceremony and protocol. Many senators, led by patrician Richard Henry Lee as well as by the first Senate president, Vice President John Adams, were inordinately concerned with prestige and titles. They decided that the Senate sergeant at arms should be called the "Usher of the Black Rod," that senators should be addressed as "Right Honorable," and that the title of the nation's president should be "His Highness, the President of the United States of America, and Protector of their liberties."†

Such senatorial pretensions to British titles were a sure sign, contemporary critics claimed, of a Senate aim to convert the American government into a system of "king, lords, and commons." But the titles agreed to in the Senate were soon rejected by the more plebeian House of Representatives and, after considerable public ridicule, with John Adams being lampooned as "His Rotundity," the Senate had to abandon them.

The early senators were also anxious that the Senate be recognized

---

* A quorum had been achieved in the House of Representatives just five days earlier. Ironically, when Congress met in joint session on March 4, 1989, to commemorate the bicentennial of its origin, only one-third of the representatives and senators attended.

† In commenting on this proposed presidential appellation two hundred years later at a bicentennial commemoration, Republican Senate leader Robert Dole of Kansas, an earlier, unsuccessful candidate for president, remarked wryly that it was the very title he had intended to use had he been elected chief executive.

as the superior of the two houses, and they chose ceremony and salary as the means to achieve this. They mandated, for example, that bills passed by the Senate could be delivered to the House by a mere employee of the Senate, the Senate secretary, whereas bills passed by the House had to be delivered to the Senate by two House *members*. The exasperated House brushed this pomposity aside and simply sent bills and messages to the Senate by the House clerk, and there was nothing the rebuffed Senate could do about it.

When the House passed a bill establishing the same salaries for both senators and representatives—$6 a day as salary, plus $6 a day for each twenty-five miles of travel to and from sessions—the Senate amended it to raise senatorial compensation to $8 and $8. The House opposed this unfavorable differential on the grounds that the Constitution intended the two chambers to be fully equal. When the dispute persisted and threatened to block passage of any salary bill at all, House members finally agreed to a $1 Senate advantage—but only to go into effect six years later, after the term of all then-sitting senators. By that time, though, the Senate had given up the fight for salary superiority.

The United States Senate was not intended to be an American House of Lords, but neither was it intended to be an American House of Commons. The latter was more or less the role assigned to the U.S. House of Representatives. The Senate was different.

A Peculiar Institution

Ours is a democracy, or, as Thomas Jefferson preferred, a republic. But the Senate of the United States is peculiar in that it is not grounded constitutionally in the principles of popular sovereignty and majority rule. Nor are some aspects of senatorial rules and procedures. And the Senate, though primarily a legislative body, exercises executive and judicial powers as well.

Delegates to the Constitutional Convention decided with little difficulty that the national legislature should be bicameral, thus following the British model as well as the framework already adopted in most of the new state constitutions. The Convention provided for major differences between the two houses in method of election, representation, term of office, and function. James Madison had thought that both houses should derive their authority directly or indirectly from the people, not from the state governments, as had been true for the weak national government under the Articles of Confederation. He had also felt that representation in both houses should be based on a state's population (a proposal that was highly favorable to Madison's Virginia, the most populous of the states). States' righters of that time had preferred the continuation of a national government whose sovereignty came from the states, rather than from the people, and the small states had demanded that each state enjoy equal representation in both houses.

The Great Compromise of the Convention—adopted by a one-vote margin of 5 to 4, with the delegation from Massachusetts divided and unable to vote and with three states absent—was probably not enthusiastically supported by anyone, and it was opposed by Madison. He vehemently objected to the idea of equal state representation in the Senate. One of his strongest arguments was that this would mean that the "minority could negative the will of the majority of the people."[4]

Madison was correct, of course. But he and a majority of the other delegates came at last to acquiesce in the compromise because it was a political necessity: no constitution could be agreed to, or ratified, without it. The House of Representatives would be popularly elected, with representation in it based on state population; there would be equal state representation in the Senate, and senators would be elected by the state legislatures.

Thus the House of Representatives would represent "the people" (not including women or blacks, of course). House members would be held immediately accountable to the people as a result of the fact that their term of office was limited to two years and because the whole membership of the House would stand for election at the same time.

Senators, on the other hand, would, as John Dickinson put it at the time, be "the most distinguished characters, distinguished for their ranks in life and their weight of property."[5] They would come from the propertied, educated, and refined class. They would not be chosen by the people nor be subject to their momentary whims and passions, but would be elected by the state legislatures for six-year terms. Only one-third of the whole Senate would be elected each two years. Property rights would be protected. States' rights, too: each state, no matter how small or large its population, would have two senators.

At first, some thought that senators were to represent their state *governments*, like ambassadors to the national government. But this was not borne out in practice. State legislatures had not been given the power to remove or recall senators. The Constitution was silent on the question of whether a state legislature could instruct its U.S. senators how to vote. Although this was tried by some state legislatures for a while, there was considerable argument about whether such instructions were really proper and binding. The first Senate decided that, during voting, senators' names should be called alphabetically rather than by states, as had been the practice in the Confederation Congress. From the beginning, the two senators from the same state often voted differently. It soon became clear that senators were individuals, not just state delegates. Too, governors found that they could deal directly with the national government without having to go through supposed senatorial "ambassadors."

If any doubt remained that senators were to represent the "people" within their states rather than state governments, it was removed in 1913 with ratification of the Seventeenth Amendment, which provided for the direct election of senators. Even with such direct election, however (and as Madison foresaw), there is still today no assurance that the Senate as a body will represent the popular or majority will of the whole country.

Originally, some had thought that the Senate would serve as a kind of "council of revision," like the British House of Lords, that its legislative function would be primarily to review and improve on measures previously agreed to in the House of Representatives. The initial terms of the Great Compromise at the Constitutional Convention gave to the House alone the power to "originate" revenue and appropriation bills and provided that these measures could not be altered or amended in the Senate. The final version of the Constitution, however, limited this House origination power to revenue, or tax, measures (although the House later assumed the power to originate appropriation bills, too) and, significantly, permitted the Senate the full power to amend tax and money bills, the same as any other legislative measures. The Senate is able, in effect, then, to originate tax and money measures itself by amending House bills (subject, of course, to House concurrence). Of course, there was never any question that the Senate could initiate legislation of all other kinds. Thus the Senate is a fully equal partner with the House in lawmaking, not a secondary council of revision. Nothing becomes law until both houses concur—and precisely so.

Old-time coffee drinkers used to "saucer and blow" their coffee (to cool it by pouring it from the cup into the saucer and then sipping it from the dish) and that is the analogy George Washington reportedly resorted to when asked to describe the principal role of the Senate.[6] The upper body was meant to be a saucer for cooling the hot coffee of the House—in other words, a restraint on public fever, an institution that could deter hurried and ill-considered action. James Madison agreed with this characterization, saying that "the use of the Senate is to consist in its proceeding with more coolness, with more system, and with more wisdom, than the popular branch."[7]

The founders figured, too, that the Senate's smaller size alone would make it more cautious and less facile in decision-making than the House. As Constitutional Convention delegate James Wilson of Pennsylvania put it, the framers believed "that the least numerous body was the fittest for deliberation—the most numerous, for decision."[8] And, even though the Senate grew in size as new states were admitted to the Union, its one hundred members today still number fewer than one-fourth the membership of the 435-member House.

The Constitution not only made the Senate more deliberative than the House of Representatives; it gave the Senate more to deliberate over. In addition to its law-making powers, the Senate has special executive and judicial powers—executive power over the confirmation of appointments and the ratification of treaties, judicial power in the confirmation of judges and as the court for all impeachment trials. The Senate apparently was expected to operate not only as one-half of the legislative branch, but also as a kind of "executive council," like similar bodies that had existed in the American colonial governments.

In fact, the Constitutional Convention first lodged in the Senate alone the powers to appoint ambassadors and Supreme Court justices and to negotiate treaties with foreign governments. But the delegates later reconsidered. They finally decided that it would not be proper to allow the Senate to fill the very positions that it could make vacant through the exercise of the judicial function of trial and conviction on impeachment charges. Thus, after the delegates had agreed on the broad outlines of the presidency and were reasonably certain in their own minds that the towering American hero General George Washington would be the first selected, indirectly through the Electoral College, to occupy that office, they concluded that it was the president who should have the authority to nominate executive and judicial officials and to negotiate treaties. To check and balance this authority, though, the delegates decreed that it could only be exercised "by and with the advice and consent" of the Senate—through confirmation and ratification.

The power to confirm appointments and ratify treaties was vested in the Senate exclusively.* The Constitution not only gave the Senate extra work, then, but it also in effect pitted the Senate and the president against each other in ways not true for the House. In addition, it made Senate decision-making extra difficult in regard to treaties: while a simple majority vote of those senators present and voting could achieve confirmation of an appointment, a two-thirds vote was necessary for ratification of a treaty.

Senate-presidential disagreement in regard to confirmation of appointments was not long in coming: the Senate rejected one of President George Washington's first nominees. And the "advice" aspect of the Senate's "advise and consent" power concerning presidential appointments never operated as well as the Senate would have liked. From the beginning the Senate tried to exercise more than reactive power in regard to appointments. As soon as they received President Washington's first list of appointments in writing, the Senate formally notified the president that they thought such presidential communications should be oral, that the Senate and the president should meet together in person, perhaps in a special chamber set aside for that purpose among the executive offices, whenever he had appointments to make. In that way, senators thought, they could actually advise and not simply consent or reject. But Washington stuck with the practice of submitting appointments in writing, for approval or rejection. He generally

---

* The Twenty-Fifth Amendment, ratified in 1967, made a lone exception to the Senate's otherwise exclusive confirmation power. A nomination by the president to fill a vacancy in the office of vice president was required to be approved by *both* the House and Senate.

sought advice elsewhere, although individual senators were able to exert some influence on appointments, especially in their own states.*

In regard to treaties, President Washington *did* at first seek prior Senate advice in person, but he soon wished he had not done so. The first chief executive went to the Senate in August of 1789 to discuss a treaty he wanted to enter into with the Creek Indian tribe. The noise of street traffic outside the open windows of the uncomfortably warm chamber made it difficult for senators to hear the details of the proposed treaty when it was read aloud by Vice President John Adams. Somewhat in awe of Washington, and wanting more information, senators sat in awkward silence for a while after Washington asked for their approval. Then they decided to refer the proposed treaty to a committee and delay action until the following day. An obviously exasperated Washington said, "This defeats every purpose of my being here."[9] But he came back the next day and sat, however impatiently, through the Senate debate that eventually resulted in approval of the treaty, although he said afterward that he would be "damned if he ever went there again," and he never did.[10]

For a time, other officials of Washington's administration continued to seek in person the formal advice of the Senate, as a body, on treaties that were being considered. But this practice was eventually abandoned. Legally the Senate's collective power in regard to treaties, like its confirmation power, became a reactive one, a power of ratification or rejection, not formal prior advice. However, presidents who have neglected to involve at least key senators in treaty negotiations have done so at the peril of later ratification defeat. And the bitter Senate debate and close vote on a treaty that Washington's secretary of state, John Jay, negotiated with Great Britain in 1794 showed from the very first days that the Senate would exercise its ratification power, like its confirmation power, quite independently, that it could never be counted on to shy away from confrontation with the president.

It was not only the Senate's lack of a popular or majoritarian base, though, or its exercise of executive and judicial powers in addition to its basic legislative powers, that made it a peculiar American legislative institution from the first and clogged its decision-making channels. Its internal workings were also peculiar.

---

* Today, the practice of "senatorial courtesy" means that the Senate will not usually confirm a presidential appointment, especially of a federal district judge, to an office in the home state of senators of the president's own party if these senators (or one senator, if only one is of the president's party) refuse to endorse the nomination. The result, in this narrow case only, is that the senators can usually not only advise concerning nominations, but actually, in effect, make the nominations.

Not a Majoritarian Body

Aspects of the Senate's internal workings seem odd in a democracy: in its earliest days, secret sessions; the development of the present seniority system for selecting its committee leaders; and the right of a minority of senators to use "extended debate," or a filibuster, as well as other dilatory tactics to prevent a Senate majority from acting.

From the first, the House of Representatives was an open body. Its freewheeling debate was fulsomely reported in the press and closely followed by the public. This and the fact that the House soon came to be dominated by the democratic agrarian Republicans of Jefferson and Madison put it more in tune with the prevalent spirit of the country.* The House of Representatives enjoyed greater national popularity than the Senate. It was also considered the more important of the two houses.

The members of the first Senate were quite distinguished, half of them having been delegates to the Constitutional Convention.[11] But the fact that Senate debate initially took place in closed sessions reduced the degree of public attention—and prestige—that might otherwise have been accorded to the upper chamber. This secrecy, although merely continuing a practice followed by both the Congress of the Confederation and the Constitutional Convention, fueled public charges that senators were inclining toward aristocracy. One magazine article of the day declared: "Upright intentions, and upright conduct are not afraid or ashamed of publicity. The spirit of the *Venetian senate* suits not, as yet, the meridian of the United States; neither does the conduct of a *conclave* or a *divan* comport with the feelings of Americans."[12]

The Federalist followers of Alexander Hamilton and John Adams were dominant in the early Senate. Most of them represented urban, commercial, and financial interests, and they lacked popularity in the country at large. Opponents of Hamilton's Bank of the United States made much of the fact that this Federalist scheme to make the rich richer and the poor poorer, as the Republicans saw it, was hatched in the secret Senate.

The Senate finally abandoned most of its secrecy in 1795, although confirmations and treaty ratifications were still handled in closed "executive sessions" for some time thereafter. The move to openness came partly in response to popular pressure and partly as a result of a certain amount of senatorial envy concerning the focus of public attention on the House of Representatives.†

Following its abandonment of secret sessions, the Senate still con-

---

* James Madison, himself, was elected to the first House of Representatives after having been denied election to the Senate by the Virginia legislature.

† Both of these reasons probably played some part again when, two hundred years later in 1986, the Senate tardily decided to follow the lead of the House and permit Senate sessions to be televised.

tinued to receive less press coverage than the House. First of all, when Congress moved into the new capitol building in Washington, D.C., in 1800, after a year in New York and ten years in Philadelphia, the Senate confined journalists to a remote public gallery in its chamber where it was difficult for them to hear. There were no official Senate stenographers in those days, either, for the reporters to rely on. Even after the Republicans gained a majority in the Senate in 1802 and gave journalists direct access to the Senate floor, the press and the public continued for a good while to find the less dignified and less stuffy debate of the House much more interesting. Senator William Plumer of New Hampshire, confiding to his diary that Senate speeches had few Senate listeners and little influence on senators' votes (an observation often made today, too), wrote:

> When a senator is making a long set speech the chairs are most of them deserted & the vote is often settled in conversation at the fire side. The conversation is there often so loud as to interrupt the senator who is speaking. . . . Under these circumstances it is often difficult for a man, who knows he is not attended to, to deliver an able & eloquent argument. . . . To this add we have no stenographer, & seldom any hearers in the galleries. . . . In the other House it is different—galleries are usually attended, frequently crouded [*sic*], with spectators—Always one, often two, stenographers attend, & their speeches are reported in the gazettes.[13]

Still, the abandonment of secret sessions by the Senate proved to be a step toward raising the prestige and importance of that chamber. In 1805, when the enormously talented but ill-starred Aaron Burr, himself a former senator, delivered his moving farewell address to the Senate as Vice President, an occasion that caused open weeping among some senators, he called the Senate "a sanctuary; a citadel of law, of order, and of liberty."[14]

The eminence and influence of that "citadel" continued to grow. A four-decade period preceding the Civil War is considered to have been the early Senate's "Golden Age." It was in the forty years from 1819 to 1859, Senate historian Richard Allan Baker has written, that the Senate "moved from a position of relative equality with the House of Representatives and the presidency to a preeminent position," as it forged compromises to deal with the "growing national tensions associated with the issues of slavery, economic development, and sectional parity."[15]

More and more men of the caliber of the Golden Age's Great Triumvirate—Daniel Webster, Henry Clay, and John C. Calhoun— sought the increasingly prestigious mantle of the Senate. By 1832, the perceptive French observer of political America Alexis de Tocqueville found the Senate much superior in quality to the House of Representatives. Tocqueville wrote that the Senate of that day contained a "large proportion of the celebrated men of America," and added:

Scarcely an individual is to be seen in it who has not had an active and illustrious career: the Senate is composed of eloquent advocates, distinguished generals, wise magistrates, and statesmen of note, whose arguments would do honor to the most remarkable parliamentary debates of Europe.[16]

The Seniority System

Secrecy as a peculiarity of the Senate gave way. But another peculiarity, the seniority system, developed (and still persists). This nonmajoritarian method for choosing the Senate's committee (and subcommittee) leaders came only after the establishment of standing committees and political party organizations in the Senate.

The Senate did not have standing committees at first. Ad hoc committees were appointed to work out the details of each particular legislative measure as it came up, the same as is still done in the British House of Commons. This system gave power to a few leading senators who were named again and again to serve on such select committees. Then, in 1816, the Senate set up its first standing committees and gave each one permanent jurisdiction over a particular class of legislation. The new system spread authority around and fostered development of specialized subject-matter expertise among the various senators.

Initially, there were twelve legislative standing committees in the Senate, although that number was later to grow.* The first twelve were: Foreign Relations; Finance; Commerce and Manufactures; Military Affairs; Militia; Naval Affairs; Public Lands; Claims; Judiciary; Post Office and Post Roads and Pensions; and District of Columbia.[17]

Then, some thirty years after the advent of Senate standing committees, political parties in that body took over the naming of committee members. America's founders had hoped that political parties, which are not even mentioned in the Constitution, would never develop here. It was a vain hope. From the outset, the issues facing the new government—for example, the establishment of a national bank—were serious ones. Officials profoundly disagreed about these issues, and they naturally chose sides. Sides became factions. Factions soon became political parties, intent on electing people of like mind to public office.

Throughout America's history, there have been only two principal national political parties at any one time, with their corresponding arms in the Senate. First there was the era of the Democratic Republicans and the Federalists. Next came the period of the Democrats and the Whigs. Then, by the time of the Civil War, the Democrats were squared off against the Republicans of Lincoln. Along the way, Andrew Jackson's

---

* In fact, the history of the Senate demonstrates an almost irresistible pressure for expansion in the number of committees—and, in modern times, of subcommittees, too. By the end of the Golden Age in 1859, for example, the number of the Senate legislative standing committees had already grown from twelve to twenty-seven.

Democrats had assumed the agrarian and democratic mantle of the Jeffersonian Democratic Republicans, but with more faith in a strong executive than the Jeffersonians had been willing to indulge. The Federalists had been superseded by the Whigs, who were equally representative of urban, commercial, and banking interests but more willing to champion the power of the national government's legislative branch.

By the end of the Golden Age, the slavery issue had begun to split the Democratic party, north from south, and the Whigs had begun to fade, to be superseded by a Republican party that was born out of opposition to the spread of slavery. But before these latter party changes were completed, southern Democrats and the Whigs had given to the Senate three of its most illustrious members of all time, Whigs Daniel Webster of Massachusetts and Henry Clay of Kentucky and Democrat John C. Calhoun of South Carolina.

When control of appointments to the Senate's standing committees was turned over to the party organizations within the Senate in 1846, the power and influence of the Senate parties grew accordingly. So did the interparty conflict in that body. Committee membership became stable, however, no longer changeable with each new Congress or dependent on either vice presidential appointment or Senate balloting, both of which methods had been tried. It was then that seniority within each Senate committee, length of service as a member of the committee, came to be the standard for determining committee rank, and thus the automatic method for choosing the most senior committee member of the Senate's majority party as the committee chair.

Mounting objections to the seniority system paralleled its rise. One critic charged that by 1859 this "intolerably bad" practice "operated to give to senators from slave-holding states the chairmanship of every single committee that controls the public business of this government. There is not one exception."[18]

To this day, although it was never written into the formal rules, the seniority system is a powerful Senate norm. It determines who will chair each Senate committee and subcommittee as well as who will serve as the minority leader—"ranking minority member" is the title—of each committee and subcommittee, although the party conferences must ratify these selections. And the mostly honorific office of president pro tempore of the Senate is still always filled by the most senior majority-party member of the Senate.

The Filibuster

The act of speaking on a legislative measure at length, as a means of preventing Senate action on it, is called a filibuster, although senators prefer the polite term "extended debate." It is a product of the formal Senate rules. A minority of senators may use the filibuster and other dilatory tactics to thwart the will of a Senate majority, by talking a measure to death or by blocking a vote on it until changes in it are forced.

"Filibuster" is an exotic word for a peculiar practice. As one historian of the Senate filibuster has pointed out, the term once referred to piracy, but by 1863, at least, was being used "as a term of reproach signifying flagrant legislative obstruction—though legislators knew how to employ the tactics long before they were favored with a satisfying name for them."[19]

Institutionalized during the Senate's Golden Age, the filibuster actually originated much earlier. In fact, in the very first session of the Senate in New York, two senators talked at length in order to waste time and prevent a vote on the question of a permanent home for Congress.[20] Extended debate, which was also once used in the House and then abandoned there, came to full bloom in the Senate with the 1825 election of John Randolph of Virginia, of whom a caustic Virginia editor wrote at the time:

> [I]t is generally felt, and pretty freely acknowledged by many of the senators themselves, that their body has lost a large portion of their own respect for it, and of the respect of the people, through Mr. Randolph's incessant talking. If every gentleman spoke as long as he does . . . a three years perpetual session would not do the *business* of a week. . . .[21]

Early senators claimed that "freedom of speech" was the legal basis for their right to talk as long as they wanted to. This was also the claim that John C. Calhoun later used in defense of a filibuster by the Democrats when Senate Whigs tried to cut off debate and get a vote on Henry Clay's controversial bank bill. It was the issue of slavery, northern opposition to it and the southern effort to preserve it, that occasioned the greatest use of the filibuster during the Golden Age of the Senate. After the Civil War, the filibuster was employed to block action on other matters.

Rising obstructionism was one reason why the Senate's public standing slumped during the last part of the nineteenth century and into the first part of the twentieth. The Senate was larger in size by then, more states having been added to the Union. Other changes had taken place also. The number of standing committees had exploded to seventy-four. The influence of lobbyists and the special interests had grown to such an extent that many observers felt that they were corrupting the Senate.

Party government had become entrenched. Each party had an elected leader, and major Senate decisions were made beforehand in closed party caucuses, or conferences, rather than openly on the floor. Senate debates, conducted in the barnlike and unacoustical (and as yet unmicrophoned) present chamber in the Capitol's north wing to which the Senate had moved in 1859 were once again tedious and of little moment. This caused the tart-tongued Senator George F. Hoar to comment humorously in 1897: "It would be a capital thing to attend Unitarian conventions if there were not Unitarians there, so too it would be a delightful thing to be a United States Senator if you did not have to attend the sessions of the Senate."[22]

The filibuster—and variations on it, including the offering of numerous dilatory amendments and motions and the demand for time-consuming roll call after roll call—flourished. When a filibuster killed his Armed Neutrality Bill in 1917, President Woodrow Wilson declared angrily: "The Senate of the United States is the only legislative body in the world which cannot act when its majority is ready for action. A little group of willful men, representing no opinion but their own, have rendered the great Government of the United States helpless and contemptible."[23] Rising public disapproval of this 1917 filibuster forced the Senate that same year to amend its rules and provide for "cloture," a means of cutting off debate by an extraordinary Senate majority.*

Despite its constitutional and operational peculiarities, the Senate became a powerful American institution, "the world's greatest deliberative body," as some, particularly senators, liked to call it. Senate membership became a matter of great prestige, "the world's most exclusive club," it was said.

The Senate was designed by the founders as a check on presidential power and public passions, an institution that could deter precipitous action. But it was also intended to be able to act when there was public, and senatorial, consensus and, in fact, to help engineer consensus.

Some Senate peculiarities could be used to block consensus-building, though, and to make timely decision-making in that body more difficult than in the House. Over the years, the Senate developed into an organization with powerful leaders and committees and with strong traditions and norms of senatorial behavior that could often bypass the blockades and enhance the Senate's ability to act in concert. This was especially true of the Senate of the 1950s.

The Senate of the 1950s

If the decades before the Civil War were the Senate's Golden Age, the first full decade after World War II might be called its Silver Era. Two best-selling books of the time, as well as a classic academic study, celebrated and spotlighted the Senate of that period.

The two popular books were written by respected journalists who covered the Senate every day. William S. White's *The Citadel: The Story of the U.S. Senate*, published in 1956, characterized the Senate as a place of tight collegiality and towering prestige, where giants like Democrat Richard Russell of Georgia and Republican Robert A. Taft of Ohio,

---

* Even with this and later rule changes, the filibuster and related dilatory tactics are still used by Senate minorities—and today more than ever—to obstruct action by the majority.

members of an "inner club," ruled its chamber and dominated its committee rooms.[24] Allen Drury's 1959 novel, *Advise and Consent*, excitingly portrayed the Senate of that same era as America's most important national forum and a place of great drama and history.[25]

The landmark scholarly work on the 1950s Senate was done by political scientist Donald R. Matthews. He went inside that body and studied it as an organization. Matthews's *U.S. Senators and Their World* described the Senate as a close-knit society with strong internal pressures for conformity to certain established folkways, or norms. He found that these norms were carefully followed by "the most influential and effective members."[26]

Among the six Senate norms that Matthews identified was the norm of "apprenticeship." The seniority system was one aspect of this unwritten rule, but it also meant that "the new senator is expected to keep his mouth shut, not to take the lead in floor fights, to listen and to learn." A "legislative work" norm, Matthews said, required senators to devote themselves to the "highly detailed, dull, and politically unrewarding work" of the Senate, to be a "work horse" instead of a "show horse."

A "specialization norm" called for a senator to refrain from trying to be active on a wide range of subjects and instead "to specialize, to focus his energies and attention on the relatively few matters that come before his committees or that directly and immediately affect his state." This particular norm "helps make it possible for the Senate to devote less time to talking and more to action."

A "courtesy" norm, constraining senators to refrain from personal attacks on each other and to avoid excessive partisanship, Matthews said, "permits competitors to cooperate." A norm of "institutional patriotism" meant that "Senators are expected to believe that they belong to the greatest legislative and deliberative body in the world. They are expected to be a bit suspicious of the President and the bureaucrats and just a little disdainful of the House."

Finally, a norm of "reciprocity" was a "way of life" in the 1950s Senate, according to Matthews, and it meant that senators should keep their word; should help a colleague when they could, with a right to expect repayment in kind, and should refrain from pushing their formal powers to the limit. Matthews wrote that a senator who ignored this last aspect of Senate reciprocity, by obstreperously objecting to unanimous consent requests and by engaging in filibusters too often, could almost slow the Senate to a halt, and he added:

> While these and other similar powers always exist as a potential threat, the amazing thing is that they are rarely utilized. The spirit of reciprocity results in much, if not most, of the senators' actual power not being exercised. If a senator *does* push his formal powers to the limit, he has broken the implicit bargain and can expect not cooperation from his colleagues, but only retaliation in kind.

Matthews found that the observance of these Senate norms facilitated Senate consensus-building, decision-making, and action.

But political scientists, journalists, and novelists were not the only ones who thought of the Senate of the 1950s as a very special place. So did senators themselves. In a kind of proud self-consciousness, they adopted a 1955 resolution offered by their powerful Democratic majority leader, Lyndon B. Johnson of Texas, then convalescing from a heart attack in Bethesda Naval Hospital, to create a Special Committee on the Senate Reception Room. The job of this new committee would be to "select five outstanding persons from among all persons, but not a living person, who have served as members of the Senate since the formation of the government of the United States, whose paintings shall be placed in the five unfilled spaces of the Senate Reception Room."[27]

The Five Outstanding Senators

The U.S. Capitol, the present massive and gleaming-white symbol and seat of America's national government, was at first only a four-story sandstone and granite box on Washington's "Capitol Hill" to the east of the White House. Next it was composed of two such boxes, first connected by a wooden walkway and later joined in stone. Finally, construction started just before the Civil War on the cast-iron dome in the middle and new wings on each side—the present south wing for the House of Representatives and north wing for the Senate. The Senate reception room is located on the second floor of the Senate side of the Capitol, just off the Senate chamber. Senators come here to meet with visitors who have called them off the floor.

In this ornate room, with its elegantly tiled floor, massive crystal chandeliers, two nine-foot, heavily draped windows, massive marble fireplace, and heroic wall painting of an imagined meeting between Washington, Hamilton, and Jefferson, are the color portraits of five outstanding senators. These are painted directly on the walls in oval panels, each topped by a raised, gold-leaf eagle and other embellishments.

When artist Constantino Brumidi finished this room in 1874, he left the oval panels vacant for the portraits of five distinguished Americans yet to be selected. Lyndon B. Johnson's resolution came some 81 years later.* Johnson saw to it that Senator John F. Kennedy of Massachusetts, who had recently authored a best-selling book about Senate heroes, *Profiles in Courage*, was named to chair the committee.

The five senators selected by the Kennedy committee afford us an

---

* At the eventual 1959 ceremony to dedicate the five portraits, Vice President Richard Nixon drew considerable laughter when he said that the first suggestion concerning the portraits had been made in 1870 but that "no action was taken. Or should I say that the Senate acted in its usual, very deliberate way."

indication of the kind of institution the Senate of that era aspired to be. The committee wrestled unsuccessfully with a definition of senatorial greatness. Kennedy wrote that a test of greatness based on legislative accomplishment was too difficult to apply. Some senators are best remembered for their "courageous negation," he said, while the legislative efforts of others failed in their own time and only much later led to action; further, history has sometimes incorrectly credited important legislation to senators who were not really responsible for it.*

Kennedy also wrote that "length of service" was a poor test of senatorial greatness, indicating nothing more than political success. He said that not even "personal integrity," taken alone, was a sufficient test of greatness in a senator, nor was "national leadership" or "contemporary popularity." He wrote:

> If personal integrity is the test, we must exclude Daniel Webster who saw to it that his "usual retainer" was "refreshed" by the National Bank when its charter came up for Senate renewal. If national leadership is required, we must exclude John C. Calhoun, who lived—and died—for Southern principles. If contemporary popularity is essential, we must exclude George Norris upon whose name (because of his filibuster against Wilson's armed-ship bill in 1917) the *New York Times* once editorialized "the odium of treasonable purpose will rest forever. . . ."

Kennedy felt that the fact that a senator was a "politician" should not be a disqualification for greatness:

> Finally, if we adopt the suggestion of many that we select those who were statesmen and *not* politicians, whom can we include? Not Clay, scheming for the Presidency; not Calhoun, embarrassing Jackson and conspiring against Benton; not Webster, earning Emerson's epitaph, "A great man with a small ambition"; not presidential candidates like Douglas or Taft or La Follette or Houston; not party spokesmen like Vandenberg or Benton. I know of no man elected to the Senate in all its history who was not a "politician," whether or not he was also a "statesman."

Kennedy said that he and his committee had concluded, then, that no standard "greatness" test for senators existed. Indeed, the committee decided that they would choose five "outstanding" senators, not necessarily the five *most* outstanding.

After polling former presidents, sitting and former senators, and 150 prominent historians and scholars, the Kennedy committee finally met to decide.† Agreement was reached quite easily on the selection of

---

* In this connection, Kennedy quoted a contemporary of Senator John Sherman of Ohio, Senator George F. Hoar, who said concerning the landmark antitrust law that bears Sherman's name: "I doubt very much whether he [Sherman] read it. If he did, I do not think he ever understood it."

† Former President Harry S. Truman, characteristically, sent the Kennedy committee a heavily annotated list of thirty-nine names, while former President Herbert Hoover first claimed insufficient competence to make a choice and then volunteered that Clay and

the Great Triumvirate—Webster, Clay, and Calhoun. Then the committee made a kind of compromise in their last two selections: they balanced their choice of a conservative, Republican Robert A. Taft of Ohio, with the selection of a progressive, Republican Robert M. La Follette, Sr., of Wisconsin. As Kennedy later reported, Clay, Webster, Calhoun, and La Follette were among the top five receiving the most endorsements from both the panel of scholars and the participating senators. Taft, he said, had been among the highest ten recommended by the scholars and was the first choice of sitting senators.[28]

At the dedication ceremony held in the Senate reception room, the record reveals, Senator Kennedy told the audience that each of the five senators chosen had been the subject of controversy during his own time:

> We are more familiar with the controversies that surrounded Taft and La Follette. But let us also remember that it was said of Henry Clay that "he prefers the specious to the solid, and the plausible to the true. He is a bad man and an impostor, a creator of wicked schémes." Those words were spoken by John C. Calhoun. [Laughter.]
>
> On the other hand, who was it who said that Calhoun was a rigid fanatic, ambitious, selfishly partisan and sectional "turncoat," with "too much genius and too little common sense," who would either die a traitor or a madman? Henry Clay, of course. [Laughter.]
>
> When Calhoun boasted in debate that he had been Clay's political master, Clay retorted: "Sir, I would not own him as a slave." Both Clay and Calhoun from time to time fought with Webster; and from the other House, the articulate John Quincy Adams, with old fashioned New England courtesy, viewed with alarm "the gigantic intellect, the envious temper, the ravenous ambition, and the rotten heart of Daniel Webster." [Laughter.][29]

Kennedy declared that the five former senators had been chosen to "call attention in these critical times to the high traditions of the Senate, and its significant role in our history" and because in the senators selected "those traditions are best exemplified."[30]

Henry Clay was "the great compromiser," risking the wrath of both sides to forge three great compromises on the slavery issue, in order, as Kennedy put it, "to save the Union until it grew strong enough to save itself." Clay was, Kennedy said, "probably the most gifted parliamentary figure in the history of the Congress, whose tireless devotion to the Union demonstrated that intelligent compromise required both courage and conviction."

Webster probably should be included, and maybe Sherman. One sitting senator, liberal Democrat Paul Douglas of Illinois, objected to the selection of Webster because he was "undoubtedly venal" and of Calhoun because of his "defense of slavery and inequality." Another sitting senator, conservative Republican Carl Curtis of Nebraska, virtually vetoed the selection of his fellow Nebraskan, progressive Republican George Norris, although Norris had tied for first, with Webster and Clay, in the rankings by the panel of scholars.

Daniel Webster was the eloquent and articulate champion of "Liberty and Union, now and forever, one and inseparable," knowing that he would be crucified by his abolitionist supporters, and he was largely responsible for the adoption of Clay's 1850 compromise, which delayed secession and Civil War for another decade.

John C. Calhoun was, Kennedy said, the most notable political thinker ever to sit in the Senate and "the intellectual leader and logician of those defending the rights of a political minority against the dangers of an unchecked majority."

Robert M. La Follette, Sr., Kennedy pointed out, was "a ceaseless battler for the underprivileged in an age of special privilege, a courageous independent in an era of conformity, who fought memorably against tremendous odds and stifling inertia for the social and economic reforms which ultimately proved essential to American progress in the 20th century."

Finally, Robert A. Taft was a senator who was not afraid to go it alone, against his party and against a majority in the Senate, when he felt principle demanded it—opposing with equal fervor, for example, President Truman's military draft of striking railroad workers as unconstitutional, and the Nuremberg trials of Nazi war criminals as unjustified under international law—and was, as Kennedy said, "the conscience of the conservative movement and its most constructive leader, whose high integrity transcended partisanship, and whose analytical mind candidly and courageously put principle above ambition."

The Senate of the 1950s that chose Webster, Clay, Calhoun, La Follette, and Taft as its heroes, viewed itself—and was viewed by those who observed it—as a great national forum, a citadel of free expression and new ideas, a deliberative body, open to the people and responsive to them, and in addition, a responsible decision-making body, capable of forging and acting on national and senatorial consensus.

Today, among many of those who watch the contemporary Senate of the United States, and indeed among many senators themselves, there is a kind of homesickness for that Senate of the 1950s, a feeling that the Senate has changed a great deal since those days, and not for the better.

A More National and Individualistic Senate

With the advent of television and other great changes in American society after the 1950s, along with the nationalization of Senate elections and campaigns and an "advocacy explosion" of interest groups, as we will see in later chapters, the Senate changed. It was nationalized: the Senate became more national and outward-looking. Senators increasingly felt the influence of powerful national interest groups. At the same time, internal power in the Senate became fragmented and more individualistic.

A careful observer of the Senate, political scientist Randall Ripley, once posited three models for distribution or dispersal of power in the Senate: centralization, decentralization, and individualism.[31] During most of the 1930s, Senate power was "centralized," that is, concentrated in the party leaders, principally the majority leader and the minority leader. From the 1940s through the 1950s, Senate leaders increasingly found themselves having to share power with the committee chairs, which meant that Senate power was more "decentralized," though still concentrated. But, beginning in the 1960s, the rules and norms of the Senate changed, so that power became fragmented and "individualized" among senators and the chairs of greatly expanded numbers of increasingly autonomous Senate subcommittees. Harking back to an earlier time in the Senate, when senators were less likely to disagree with committee recommendations, one senator said:

> When I first arrived here, . . . you had a situation where committee chairmen regularly supported other committee chairmen, . . . and members supported the chairmen more regularly—it was pretty much a practice. . . . You had the majority leader [Lyndon Johnson] supporting all the chairmen, the chairmen supporting the majority leader, and it therefore gave the chairmen of each committee more power.[32]

From the 1960s on, the Senate became an increasingly open and democratic place and one in which power was more widely shared. A huge growth in staff assistance for senators and their committees and subcommittees both resulted from and enhanced the trend toward more individualism. Individualistic senators began to use the filibuster and other dilatory tactics much more on the Senate floor. The Senate found it harder to take timely action. Democratic Senator David Pryor of Arkansas noted in 1987 that the Senate of the Ninety-ninth Congress (1985–1986) had spent five hundred *more* hours in session, with four hundred *fewer* roll calls, than the Senate of the Ninety-third Congress twelve years earlier (1973–1974). "In other words," Pryor concluded, "as of last year, we were churning out 35 percent less work product while lengthening the amount of time in session by 20 percent."[33]

Many observers of the Senate, and indeed many senators themselves, began to feel that "teamwork in the Senate [had] given way to the rule of individuals,"[34] that this individualism had "dragged this proud institution into a slow-motion system of inefficiency and procedural imprisonment,"[35] and that the Senate had become mired in a tar pit of obstructionism, stalemate, policy gridlock, and frustration.[36] "There is a lack of accomplishment, or maybe more accurately a sense that the whole system is breaking down," one senator declared after announcing his retirement in 1988, and another, also retiring that year, asserted that "the Senate has become difficult to lead, consensus is illusory and the whole policy-making process stands on the brink of incoherence."[37]

Political scientist Nelson Polsby summed up the results, as he saw them, of all the changes which the Senate underwent after the 1950s when he concluded that the contemporary Senate has become "an individualistic, staff-saturated, publicity-hungry, and public-regarding institution serving the ambitions of many of its members to achieve star billing in connection with national issues. . . ."[38] Perhaps that is too harsh a description, but it certainly captures the feeling that many observers have about the present United States Senate.

As the Senate became more nationalized and more open and democratic, it became *more responsive*, more representative of all the people and more accountable to them, more a generator and popularizer of new ideas, but it also became *less responsible* and efficient, less decisive as a legislative body. As one authority on congressional change has stated:

> Responsiveness implies permeability; more openness and participation allow more interests access to Congress. This, in turn, suggests a diminution in responsibility, a decline in efficient production of policy. . . . Responsiveness and accountability expose the assembly to multiple pressures and make it accessible to many points of view. Permeability undercuts decisiveness, however. . . .[39]

There are those who feel that the way the Senate is organized, the way power is distributed and dispersed, is governed by internal-change cycles: that rising Senate opposition to concentrated power causes centralization (party-leader power) to give way to decentralization (committee-chair power); that decentralization is eventually replaced by individualism (senator and subcommittee-chair power); and that, finally, as senatorial stalemate increases and senatorial and public frustrations rise, especially when accompanied by electoral upheaval and political realignment, reorganization occurs to centralize power again, thus completing the cycle.[40]

Indeed, as we shall see, there have been some recent developments—particularly, strengthened party conferences and the use of the summit device—that indicate some movement toward arching over the Senate fragmentation of power. But the Senate of today is considerably different from the Senate of the 1950s. It is less responsible and efficient. Senators' behavior is different. The purpose of this book is to examine the changes that have taken place in the Senate and the reasons for them, as well as the problems these changes have caused, and to consider ways by which the Senate might be made more responsible and efficient again as the great national deliberative and decision-making body it was intended to be.

II

THE SENATE'S NATIONALIZED ENVIRONMENT

2

The Nationalization of America
and of Senate Elections

J. William Fulbright was thirty-nine years old in January 1945 when he boarded a train in his hometown of Fayetteville, Arkansas, to go to Washington to be sworn in as a newly elected member of the Senate of the United States. An urbane former Rhodes scholar, he had just completed one term as a member of the U.S. House of Representatives.[1]

Fulbright was beginning what would eventually stretch to thirty years of service in the Senate. With time and the help of Senate Majority Leader Lyndon Johnson, he would in 1959 become chairman of the Senate Foreign Relations Committee. With more time and despite the bitter displeasure of *President* Lyndon Johnson, he would in the 1960s become a major and outspoken critic of America's war in Vietnam. The United States Senate that Fulbright joined in 1945 would change considerably by the time he was defeated for reelection in 1974. Arkansas would change. America would change, too—change in a way that would amount to a nationalization of American society. And the change in America and Arkansas would prove to be a principal cause of Senate change.

By today's standards, transportation and communications were slow in America during Fulbright's first years in the Senate, and into the 1950s. Back then, for example, it took Fulbright two entire nights and a day, with a train change in St. Louis, to get from Fayetteville to Washington's Union Station, as he was to recall in an interview in his elegant Washington law office more than forty years later.[2] Senator Fulbright was at first reimbursed by the Senate for the cost of only one round trip per year between Arkansas and Washington—and that was the only round trip he made. He went to Washington each January when the Senate session opened. And there he stayed, without even one home visit, until the Senate adjourned sine die, finally for the year, sometimes as early as June in those first years (as contrasted with today's practically year-round sessions).

Did many people from Arkansas come to Washington to see him in the early days? "Oh, hell, no!" Fulbright answered spiritedly. Not many Arkansans telephoned, either, he added, and the mail from home was very light. The pace of the Senate was almost leisurely; there was time to think, time to spend with colleagues. "I rather liked the occasional night sessions we had, because you could sit and visit with your colleagues," Fulbright recalled.

What changed things? "The advent of the airlines, first of all, in the 1950s," Fulbright responded. "And, then, the greatest change was television, the most dangerous development, a curse, a dangerous instrument. It changed the relationship between a senator and his constituents."

The Nationalization of American Society

Airline travel and television—these were, indeed, major changes in the Senate's external environment that amounted to a kind of nationalization (as opposed to localization) of American transportation and communications. Take air travel: in 1950, Americans flew a total of a little over 10 million miles. Ten years later, in 1960, airline travel in the United States had climbed to 39 *billion* miles! By the beginning of the next decade, Americans were racking up 132 billion miles of airline travel annually, and 250 billion miles by 1980.[3] Many more people could go to Washington personally, where as a part of their trip, they could observe their senators in action.

Television came to America in a rush. Dramatically and almost at once, it altered lives and politics. Consider the fact that in 1950 only 9 percent of American families had television sets, while just five years later 65 percent did! By 1965, 95 percent of American families could watch television at home.

Most people began to get their news from television, including news about what their senators were doing in Washington. Television news helped to focus their attention at the national, rather than the local or state, level. William M. Lunch, in his book *The Nationalization of American Politics*, made this point: "When the news was transmitted primarily by newspapers, political coverage was primarily local. . . . On television, however, political news is defined almost exclusively as national politics and government, most of it from Washington."[4]

One reason why Senator Fulbright received so few phone calls from Arkansas during his early years in the Senate was that not many people had telephones then. In 1950 America was a country of 150 million people but only about 26 million residential telephones (about 140,000 in Arkansas, then a state of nearly 2 million people). The number of home telephones in the country had more than doubled by 1960 (and about half the residences in Arkansas had telephones that year). By 1970, 92 percent of America's homes, and 77 percent of those in Arkansas, had telephones, and by 1980 the figures were 96 and 87 percent, respectively. People could place calls from home to their senators—and they did.

Americans—and Arkansans—were changing rapidly, too. They were becoming more urban, better educated, and more mobile. Their standard of living was improving as America's economy became increasingly national (rather than local or regional), and a rising national tide raised nearly all boats. In 1950, 64 percent of Americans lived in urban areas, and a third of Arkansans did. By 1960, 70 percent of America's population, and 43 percent of Arkansas', was urban. In 1970, the figures were 73.5 and 50 percent, respectively, figures that were still pretty much the same in 1980, except for a slight increase for Arkansas (51.6 percent urban).

The standard of living rose markedly through the years,[5] and with it education levels. Slightly over 34 percent of American adults were high school graduates in 1950; by 1980, 74 percent were. Less than 8 percent of the country's adults had a college degree in 1950; by 1980 over 19 percent did.[6] During the same 1950 to 1980 period in Arkansas, the percentage of college degrees among those over age 25 (a smaller group than "adults") more than tripled, rising from 3 percent to nearly 11 percent. Better-educated people are more politically active, more attentive to what is going on in their government, than are less well-educated people.

Not only through airline travel, but more importantly for most Americans—and Arkansans—increased mobility came for more people through automobile ownership. In 1952, for example, there were 43.6 million cars in America, 333,000 of them in Arkansas. These numbers grew sharply and steadily each year thereafter, so that by 1986 both figures had approximately tripled: Americans owned 135.6 million cars, Arkansans almost a million. A lot of these people with automobiles began to use them to drive to Washington, among other places.

Growth in the National Government

As American society changed, so did our national government. It grew enormously, especially after the 1950s. This was partly an effect, partly a cause, of the changes that were taking place in the country. America had become a great world power, and that also meant big government. Fulbright came to the Senate toward the end of World War II. In his first year there, 1945, President Franklin Roosevelt died and Vice President (and former Missouri senator) Harry S. Truman succeeded to the office. The war ended the same year, and America began a transition to peace-time. But then and through most of the 1950s, federal spending contin-ued at levels that now seem very low indeed, even minuscule.

The nation's first full year of peace was 1947, Fulbright's third year in the Senate. That year the entire federal budget amounted to only $35 billion. (John Stennis of Mississippi, later to be president pro tempore, came to the Senate in 1947, and that $35 billion federal budget of his first year stuck in his mind ever afterward.)[7] Even ten years later, in 1957, the total federal budget amounted to only $80 billion. (That was the year, incidentally, when a future Senate maverick, William Proxmire of Wisconsin, finally won a Senate seat after more than one try; for Prox-mire, a special foe of wasteful government spending, the $80 billion 1957 federal budget was a benchmark that he often referred to.)[8] A fed-eral budget that had been $35 billion in 1947 and $80 billion in 1957 had almost unbelievably mushroomed by 1988 (a year, incidentally, when both Stennis and Proxmire retired from the Senate) to $1 trillion!

The business of the Senate also expanded greatly. There were only seventy-eight roll-call votes on the Senate floor, for example, during all of 1947, whereas in 1988 there were ten times that number (as Senator Stennis, comparing his first and last Senate years, later related to a col-league).[9]

The numbers of Senate staff members repeatedly doubled and dou-bled again through the years. The average senator had only six personal staff members when Fulbright came to the Senate in 1945.[10] Even nine years later, senatorial staff numbers had not grown much. For example, when future Senate President Pro Tempore Strom Thurmond came to the Senate from South Carolina in 1954,* he had only seven people on his personal staff, three men and four women; by 1988 he had fifty-two personal staff members.[11] The average senator in 1988 had a personal staff of thirty-nine.[12]

Change in the Washington Media

Media attention to, and coverage of, the Senate multiplied too. In the late 1940s and into the 1950s, senators made little news back home, since few local news organizations maintained bureaus in Washington. The majority of senators made no *national* news at all (although it should be noted that Fulbright was something of an exception, ignoring from the first, especially on foreign affairs matters, the Senate norm of the time that new senators should avoid the limelight).[13]

Until the 1970s, when new "sunshine" rules were adopted, "mark-

---

* Thurmond was first appointed to fill a vacancy. After resigning he was elected in his own right in 1956, initially as a Democrat, then as a Republican after changing because of the Democratic party support for black civil rights.

up" sessions in Senate committees and subcommittees—where final decisions are taken on amendments to pending legislation and on whether to report the legislation to the full Senate—were regularly closed to the press and the public. Most often the actions of a senator in committee or subcommittee went completely unnoticed.

Capitol reporters were much less aggressive in those earlier years. "In the old days," Floyd Riddick, former Senate parliamentarian has said, "no reporter would dare stop a senator on his way to or from the Senate and ask for a comment. The senator would have said, 'See me in my office.' Senators are now more accessible."[14]

In earlier times, too, and as late as the 1960s, there was an unwritten journalistic code in Washington that placed the personal habits and behavior of senators and other national political figures more or less off limits to reporting. For example, during a late-night Senate debate in the 1960s on President Lyndon Johnson's antipoverty program, an influential southern senator then known for drinking too much on occasion (he later reformed) came onto the Senate floor almost falling-down drunk and lunged into the argument. The pending question was whether state governors should have a veto over federal antipoverty programs planned for their states. Republicans favored veto power; Democrats were generally opposed to it. Two Republican senators, Peter Dominick of Colorado and John Tower of Texas, began to bait the incapacitated senator. Word spread immediately, and senators cascaded back into the chamber from the cloakrooms where they had been killing time while awaiting the final vote. The press gallery filled up almost at once, too, and soon all other Senate galleries were packed as well. Spectators, glued to the scene by an almost morbid fascination, gasped audibly when the drunken man started to quote a personal conversation with President Johnson, something then thought improper in Senate debate unless specifically authorized by the president.

"I talked to Lyndon this very afternoon," the senator said, his words slurred, "and he said, '——, if you let those governors have a veto over these programs, they'll kill me politically and they'll kill you, too!'"

Senator Dominick rose. "Mr. President, will the senator yield?" he asked. A grave and ashen-faced Senator Stennis, always concerned about the Senate's reputation, was seen to pull at the senator's coattail and quietly urge him to sit down.

"Yes, I'll be glad to yield to my distinguished colleague from Colorado," the impaired senator said, ignoring Stennis and, in fact, moving a couple of seats to get away from him. Senators, especially those on the Democratic side of the aisle, looked at each other with dread.

"The Senator would not say that his own governor, Governor ——, could not be trusted with this veto power, would he?" Dominick asked.

"That's exactly what I *do* say," the Southern senator bellowed, heedless of caution or consequence. "Who did the governor appoint to head the poverty program in my state? A hot check artist! How do I know he's a hot check artist? Because he gave *me* a hot check! Who did he appoint as deputy director of the poverty program? A member of the Ku Klux Klan!"

The reporters in the press gallery were writing and laughing at the same time. On the floor, the drunken senator and senators Dominick and Tower continued to go at it for some time, until, at last, Senator Stennis succeeded in getting him to quit.

A good many persons went home that night thinking that the man in question had ruined himself, that his career surely could not survive. But the next morning they found, much to their surprise, that neither the Washington newspapers nor those in the senator's home state carried even one word about the episode. There was no report on television or radio, either. And, since senators have a right to edit their remarks in the *Congressional Record* before it goes to press each night, there was nothing to embarrass the senator in the official transcript. "It took my staff two hours last night to clean up the *Record*," the senator in question, seeming much chastened, remarked to one of his colleagues when the Senate convened the next morning.[15]

Journalistic codes changed after the Watergate scandal. Such an episode not only would be fully reported today, it would surely now cause what has come to be called a press "feeding frenzy" and a spate of additional and related investigatory stories about the senator involved. Political scientist Larry Sabato has written that the "lapdog" Washington journalism of post–World War II changed to "watchdog" journalism in 1966 and, after 1974, to "junkyard-dog" journalism.[16]

The practices of Washington reporters became much more adversarial at the same time that their ranks expanded enormously. Between 1960 and 1976, the number of radio and television journalists accredited to cover Congress more than tripled (to six hundred), and the number of print journalists went up by more than a third (to eleven hundred). By the mid-1980s, around ten thousand journalists were working in Washington. Local newspapers and television stations opened Washington news bureaus or contracted for regular capital reports by independent stringers. Television networks expanded their nightly news programs in 1963 from fifteen minutes to thirty minutes. The number of network news programs soon increased, and new cable networks appeared. Continuous, twenty-four-hours-a-day radio and television news programs became available almost everywhere in the country. Senate sessions began to be telecast live. Thus today, what senators do—or do not do—is much more intensively covered by the media. And senators themselves have become adept at using the press and all the modern means of communication to stay in touch with their constituents.

New Activist Groups Spring Up

As voters became more urban, more mobile, more affluent, better educated, and better informed, they became more attentive to public affairs. Ethnic and minority and other groups began to become much more assertive in the late 1950s and afterward.

It was in 1957—in Little Rock, Arkansas, as a matter of fact—that troops were first used (by President Dwight Eisenhower) to enforce the right of African-American students to attend a previously all-white school. Throughout the South, schools had been rigidly segregated before implementation of the Supreme Court's *Brown* v. *Board of Education* decision of 1954.[17]

The black civil rights movement gained its greatest strength in the mid-1960s. The landmark Civil Rights Act was adopted by Congress in 1964, and a highly effective Voting Rights Act was adopted in 1965, over the determined opposition of southern Democrats (including Senator Fulbright; from the start of his Senate career, Fulbright had justified his opposition to civil rights legislation by declaring that public officials have no alternative but to vote in accord with strongly held constituent views on domestic issues, in order to avoid defeat and be able to lead public opinion, if need be, on world issues).[18] Large numbers of new activists were introduced to politics through the civil rights movement. The Voting Rights Act of 1965 caused an immense rise in the number of black voters throughout the South. In Arkansas this meant that Fulbright's longtime opposition to civil rights legislation became increasingly troublesome for him politically.

A very large and activist U.S. peace movement began to develop in the last half of the 1960s in opposition to America's war in Vietnam. Fulbright, who had broken with President Johnson on the war issue, was not only in step with this movement but was one of its heroes. Even he must have been astonished, however, as others were, by the volcano of citizen activism that the peace movement energized.

After the peace movement subsided somewhat, a vigorous environmental movement developed, with "earth day" rallies and other mobilizations. Some conservative politicians called environmental activists "tree huggers," but they ignored the growing environmentalist ranks and efforts at their political peril.

Impact on the Senate

What conclusions can be drawn concerning the impact on the Senate of the great societal changes, as well as of the growth in government, that occurred in the late 1950s and afterward? Americans became more aware of, and attentive to, public—and senatorial—affairs. Senators became more accessible, more exposed to public scrutiny and personal contact.

By 1988, the average senator received nearly seven hundred letters a day, talked on the telephone more than twenty times daily, and kept personal appointments with six people or groups in the office (not counting the much larger number of people who were taken care of by senatorial staff members).[19]

Communications systems work both ways, of course. Not only were senators bombarded with information and contacts from their constituents, but senators, themselves, began to bombard the folks back home almost as much, and this generated a great deal more bombardment of senators in return. Transportation systems work both ways, too. Not only did it become easier through the years for the voters to come to Washington to see their senators; it also became easier for senators to go home—and because they could, they had to. While in the early years of Senator Fulbright's service, for example, most senators never went home during a session, by 1988 the average senator went home more than twice every month.[20] Senators began to be preoccupied with "staying in touch."

Senators found it politically more and more difficult to go to Washington and serve quietly as an apprentice for a time, being seen and not heard, attending only to the nitty-gritty of legislating, not feeling pressed to make much news—or a name. They found it less and less likely that they could specialize in as few as two or three subject areas. They found it harder and harder to "go along in order to get along" in the Senate; to play follow the leader, to more or less automatically back committee recommendations on various legislative measures that came before the Senate; to help another senator out, expecting similar help in return; or to refrain from using all the parliamentary tools available to them to aid in their cause. Once senators had done all of these things without having to worry about public opinion back home, if there was any. They could no longer do so.

The times changed, the country changed, and electoral politics and campaigns changed. Senators who did not change found it increasingly difficult to get reelected.

The Nationalization of Senate Elections

"Rising educational levels and enhanced communications and travel no doubt" gave rise in Congress in the late 1960s to "a heightened sensitivity to constituency concerns" and an increasing preoccupation with reelection—so said an expert observer of Congress, Roger Davidson.[21] That was especially true in the Senate.

The late Democratic Senator Richard Russell of Georgia once joked that you could watch senators walk down a hall and immediately tell what part of their six-year term they were in. Those recently elected or reelected, with a full six years ahead of them, Russell said, walked along looking up, thinking lofty thoughts. Those with four years remaining walked looking straight ahead. But senators with only two years left,

Russell said, looked down as they walked, carefully watching where they stepped.[22] That was, indeed, the old Senate style before so many major societal and electoral changes—Washington orientation for four years, "home style," in political scientist Richard Fenno's words,[23] for the last two. But the old way became increasingly unreliable. Russell himself, in planning for a 1970 reelection campaign (that ultimately turned out to be a walkover), worried aloud about having lost touch back home over the years. "Most of my best supporters are in the graveyard now," he said.[24]

Fulbright had the same problem when he decided to run for a sixth term in 1974. He had always followed the old four-year, two-year style, and, as with Russell, it had been successful for him. Six years earlier, when some observers thought Fulbright might be defeated for reelection to a fifth term, especially because of his opposition to the Vietnam War, Fulbright had turned over the gavel of his Foreign Relations Committee to the next most senior Democrat, John Sparkman of Alabama, and had returned to Arkansas for a full-time, daily regime of handshaking and local "speakings," and Main Street and county fair, shirtsleeves and galluses, down-home campaigning. It had worked again.

But that was the last time. Late-term home style and last-year campaigning did not work for Fulbright in 1974. The outcome was reported by a Fulbright biographer:

> Fulbright, it turned out, had not done his homework on Arkansas politics. He was, quite simply, out of touch, belonging more to the urbane elegance of Georgetown than the bucolic homeliness of Arkansas. . . . Dale Bumpers, the state's popular young governor, sensed Fulbright's vulnerability and joined the primary race. When the votes were counted on May 28, Fulbright had been defeated by a humiliating margin of nearly two-to-one.[25]

An old Senate giant had been replaced. Old style had given way to new style. In fact, such Senate replacements had begun to take place much earlier than 1974. The year 1958 had perhaps marked the beginning of the turnover. That year, thirteen new Democratic senators and three new Republican senators had been elected—including, for example, Democrats Philip Hart of Michigan, Edmund Muskie of Maine, and Eugene McCarthy of Minnesota and Republican Hugh Scott of Pennsylvania. The class of 1958 was the vanguard of the eventually successful challenge to the "Senate establishment" and its old ways of doing things.[26] These newcomers represented the first of several succeeding waves of Senate change, waves that themselves had developed from societal and electoral changes.

Television and Senate Campaigns

Senator Fulbright called television a "curse, a dangerous development." Whether or not that assessment is justified, there is no doubt that television has had great impact on society and on Senate campaigns as well. Senatorial campaign advertising now is overwhelmingly television advertising, particularly in one-minute and thirty-second spot commercials. And most plans for using *free* media in a Senate campaign—demonstrating support for day care legislation by appearing at a local day-care center, for example—are aimed at getting the candidate on local television newscasts. Television has been a big reason for the appearance of a burgeoning new campaign industry of consultants that no successful Senate campaign, whether of incumbent senator or of challenger, would today think of being without.

Senate campaigns of the 1960s were still pretty much local operations, with the candidates often acting in reality as their own campaign managers. By the 1970s, the general practice was to hire one of several "turn-key" campaign firms—like Matt Reese and his group who worked for Democrats—to manage the whole campaign, from polling to advertising. Since the beginning of the 1980s, Senate campaigns usually do not hire an overall consultant or grand strategist. Instead, particular kinds of consultants—and a lot of them—are hired for particular campaign tasks.[27]

There are "name" Democratic and Republican consultants in all the various categories, and they do not come cheap. An announcement by a Republican Senate campaign that it has hired a certain nationally known media consultant, the Robert Goodman Agency, for example, and a polling consultant like Tarrance and Associates, or by a Democratic Senate campaign that it has employed a firm like Greer, Margolis, Mitchell and Associates to handle media and a group like Garin-Hart Strategic Research to handle polling, is a strong signal—to the journalists who cover campaigns, as well as to interest groups, their lobbyists, and their campaign-financing arms, the PACs (political action committees), and to others whose opinions count or who contribute big money—that this is a "serious" campaign with a real possibility of winning.

During each election cycle, one of two monthly magazines of the campaign industry, *Campaigns & Elections*, runs four pages of Senate campaign lists, two pages for the Democratic candidates and two for the Republican candidates, giving the name, address, and telephone number of the campaign manager, treasurer, and press secretary for each campaign as well as the media consultant, the former media consultant, the polling firm, the former polling firm, and the campaign budget (in millions of dollars).[28]

As late as the 1960s, a good many Senate campaigns were successful without ever having had the benefit of even one paid public opinion poll. These campaigns relied on the "feel," or intuition, of the candidate and others, and on newspaper polls, when they were available, for finding out what people were thinking and how the race was going. Not any more.

Polling and Pollsters in Senate Races

Today a good pollster is considered crucial for a winning Senate race. The pollster is selected early and is typically paid between 4 and 8 percent of the total campaign budget. A first "benchmark" poll is generally followed by one or more similar overall polls, to check for trends. In a Senate campaign, especially the campaign of an incumbent senator thought to be vulnerable, the beginning benchmark poll is usually taken at least eighteen months before the actual vote. Democratic Senator Tom Harkin of Iowa, facing a November 1990 reelection vote, was already on his second poll by February 1989. In most benchmark polls, which cost around $25,000, six hundred or more representative people in the state are interviewed by the polling firm, each interview requiring about twenty-five minutes. In such a poll, the normal range for name recognition for a sitting senator is between 89 and 96 percent (somewhat lower in large states). A key question in the interviews is something like: "Thinking ahead to the next Senate election, would you vote now for Senator ——, would you consider voting for someone else, or would you vote to replace Senator ——?"

In these beginning benchmark polls, the average senator scores 40 to 46 percent for reelection; this is called the "core support." Forty-eight percent or better core support means—and is taken to mean by prospective contributors and prospective opponents alike—that the senator will be hard to beat. Against a senator like that, opposition money is hard to raise, and it is hard to get a good challenger to run. Forty percent core support or below means that a senator is clearly vulnerable; opposition and PAC vultures will begin to circle. Respondents in the benchmark poll who say that they would vote to replace the senator are called the "core opposition"; if this group runs as high as 20 percent or more, the senator is in trouble.

Interviewees in these early polls are asked to rate the job performance of the incumbent senator. Most senators have a job-approval rating in the mid-fifties (combining those who answer "excellent" with those who say "good"). If job approval is 60 percent or better, the senator is said to be in pretty solid shape.

Trial heats are also run, pitting the incumbent senator against likely challengers. A vote of 57 percent or more in these match-ups indicates that the senator will be tough to beat, while a vote of 52 percent or under shows serious vulnerability.

Polls are one reason Senate campaigns have been starting earlier and earlier. A poll that looks good when it is announced, or leaked, can be a crucial factor in building early financial and other support for a campaign. Peter Hart Research conducted a Kentucky poll in July of 1988 for County Judge Harvey Sloane, a Democrat, in preparation for his race against incumbent Republican Senator Mitch McConnell. Sloane's press release about the poll was well reported in the local media, together with his statement that put a Sloane-oriented "spin" on McConnell's standing: "For an incumbent to only have 44 percent more than two years out tells us he's basically weak. He's very vulnerable. He's very beatable."[29] (Unfortunately, from Sloane's viewpoint, McConnell did not prove to be all that beatable at election time.)

Media advertising is sometimes started early in a Senate campaign for the primary purpose of improving the candidate's standing in the polls. "The thing that's really shocking from the perspective of 10 years ago is how early it all starts," Garin-Hart Strategic Research president Geoffrey D. Garin has said. "The releases of polls become important points in the campaign. You do early media [advertising] to get early poll numbers that look good. And the back-and-forth [attacks between candidates] starts early."[30]

Senate campaign television ads used to begin running (or "go up," as they say in the trade) no earlier than the late summer or the early fall of the election year. Not any more. Senate campaign commercials are now quite commonly run in the spring—as, in 1988, in California, Rhode Island, Minnesota, Ohio, New Jersey, and Mississippi—and it is highly risky for the other side to let them go unanswered. Some Senate campaign commercials now commence even earlier. Octogenarian Democratic Senator Quentin Burdick of North Dakota, facing reelection in 1988, began running television commercials back home in the summer of 1987 that featured Burdick saying, "You and I know there is a young Congressman who would like my job. He can have it. Someday." The message was that it was all right for fellow Democrat U.S. Representative Byron Dorgan to aspire to be a senator, but that he should have the patience to wait until after Burdick could finish one more term. Dorgan decided not to run, and Burdick was reelected (though he was to die in office before his term was up).[31]

Polling consultants do far more today, though, than simply measure the standing of the candidates and what public opinion is on key issues. Pollsters have become major campaign strategists.[32] As one report puts it:

> Pollsters seek issues that capture public attention and should be stressed in campaign ads, speeches and debates. Polls also attempt to monitor daily shifts in public opinion, and the pollster suggests adjustments in campaign strategy to respond to such shifts. And pollsters analyze candidates' strengths and weaknesses and advise how appearances, money, advertising and staff ought to be deployed.[33]

The Role of Focus Groups

Polling consultants, then, look for so-called "hot button" issues that can be used in a campaign. To find them and to "flesh out the bare bones" of regular polling information, polling consultants, and the campaigns they work for, now regularly use the device of "focus groups."[34]

A focus group comprises about twenty selected likely voters who are each paid $25 to $50 to take part in a free-flowing, two-hour discussion about the candidates and issues in an upcoming campaign. Usually the pollster will attempt to get a cross section of voters, although the group may sometimes be overweighted in favor of swing voters, independents, or undecideds.*

The focus group session is frequently held in a room with a one-way mirror, so that campaign staff and consultants can watch and take notes. It is led by the pollster or a trained facilitator, often a group therapist.

A typical session begins with "mood" questions that seek to find out how people feel generally about the future of their country and state, and the pollster or facilitator may suggest that focus group members use weather forecasting terms, for example, like "stormy" or "cloudy," to encourage free assessments. Next may come questions on issues, to get at what the people in the group think are the most important public problems and why. Following that, members of the focus group may be asked to say what word comes to their mind when they hear various political names, which are then thrown out starting with the president and working on down, including both Democrats and Republicans and, of course, the candidate in the race that the pollster is really interested in (without revealing this).

Discussion may then be steered to the question of what sort of person members of the group think should be sought to fill the particular office involved—how, for example, they might write a want-ad for the office. The pollster, or leader, may next ask focus group members to react to key candidates and their personalities, with questions like: How would you describe —— to a person who is supposed to meet her in a hotel lobby? What actor would play the part of —— if they were making a movie about his life? What do you think —— would do if he got out of politics?†

Members of a focus group may also be asked to rate the candidates in respect to a list of specific personal qualities and in regard to arguments that might be used for and against them—arguments such as: He fights for the farmers. He opposes waste in the Pentagon. She stays real close to the people. He's weak on defense. He favors abortion. The pollster or facilitator may show the group certain television spots that have been aired or that are soon to be run, or a tape of a news interview, and ask for reactions.

When the focus group session is finished, the pollster will have

---

* On occasion, a particular focus group may be made up solely of certain types of target voters, such as blue-collar workers or independents.

† To the last question, involving one pollster's own candidate, a person in a focus group said, "He would probably become a Philadelphia lawyer."

gained some idea about what the subjective, emotional reactions of voters (or selected voters) can be expected to be toward the candidates and toward possible campaign themes.

It was in such a focus group that the 1988 Bush for President campaign discovered how powerful a negative emotional reaction there would be to the "Willie Horton" issue—the temporary furlough that Massachusetts Governor Michael Dukakis, the Democratic candidate for president, had approved for a state prisoner who had then committed additional crimes. This moved Bush campaign manager Lee Atwater to predict that, despite the fact that Bush was then trailing Dukakis badly in early polling, "If we can make Willie Horton a household name by election day, Bush will win." It worked.

Media Consultants and Tracking Polls

From the first day in a Senate campaign until the last, from the benchmark poll to focus groups to "tracking polls," the pollster works in tandem with the media consultant. Media consultants are the "big boys" of Senate campaigns in terms of costs, importance to the campaign, and the signal they give to journalists and to PACs and other contributors about the seriousness and credibility of a campaign.

Media consultants produce the television and radio ads and buy the air time, and they are paid retainer fees of between $75,000 and $150,000, plus a commission of 15 percent of the cost of the media buy. The 1990 Senate reelection effort of Tom Harkin of Iowa, for example, had a budget of $6 million, half of which was earmarked for television advertising; it was reported that the media consultant in that race was paid a total of $600,000.

The themes and content of the media advertising campaigns for a Senate race are taken by the media consultants from the polls and focus group results, and their effect is monitored, these days, by tracking polls that measure public opinion on a daily (or more correctly nightly) basis and look for trends. Tracking polls were first used in Senate campaigns in 1980. Four years later every competitive Senate race was using them, at least during the last month before the election. Tracking polls work this way: each night, quick telephone interviews are made with about 150 representative people. Rolling averages are kept for the latest three nights, the results of an earlier night being dropped when the new third night's results come in.* By means of tracking polls, campaigns can get a quick snapshot of opinion, and trends can be spotted almost at once.

---

* A campaign with money to spare may do tracking polls with as many as three hundred interviews a night. Democrat John D. Rockefeller IV spent several million dollars of his personal fortune in first getting elected to the Senate from West Virginia in 1984, and it is thought that his lavish campaign regularly paid for as many as 750 tracking-poll interviews a night!

Quickness is the reason that tracking polls and media advertising are used in conjunction with each other. "In the last month of a campaign," Democratic media consultant Joseph Rothstein has said, "both sides will poll nightly, test how that day's media and campaigning has played, and trace the results."[35] Rapid changes can be made. The campaign can react quickly, too, to the media advertising of the opposition. "Just having the media is not enough," Democratic pollster Harrison Hickman has said. "You have to know if your media is working, and if your opponent's media is working. Information is now a much more important commodity."[36]

Other Campaign Consultants

The media consultant and the pollster are the most important among Senate campaign consultants, but there are others. It is quite common in contemporary Senate campaigns for an "opposition research" consultant to be hired. These persons get the dope on an opponent—attendance, voting, tax, and credit information, as well as writings, speeches, and the like, things that can be used to negative advantage—at a cost to the campaign of between $10,000 and $20,000.[37] "What they do is follow someone's record," one Republican operative has said concerning opposition research. "They delve and dig and dig and dig in the damnedest places you've ever seen, where you wouldn't think to automatically look. . . . They will go back 20 years on a guy, and if you've lied on your resume, we're going to find out about it."[38] Opposition research thus provides the basis for a lot of the negative advertising that has become so standard in recent years.

Direct-mail advertising has long been a commonplace in lesser races in big states like California and New York. Specialized consulting firms have been growing more and more expert and effective in this means of persuasion—and attack. Now, increasingly, Senate campaigns are using direct-mail advertising, too, and thus direct-mail consultants.

More and more Senate campaigns hire national fund-raising consultants. There are two types. One kind specializes in the highly developed and well-worked field of direct-mail solicitations—renting lists, testing fractions of them, mailing heavily to lists that show greatest promise, and continuing with later, additional appeals to earlier responders. The other kind of national fund-raising consultant increasingly seen in Senate campaigns specializes in dealing with the thousands of PACs that contribute money on behalf of their special interest groups, companies, or firms. There are about two dozen such consultants in Washington.[39] They choose which PACs to solicit, ply them with poll results and news clippings, set up meetings with them for the candidate and advise the candidate on the correct approach, and invite PAC officials to fund-raising events.

Some Senate campaigns hire debate coaches, since candidate

debates have become so common. These consultants help prepare the candidates for handling various kinds of issues and questions and even advise concerning the lighting and physical aspects of debate staging. Very occasionally, a Senate candidate or campaign may hire a speech coach, as Democratic Senator Bill Bradley of New Jersey was reported to have done before his 1990 reelection campaign.

National pollsters, media consultants, and other Senate campaign consultants have helped to nationalize Senate campaigns.

Electronic Politics and Senate Campaigns

Senate races have also been changed by what is called "the new electronics politics"—computers, satellite feeds, fax machines, laser printers, and more (such as hand-held meters that can record the almost second-by-second reactions of individuals in pollster-selected groups to television ads, debates, and other appearances). For example, a campaign industry magazine has reported:

> By 1988, it was standard procedure in most campaigns to share data files between laptop and portable computers in the field and headquarter hard-disks. Campaigns transmitted speech texts and memos to candidates on the road, advance staff raced ahead with data disks to load remarks on tele-prompters, and press secretaries utilized portable printers to distribute speeches to the press well ahead of deadlines. . . . Personal computers also sorted and organized lists of supporters, press contacts, fundraisers and friends—while researchers arranged data on the opposition for instant retrieval, loading up an opponent's record into user-friendly databases.[40]

Senate (and other statewide) announcements of candidacy, which used to be made in the capital city and then followed up with repeat announcements in other state media markets as rapidly as the candidate could be flown from one to the other, can now, by satellite, be made in all the media markets simultaneously. For example, when Republican U.S. Representative Trent Lott of Mississippi opened his successful 1988 campaign for the U.S. Senate, his announcement garnered free news coverage in all parts of the state at once, live via satellite. As Lott's press secretary, Tom Bagwell, has reported: "We planned the announcement for 6:01-and-30-seconds p.m. We started with an empty podium and told the local [television station news announcers] that when Trent Lott and his wife went up, they could say 'Let's go live to Jackson'—and 70% did!"[41]

New technology provides today's Senate campaigns the capability for what is called "quick turnaround"—rapid execution of tactical and strategic campaign decisions, as well as rapid reaction to the other side's actions. After being appointed to fill a vacancy, Republican Senator David Karnes of Nebraska was defeated in a 1988 election bid by a more attractive and effective Democratic candidate, former Nebraska Governor Robert Kerrey. But Karnes's campaign profited from the quick turn-around afforded by technology, as, for example, when his consultants wanted to make the most out of a late endorsement of Karnes by the *Omaha World Herald*. Baltimore-based media man Adam Goodman reported how they did it:

> *12:00 Midnight*: We're tipped off about the endorsement. I alert our creative crew, persuade an announcer to cancel an appointment for the next morning, and set up a conference call for 7:30 a.m.

> *7:30 A.M. Thursday*: We begin hammering out our ad copy by phone, with the campaign director in Nebraska, general consultant Scott Cottington, and our creative people. Sign-off achieved over the phone!

> *9:00 A.M.*: Announcer tapes the spot.

> *9:30 A.M.*: We add the video, including available footage [tape of a fax of the newspaper editorial with electronically highlighted segments].

> *10:15 A.M.*: Finished, we dub it and race to the Baltimore ABC affiliate. We were lucky to get 15 minutes between 10:30 and 10:45 a.m. They uplink it, and their affiliate in Omaha downlinks it.

> *1:00 P.M., Noon in Omaha*: Our campaign workers take dubs from the ABC affiliate to the other Omaha stations—and in some cases because of their reading habits, people in Omaha actually saw our ad about the endorsement before they had read the newspaper![42]

The new technology, coupled with new techniques like tracking polls, now permits candidates to have "a conversation with the voters," according to Democratic media consultant Robert Squier. He has said: "Before it was like trench warfare in World War I. You'd get the guns ready, and three weeks before the election, you would fire off your stuff, and then on Election Night, you'd count the bodies. Today, you listen [to the electorate] and talk back and listen again and talk back."[43]

Negative Ads in Senate Campaigns

Today's new techniques and new technologies, and their use in combination, have produced something else not quite so benign: a startling, and highly worrisome, increase in the use of negative campaign advertising.

There have always been negative ads and attacks by one political candidate against another. But never to the degree that developed in the 1980s. In earlier times, most candidates worried that negative ads would boomerang and hurt the attacker; the general rule was that you tried to avoid even mentioning your opponent's name (and you even tried to ignore attacks). But in 1980, some congressional "fixtures" like Democratic Senator Gaylord Nelson of Wisconsin were the targets of rough television advertising attacks by the highly conservative National Conservative Political Action Committee (NCPAC)—and they lost. Conventional wisdom about negative advertising changed. Democratic consultant Jill Buckley has pointed out why: "People say they hate negative advertising. But it works. They hate it and remember it at the same time. The problem with positive is that you have to run it again and again and again to make it stick. With negative, the poll numbers will move in three or four days."[44]

Today's focus groups, polls, and opposition research can fine-tune negative themes that show promise of effectiveness, television commercials can be produced and "go up" with a speed that allows quick reaction to fast-breaking events or to an opponent's mistakes, and tracking polls can gauge the effectiveness of the ads, literally overnight, so that ads that are working can be continued and ineffective ads can immediately be replaced. "It's [a] clearly terrible . . . combination of the ancient art of negative campaigning with modern technology," according to Republican Senator John Danforth of Missouri.[45]

The typical Senate campaign ad of the 1970s, particularly a beginning ad, was positive, pleasant, and upbeat, showing the candidate in informal clothes, talking with a farmer, for example, or pushing a child in a swing. By contrast, the typical ad of the 1980s might be the now-famous thirty-second "bloodhound" commercial run by county executive Mitch McConnell, who was then a badly trailing 1984 Republican challenger against Kentucky's incumbent Democratic Senator Walter "Dee" Huddleston. This Roger Ailes ad has become a classic. In addition to hitting hard at reports that Huddleston had been making too many speeches around the country for money, instead of tending to Senate business, the spot also had the advantage of being funny. It showed a detective following a pack of dogs as they ran sniffing here and there in vain around the U.S. Capitol grounds and past swimming pools and other places, an announcer's voice saying at the end: "We can't find Dee. Maybe we ought to let him make speeches and switch to Mitch for Senator."[46]

An Ailes Communications associate explained why they decided to use the spot, and he noted how well it had worked for McConnell: "We simply had to go negative. We had to take some points off Huddleston very quickly. We kept racking our brains looking for the home run. We brought 'Bloodhounds' out, and it was like lighting a match on a pool of gasoline. It simply exploded."[47] McConnell moved up dramatically in the polls—and went on to a stunning victory. More and more Senate incumbents and challengers came to believe that they, too, had to "go negative" to win.

Senate races are like "little presidential races," in that techniques and tactics that are found to work in presidential races are sooner or later copied in Senate races (in recent years it has been sooner rather than later). As far back as 1964, President Lyndon Johnson's campaign used a now-famous and highly effective negative commercial that played on widespread fears that his Republican opponent, Senator Barry Goldwater of Arizona, was trigger-happy. The ad featured a sweet little girl's counting as she pulled petals from a flower suddenly turning into a countdown for, and then the blast of, a nuclear bomb explosion. There was considerable criticism of this ad in the media, and it was pulled off the air rather quickly, but not before it was clear that the ad had been effective.

Another major reason negative advertising has spread especially in Senate campaigns is that these campaigns, unlike most of those for the House, usually have sufficient money, and therefore the skilled consultants, to implement fully a negative campaign strategy. No recent Senate campaign was meaner than the 1988 Lautenberg–Dawkins race in New Jersey, which one local political scientist called "clearly the most nasty, dirty, vicious campaign that I've seen in the last 12 years."[48] Democrat Frank Lautenberg was the incumbent. Republicans had targeted him as beatable because incumbents are thought most vulnerable when facing their first reelection effort, as Lautenberg was, and because New Jersey voters seemed to have little clear impression of him. Lautenberg was a self-made millionaire who had founded a successful computer business before being elected to the Senate. There, in six years, he had built a solid, if somewhat colorless, reputation for hard work. His Republican opponent was Pete Dawkins, a former Heisman trophy winner and Rhodes scholar with a doctorate from Princeton, a retired Army brigadier general and decorated Vietnam veteran, and a successful Wall Street investment banker.[49]

Typical for today's Senate campaigns, the race featured big, big spending and the usual swarm of high-visibility consultants. Lautenberg's campaign manager, James Carville (who later managed Bill Clinton's 1992 presidential campaign), was paid $140,000, plus a bonus. Paul Maslin was the Lautenberg pollster (for $180,000), and the media consultant was Squier, Askew, Knapp (for total fees and commissions of $650,000). General consultant Roger Stone managed the Dawkins campaign for about $110,000. Alex Gage handled the Dawkins polling (for $80,000), and the Ailes firm and its associate, Greg Stevens, handled the media (for total fees and commissions of $770,000).

Dawkins's first television commercials, which started running in February of 1988, were biographical, image-building sixty-second spots that featured a sincere Dawkins, who had only recently moved to the state, saying that he had become a New Jerseyite by choice: "I moved around a lot. I lived in a lot of places. But I have to tell you that throughout all those years, in all those places, I never found a single place that had as good people or as much promise as I found right here in our Garden State." Afterward, Lautenberg's consultants found in their focus groups that people were actually "hooting down" the sincerity of Dawkins's attempt to answer the anticipated "carpetbagger" problem: "I mean, *I* like New Jersey, too, but who the hell does he think he's kidding?"

So, while Dawkins opened his fall campaign in September with a thirty-second version of his earlier, image-building spot, Lautenberg's campaign was launched with a tough television attack ad, the only time that anyone could remember an incumbent "going negative" to begin with. The ad both raised the carpetbagging question about Dawkins and questioned his sincerity and genuineness. It began with an announcer asking, "Why did Pete Dawkins move to New Jersey last year?" then cut to a clip of the Dawkins commercial itself and Dawkins's statement, and then ended with the announcer saying, "Come on, Pete. Be real!"

Pollsters are just as interested in gauging the "negatives" of their candidates and opponents—that is, the amount and intensity of opposition to them—as they are in the "positives"—the amount and intensity of support. Lautenberg's polls showed that Dawkins's negatives started to rise as soon as the "be real" negative ad appeared. A later, positive spot they ran that extolled Lautenberg's accomplishments did not produce nearly as good results as the negative ad.

Then Dawkins himself went negative, with an ad that made accusations about certain Lautenberg corporate stock transactions. Lautenberg countered with a negative spot that charged that an Army base Dawkins had once commanded had been guilty of ocean pollution and toxic chemical leaks. Dawkins came back with a highly effective anticrime ad that attacked Lautenberg for his opposition to the death penalty.

Both campaigns' polls showed Dawkins gaining, the gap between the two candidates narrowing. With eighteen days to go before the November election, Dawkins, perhaps responding to rising media criticism of the campaign's negativism, announced that he would stop his negative commercials: "He attacked me and I attacked him and he attacked me, and the result is the people of this state are confused, and they don't know where either of us stand on the issues." But Lautenberg stayed on the attack, running two more negative spots that questioned Dawkins's character and motivation.

Dawkins's funds began to run short in the last days of the campaign, but Lautenberg had hoarded his. Switching to the positive at last, Lautenberg outspent Dawkins at the end by a three-to-one margin, and on election day he defeated Dawkins by a 54- to 46-percent margin. Notably, however, fewer than half of New Jersey's eligible voters cast a ballot in that race.

Virtually everybody believes today that you have to fight fire with fire, that it is not enough simply to answer a negative attack—you must meet an attack with an attack of your own. Initially, when Republican Senator Danforth was attacked by tough negative ads during his reelection campaign of 1982, he resisted consultant advice that his campaign respond in kind. But after his opponent, Democrat Harriet Woods, pulled even with him in the polls, just two and a half weeks before the election, Danforth relented and "went negative," too. "You ask, who is in control?" Danforth said later. "Really, it's your pollster and your media person. You're polling constantly to gauge what effect your opponent's ads are having, and if he puts on something that works, you instantly air-express it to your media man, and you hope inside 24 hours, he's got an answering ad ready. It's the opposite of a debate on policy. It's all very quick—and dirty."[50]

Negative advertising has had another effect on Senate campaign strategy. Candidates, particularly Senate incumbents, now frequently start television spots a year ahead of time for the purpose of "inoculation"—that is, to immunize themselves, in a sense, against expected attacks. Republican Senator James Abnor of South Dakota expected that fellow Republican and state governor William Janklow would run against him in 1986 and attack him for being ineffective in the Senate. So Abnor started running television commercials in November of 1985 that featured testimonials from other Republican senators, saying what a great senator Abnor was. (Abnor was renominated all right, only to fall in the general election to an attractive young Democratic U.S. House member from South Dakota, Tom Daschle.) Now, as one campaign consultant has said, the conventional wisdom is that "Inoculation and pre-emption are what wins campaigns," or as another put it, "If you know what your negatives are, and you know where you're vulnerable, you can pre-empt it."[51] Inoculation is another reason why Senate campaigns are starting earlier and earlier.

Some political consultants and political scientists believe there is nothing wrong with negative ads, arguing that the press is an adequate watchdog against unfair charges or that even negative ads contain useful political information. Some people think it would be difficult to draft a law restricting negative campaign advertising in a way that did not violate the free speech and free press provisions of the First Amendment.[52]

Most observers, though, feel that something must be done about negative ads, thirty-second spots in particular. They are right. This type of advertising focuses voters' attention away from the real issues in a campaign. And there has been no letup in their use. Hal Bruno, ABC News political director, has said, "1990 politics was an open sewer—it's the rottenest campaign year I've ever covered. And I'm afraid 1992 is only going to be worse."[53]

Some see a connection between rising use of negative ads and declining voter turnout. There may be something to this, although the decline in American voting—though up to 55 percent turnout in the 1992 presidential election but still little over one-third turnout in recent congressional elections—has been too dramatic to be more than partially explained in this way. Rutgers political scientist Clifford Zukin thinks that the low turnout in the Lautenberg–Dawkins race was the result of "a very serious dysfunctional cynicism about politics."[54] Sociologist Herbert Gans of Columbia University also feels that what he calls "pernicious . . . scurrilous . . . increasingly demagogic use of commercial advertising" has contributed to the decline in American voting.[55]

There is also a belief in some quarters that the negative campaigns of recent times have produced a spiraling effect, that is, that they raise the level of public cynicism about politics and turn people off; as a result, ever more negative advertising is required to get the voters' attention back again. As Democratic pollster Paul Maslin has put it, "Unless you jolt them, you won't get results."[56] And negative advertising is what jolts them.

The rise in the use of negative ads has affected senatorial behavior. Democratic Majority Leader George Mitchell of Maine has said that it is not unusual to hear senators talking before a vote about what kind of use might be made of it against them in a negative ad. "The remark is frequently made: Watch out for this one, guys; this could really be made into an effective 30-second spot."[57]

Some Senate votes are even forced so that they can be used as fodder for future negative ads. This is not one senator or Senate candidate running against another, but an effort that is party or ideology driven. According to former Democratic Senator Paul Sarbanes of Maryland, "There is much more of a political agenda working in the Senate, now. Amendments are offered and brought to a vote for political embarrassment, not substantive intent."[58] Sarbanes said the offenders were senators from what he called "the far right." Democrat Bill Bradley spoke of the same group when he said that there are "five or six senators who poison the atmosphere" and who "choose amendments, not on their substantive merit, but solely for specific partisan use in thirty-second television ads; they already know in advance how they will frame the ad."[59]

Senators are aware, now, that any vote, no matter how early in a six-year term or how seemingly obscure, can come back to haunt an incumbent in a negative campaign ad in the next election. "Roughly half the senators on each side think much more about how any individual vote can be used against them than they ever have before," one polling consultant has noted.[60] A Republican media consultant agrees, saying, "People [in Congress] now immediately think not how it will play in Peoria, but how it will play in a TV commercial."[61]

Looking toward the 1990 congressional elections, *Washington Post* columnist David Broder urged that political journalists run regular critiques of campaign spots. Many throughout the country did, assessing the factual basis and fairness of negative ads in particular.[62] This may have had some effect, especially because some victims of attacks began to make use of the media critiques to respond to negative ads, as did Michigan's Democratic Senator Carl Levin, for example, to counter the negative ads of his unsuccessful Republican opponent, Representative Bill Schuette, that year.[63]

Still, legislative regulation of negative advertising is needed. Missouri's Senator Danforth co-authored, with Democratic Senator Ernest

F. Hollings of South Carolina and others, legislation to require that television-commercial attacks by one federal candidate on another be made by the candidate personally, not by an announcer or surrogate. "If a candidate wants to sling mud at his opponent, the public should be able to see the candidate's dirty hands," Danforth said in support of the measure.[64] The Danforth–Hollings proposal would also require radio and television stations to offer free air time for a response by a candidate who has been attacked in a negative ad sponsored by a PAC or other group independent of an opponent.[65] The Danforth–Hollings approach was largely incorporated in a campaign-finance reform bill that passed the Senate in 1990 but died in the House.[66] These restrictions on negative advertising, reintroduced in the Senate as the Clean Campaign Act of 1991,[67] should be enacted into law.

National Spotlight on Senate Campaigns

The national newsletters and reports that now service the PACs, journalists, and others concerning Senate races are another element in the nationalization of Senate elections. In the fall of 1987, for example, two campaign entrepreneurs, Republican Doug Bailey and Democrat Roger Craver, started the computer-based *Presidential Campaign Hotline,* which provided paid subscribers a wealth of digested and excerpted press reports, polling data, assessments, campaign press releases and statements, and other information about the upcoming presidential campaign on a daily basis. The *Hotline* founders were swamped by journalists, PAC officials, consultants, and assorted campaign junkies, all eager to pay whatever was required to get the service (by computer modem or in mimeographed form).

Soon, particularly to make the *Hotline* more attractive to the PACs and thus saleable year-round, not just on presidential campaign cycles, Bailey and Craver expanded its daily coverage to include Senate races, too. For example, the July 26, 1988, *Presidential Campaign Hotline* not only carried extensive information concerning the Dukakis–Bush race but also reported on polls, editorials and news stories, strategy, and campaign statements about "target" Senate races—in Minnesota, Tennessee, Texas, New Jersey, Nevada, Ohio, California, and Mississippi. That day, those interested in the Minnesota contest, for example, a race that pitted incumbent Republican David Durenberger against Democratic Attorney General Hubert "Skip" Humphrey, could read this statement from the ultimately victorious Durenberger campaign:

A MOR poll commissioned by the Durenberger committee revealed that Skip Humphrey's negatives rose a dramatic 10 points, from 19 to 29, after his spate of negative advertising (Durenberger's went up only 5, from 11 to 16). Durenberger's positives stayed at 73, while Skip's dropped from 61 in April to 52. Since his negative advertising backfired, Skip's only hope is

Dukakis' coattails. But he may prove too much of a drag for Dukakis to win in Minnesota either.[68]

Similarly, a number of other Washington political newsletters now also give their subscribers early, detailed, and continuous reports on Senate races. The weekly *GRC Cook Political Report* is one of the best; as early as February 1989, twenty-one months ahead of time, it was already providing detailed assessments of the 1990 Senate races and handicapping each one for its subscribers, rating them "Solid Dem," "Likely Dem," "Lean Dem," "Toss up," "Lean Rep," "Likely Rep," "Solid Rep."[69]

Pollster Paul Harstad, with Garin-Hart Strategic Research, has said concerning recent developments in campaigns:

> There is a new immediacy and demand for daily posturing, overnight spin control, and positioning of one's campaign for consumption of the PACs, print and television reporters, political insiders, and the conventional wisdom. There is far more visibility, now, more scrutiny, more creation and manipulation of the expectations. It almost makes Senate races mini presidential campaigns. The net result of all this—early starts, more negative ads, quicker turn around, daily spin control, more money, more visibility— is that campaigns are harder to control. Anything can happen, mistakes are more visible, there's more toe to toe, and incumbents get roughed up early.[70]

Partisan Nationalization of Senate Races

The nationalization of Senate campaigns also results from their increasing nationally partisan nature and the greatly heightened activities of the Senate Republican and Democratic campaign committees.

Partisan changes in the electorate have fostered greater partisanship in Senate elections.* The South changed. Party realignment there, with many white Southerners switching their affiliation from Democratic to Republican, first appeared with the national Democratic Party's 1948 adoption of a strong civil rights plank. It spread like Johnson grass after the civil rights conflicts of the 1960s.[71] At the same time—and related to the white Southern conversion—there was a huge increase in the numbers of Southern black voters, as a result of the Voting Rights Act of 1965, and these black voters were, overwhelmingly, Democrats.

There were national changes, too. Party identification became more closely related to income: more Democrats at the lower levels, more Republicans at the higher levels.[72] Ideological differences between Democratic identifiers and Republican identifiers grew, as Republicans became more conservative, Democrats more liberal.[73] These differences were most pronounced among party elites, the kind of people most

---

* More on this in Chapter 6.

likely to be active in Senate and other nominating contests.[74] Each party's identifiers (the "party in the electorate"), as well as its elites, then, became more internally homogeneous at the same time that they became less like those of the other party.[75]

These changes in the electorate, both in the South and nationally, produced more homogeneity and cohesion within each of the two Senate parties,[76] more partisan conflict between them,[77] and more nationally partisan Senate elections.[78]

Today, as one study has shown, "Both parties have an equal chance of winning any seat in the Senate."[79] Similarly, political scientist Alan Abramowitz has found that "two-party competition has spread to every region of the country," and "no state can be considered safe for either party."[80] Elevated party competitiveness for Senate seats is evidenced, for example, by the fact that since 1968 close to half the states, an average of 45 percent, are represented by "mixed" Senate delegations—that is, a Democratic senator and a Republican senator.[81] There were twenty-six such mixed delegations in 1979, twenty-one in 1993.

The two internal Senate political party campaign committees—now called the National Republican Senatorial Committee (NRSC) and the Democratic Senatorial Campaign Committee (DSCC)—were first established in 1916, following ratification of the Seventeenth Amendment, which provided for the direct election of senators. But it is only in recent years that these campaign committees have become such big and important operations.

Each Senate party campaign committee raises money for, and otherwise assists with, the election and reelection of its own partisans. Both now have their own national headquarters and a large staff—in 1988, 45 for the DSCC and 101 for the NRSC.[82] Each works at recruiting candidates, running campaign schools, coordinating and providing campaign consultants and services, and assisting with PAC fund-raising. The DSCC raised a total of $17.5 million in 1989–1990, the NRSC $65 million.[83]

The Democratic committee now regularly has enough money to give all of its candidates with competitive races the maximum contribution allowable for such committees by law, $17,500. The Republican committee has enough to give every one of its incumbent senators seeking reelection and all of its competitive challengers the maximum. In addition, during each election cycle, each of the committees now makes permitted "coordinated expenditures" on behalf of its candidates, that is, direct expenditures for advertising, polling, and fund-raising for them, for example—ranging from a little over $100,000 in the least populous states to around $2 million in the large ones.[84]

The campaign committees showcase their candidates before PAC groups and other prospective contributors. In the last months of an election cycle, they each sponsor weekly meetings in Washington for PAC officials, journalists, and other interested observers. At these meetings committee staff members, campaign consultants, and sometimes the candidates themselves appear and report on the status of the campaigns.

One campaign cycle barely ends before Senate campaign activities for the next one begin. Looking toward the 1990 elections, for example, the Democratic Senatorial Campaign Committee held an intensive and highly specialized three-day "Incumbents Campaign Planning Weekend" in March of 1989. It featured, for example, five consultants each reporting on the successful 1988 campaigns of Frank Lautenberg of New Jersey and Howard Metzenbaum of Ohio; a detailed analysis of the losing 1988 campaign of Senator John Melcher of Montana; other presentations by experts on such topics as "Nurturing the Involvement of Major Political Donors," "Running for the Senate in the Age of Atwater," and "Discouraging Potential Opponents"; and technical workshops on everything from opposition research, to polling, to media, to field organization. By June of 1989, the Democratic campaign committee had already sponsored a two-day "DSCC Vendor/Consultant Forum" for consultants and Senate-campaign staff members.[85]

Frequently, now, the party campaign committees in the Senate, and sometimes the national committees of the two parties as well, make the release of funds and other help for a Senate candidate contingent on that candidate's agreement to hire certain recognized consultants. Political scientist Gary Jacobson has described this practice and its impact this way:

> On occasion, party officials have insisted that candidates hire specific campaign consultants as a condition of receiving the national party's full range of assistance. In this way, the party influences [helps to nationalize] campaign strategies and messages even if party officials do not consciously try to impose a uniform approach. Schools for candidates have a similar effect, as does the party's reliance on a common pool of outside campaign professionals.[86]

Heightened Senatorial Insecurity

As American society, and Senate elections as well, have become nationalized, senators have come to feel much more electorally insecure.

Especially because of the greater prestige and power of the Senate office and its longer term, senators up for reelection, far more than House members, can expect to face quality challengers, the kind of candidates that can get money and media attention, and thus the kind that might beat them.[87]

Senators are also unlike House members, as two authorities have pointed out, in that "senators, after all, are national figures, identified with the making of national policy," while representatives are more associated with "district matters and constituency services."[88] Voters are more likely to hold senators individually responsible for national policy, and this makes them more vulnerable to attack for their views and actions.[89]

Senate campaigns, unlike most House campaigns, fit whole major media markets—a daily newspaper's circulation area, a television station's broadcast area. This makes today's mass advertising in Senate races much more cost-effective and feasible—and more necessary and costly, as well. Senators and Senate challengers are generally considered by the news media to be newsworthy, and their campaigns are usually well covered. The result is that in Senate races, today, "well-financed opponents can approach the incumbent's recognition levels"[90] and make a race of it virtually anywhere.

The point of the late Senator Russell's humorous story about his being able to tell which part of their terms senators were in by watching how they walked was well taken, back then, in the sixties: only the senators with two years to go before election looked down as they walked. Only they, in other words, were immediately concerned with reelection. Things have changed. Nowadays, senators feel that they must figuratively walk looking down all the time.

Political scientist Richard Fenno, studying Senate elections during the period 1976 to 1980, found that there was still some of that old electoral-cycle way of thinking, particularly among the veteran members of the Senate. He quoted a senator with eighteen years of service as saying, "We say in the Senate that we spend four years as a statesman and two years as a politician. You should get cracking as soon as the last two years open up."[91] But Fenno found indications that that kind of thinking had changed for the newer members of the Senate, and he noted Robert Peabody's conclusion that:

> Few Senators can afford the proverbial luxury of serving as statesmen for four years, then reverting to the political role for the remaining two years of the term. Many of the younger group of senators appear to be following the practice of running hard through most of the six-year term, not unlike the experience of the House incumbents.[92]

And well they might. As a recent study concludes, "it seems clear that for at least the last 35 years incumbent senators have become more electorally vulnerable than incumbent representatives."[93] The reelection rate for incumbent senators is not as high as that for House members, a considerably lower percentage of Senate seats are considered "safe seats," and the margins of reelection victory for incumbent senators is much narrower.[94] In the twenty-four years from 1960 through 1984, 70 percent of House incumbents won reelection by 60 percent or better of the general election vote, whereas only 44 percent of Senate incumbents did.[95]

What some called "the high mortality rate" of senators at the polls in the 1970s gave Senate incumbents a bad scare. Senator Fulbright was defeated for reelection in 1974, and of the one hundred senators who answered the next year's first roll call in January of 1975, almost a third—thirty senators—were defeated within the following six years. Senators learned that they had to be "politicians" year-round.

When Donald Matthews described the norms, or unwritten rules, of senatorial behavior in the Senate of the 1950s—apprenticeship, legislative work, specialization, courtesy, institutional patriotism, and reciprocity—he noted even then that these norms were less likely to be observed by senators who came from a "two-party, or a large and complex, constituency."[96] Increasingly, following the 1950s and 1960s, that description began to apply to *most* members of the Senate. Matthews explained the reason why such senators were less likely to conform to Senate norms: "The political insecurity of a senator of this kind of state is likely to result in a shortened time perspective, an eagerness to build a record quickly, an impatience with the slowness of the seniority system. . . . A senator whose seat is in grave danger is much more likely to 'rush it' than one who can count on re-election unless he makes a major blunder."[97]

There has been a nationalization of American society since the 1950s, then—in rapid communications and transportation; greatly increased mobility and urbanization of the people; a higher general standard of living and educational level; an enormous growth in the national government; and a great increase in media outlet numbers and the intensity of their activities.

Senatorial campaigns have become nationalized, too—"little presidential campaigns," with national consultants, strategies, techniques, partisan sponsorship, and greatly increased national attention.

As a result of these external changes—in society itself and in Senate campaigns—senators have become more national in their outlook at the same time as they have become more accessible and subject to public scrutiny, more politically exposed and limited, and more electorally vulnerable and insecure. All this has affected behavior in the Senate, causing senators to be less willing to follow the old Senate norms and, in fact, changing the norms that had facilitated Senate efficiency in legislating.

Entwined with these external-change causes and their senatorial effects was another: the rise in the numbers, influence, and activities of American interest groups.

3

The Nationalization of Issues, Interests, and Campaign Financing

In late September 1991, Senator Robert Kerrey of Nebraska stood on a platform in front of the capitol building in Des Moines, Iowa, and declared to an enthusiastic home-state crowd that he would be a candidate for the 1992 Democratic nomination for president of the United States. Kerrey had by then served in the U.S. Senate less than three years. Yet few observers were surprised by his announcement.

Even back at the start of Kerrey's 1988 Senate campaign, a visitor could talk to ten people in Nebraska and at least five of them would volunteer that they thought Kerrey would one day be president of the United States (though, as it turned out, 1992 was not to be his year). Kerrey had been a storybook candidate for the Senate, and he had run a storybook campaign.[1] The first thing he did after announcing his Senate candidacy was to hire a campaign manager who knew how to raise money.[2] It was a wise decision.

There was no doubt that Kerrey, then forty-five, was a politician with "star quality."[3] He was a Congressional Medal of Honor winner who denounced the Vietnam War as immoral, a marathon runner despite having lost a leg in Vietnam, a successful restaurant owner who had turned to politics and beat an incumbent to become governor, an effective governor whose romance with Hollywood actress Debra Winger captivated press and public alike, an eminent politician who renounced a second term in the statehouse despite a 70-percent approval rating, and a Democrat in a state where Republicans have the registration edge. He was elected to the U.S. Senate in 1988 with 57 percent of the vote.[4]

In March 1987, Nebraska's governor had appointed his fellow Republican David Karnes to fill a Senate vacancy created by the death of Democrat Edward Zorinsky. Kerrey declared his candidacy for the Senate seat in November of that same year and, in December, hired Paul Johnson to manage his campaign. Johnson, who had just run Tom Daschle's winning 1986 U.S. Senate campaign in South Dakota, went to work at once to establish a campaign budget and a detailed plan for raising the money it required.

Johnson set $3.1 million as the amount that would be needed to beat an incumbent senator in a state like Nebraska. That figure was three times what any statewide candidate had ever spent in Nebraska before. It was the equivalent of nearly $2 each for every man, woman, and child in the state. The rival Karnes organization separately decided upon its own, parallel, $3-million budget. Thus a total of more than $6 million was budgeted for the two campaigns in a state with a little over 1.5 million people. Until then, most Nebraskans would probably have thought that politicians would need to drop cash out of helicopters to get rid of that kind of money. But before the 1988 Senate race was over, both the Karnes campaign and the Kerrey campaign had actually raised and spent those enormous sums. In fact, the Kerrey campaign exceeded its budget target and ultimately was able to harvest a total of $3.31 million in contributions. But doing so was not easy. Nor was it accidental.

Typically, over half the Kerrey budget was set aside for media and media consultants (Rothstein and Company of Washington, D.C., for television and radio; a South Dakota firm for newspaper ads). Most of the media money was, of course, earmarked for television advertising (leaving $200,000 for radio spots and only $50,000 for newspaper ads). Twenty-three separate thirty-second television spots were used during the campaign, and a half-hour television profile of Kerrey was run twice, statewide. About $500,000 was budgeted for a highly paid campaign staff and for polling services, field offices, get-out-the-vote efforts, and related campaign costs. Another nearly $200,000 was apportioned for printing, computers, postage, travel, and lodging.

But campaign manager Johnson knew that you have to spend money to raise money, and from the first he budgeted also, and generously, for that purpose—$650,000 in all, about 21 percent of the amount he hoped to raise.

Johnson divided fund-raising into in-state and out-of-state categories. His plan was to raise a little over $1.2 million of the originally projected $3.1 million Kerrey campaign budget at home, in Nebraska, and he hired five full-time staff people to supervise this effort. He budgeted nearly $400,000 for in-state fund-raising costs: $130,000 for Nebraska telemarketing (where hired callers would work special telephone lists of prospects), projected to raise $250,000; $100,000 as the cost of raising $550,000 from Nebraska large givers ($100 or more) through personal calls by Kerrey himself and from fund-raising parties; $140,000 to be spent to raise $350,000 at Nebraska small-donor functions; and $28,000 as the cost of raising $110,000 through Nebraska direct-mail solicitations. All these targets, except those for small donors in Omaha and Lincoln, were eventually met or surpassed.[5]

Senate campaigns, today, raise more than half their funds out of state.[6] That was true in Nebraska in 1988, too, as Johnson, a campaign veteran, fully anticipated. He earmarked $250,000 of the Kerrey budget as the cost of raising this national money.

Johnson expected half the national funds, in turn, to come from Washington, D.C., and New York, mostly from political action committees (PACs), the campaign-financing arms of interest groups, corporations, and their lobbyists. He hired a nationally known fund-raising firm to work the lists of Washington and New York contributors and PACs (with labor, business and corporate, progressive-issue, and agricultural PACs specially targeted) and to arrange Washington and New York receptions, meetings, and telephone calls for Kerrey. To supervise efforts that would raise the rest of the out-of-state money, from donors outside Washington and New York, Johnson hired two full-time, experienced national fund-raisers—one from the Daschle campaign, the other from an earlier campaign of former Colorado Senator Gary Hart—and set them up in a special Omaha office. Both national efforts would pay off, as it turned out; in fact they would exceed expectations.

Raising money is a central focus of today's Senate campaigns, and in Nebraska in 1988 it was the "ultimate priority and the most dominant activity in candidate Kerrey's day," according to the campaign press secretary (later *Senator* Kerrey's press secretary), Steven J. Jarding:[7]

> Kerrey would typically spend on average four to six hours per day on the phone soliciting funds from individuals who were considered major donors. Virtually all Kerrey's public appearances were set up to coincide with fund-raising opportunities. . . . In any given week, Kerrey would likely spend 30–40 hours on the phone raising money and would attend 10–15 fund-raising events in his honor.[8]

Kerrey was a dynamic and immensely attractive candidate, almost a celebrity. He was winningly unconventional and forthright: when his opponent charged that he was a liberal who would vote down the line with the Democrats in the Senate, Kerrey answered, "So what?"[9] He ran the kind of highly professional and nationalized Senate campaign that is typical today. Opponent Karnes inadvertently helped Kerrey, in farm-state Nebraska, when he made a locally booed and later highly publicized statement that "We need fewer farmers at this point in time."[10] Kerrey was elected by a 100,000-vote margin, while the Democratic presidential candidate Michael Dukakis was losing Nebraska to Republican George Bush by 130,000 votes. Typically, Kerrey could not have done it without big, big money, a lot of it with invisible strings attached.

A National Advocacy Explosion

Campaign financing is one important aspect of the growth in influence of America's interest groups. There are others. Since 1960, America—and its national capital—has experienced an "advocacy explosion," as one expert observer has put it.[11] Senators today must deal with vastly more Washington-based interest groups (organizations that seek to influence public policy) and business corporations, as well as the lobbyists for both, than their counterparts of the 1950s ever imagined. There were two aspects of this explosion: the number of groups grew enormously; and so did the range, volume, and intensity of their activities.

Many, many new groups were formed to influence national policy. Of Washington-based organizations that a 1981 Schlozman–Tierney study looked at, 40 percent had been founded after 1960, 25 percent after 1970.[12] There was also a "massive immigration" to Washington of existing organizations. For example, of groups in the Schlozman–Tierney study, the estimate was that "about seventy percent had opened their Washington offices after 1960, slightly under half after 1970."[13]

Fewer than twelve hundred trade associations—organizations like the National Funeral Directors Association and the National Independent Retail Jewelers, for example—had offices in Washington in 1960. By 1986 that number had tripled, to 3,500, with a combined Washington work force of about 80,000 people.[14] Many business corporations opened new offices in Washington; one directory showed thirteen hundred Washington-based corporation offices in 1986, compared with earlier lists of only one hundred in 1968 and five hundred in 1978.[15] The existing Washington offices of many business corporations were greatly expanded. For example, in the ten years after 1968, the Washington office of General Motors grew from three staff people to twenty-eight.[16]

A mere 365 lobbyists formally registered with Congress in 1961.[17] By 1989, even with fairly loose registration requirements (only those who collect or receive money for the "principal purpose" of lobbying members of Congress, by personal and direct contact with them, have to register[18]), the number of active lobbyists registered with the Secretary of the Senate had swollen to over 5,600—a figure that comes out, of course, to a registered lobbyist-to-senator ratio of better than 56 to 1.[19]

But that is far from the whole story. The fact is that there are many more lobbyists than those who actually register. Over twelve thousand people were listed in the privately published 1989 *Washington Representatives* directory—a directory of "persons working to influence government policies and actions to advance their own or their clients' interests."[20] Just twelve years earlier, the same publication's listings had totaled only a third as many.[21]

A large percentage of lawyers licensed to practice in the Washington area are actually lobbyists. From 1961 to 1989, the number of Washington-area attorneys quadrupled, increasing from slightly fewer than 13,000 to nearly 51,000 in those twenty-eight years.[22] A large number of out-of-town law firms opened Washington offices, and a 1983 study showed that two-thirds of these Washington branches had opened in just the preceding ten years.[23]

A great increase in lobbying by governments—domestic and foreign—was another aspect of the advocacy explosion in Washington. Domestic-government organizations, like the National League of Cities, the National Governors Association, and the National Conference of State Legislatures, maintain a Washington presence. More than two-thirds of the state governments now also have formal offices in Washington. So, too, do local-government agencies like the Kansas City Board of Public Utilities, for example, and a large number of professional organizations of state and local government officials, from highway engineers to county welfare directors.[24]

The "tremendous growth in foreign lobbying," by businesses and governments, as an expert Congress watcher, Norman Ornstein, has said, "is one of the most significant changes that has taken place in the past ten years."[25] The numbers of registered agents of foreign governments in Washington has swollen to over one thousand.[26] Ornstein attributes the growth in their numbers and activities to the increased assertiveness of Congress in foreign policy and the ease of access that foreign lobbyists have to the congressional policymaking process. In 1984, Japan, for example, spent $14 million on Washington lobbying, particularly to sidetrack additional restrictions on Japanese imports into the United States.[27] In 1989, one Washington lobbying firm, Black Manafort Stone & Kelly, was being paid a total of $5 million a year by foreign clients—including $950,000 from a Philippine political party, the Union for National Action; $500,000 from the government of Kenya; $1 million from Zaire; $600,000 from Jonas Savimba and the UNITA rebels in Angola; and lesser amounts from others, including Somalia and Peru.[28] Foreign embassies in the United States—those of many Latin American countries, for example—have begun to operate like "live-wire lobbyist[s] for a domestic interest group."[29]

"Think tank" is a fairly recent Americanism for a nonprofit public policy research or study center. The "explosion in the number of think tanks," as one expert has put it,[30] as well as their significant influence on policy, is fairly recent, too. There were an estimated one hundred of them in Washington in 1991 (not to mention important think tanks outside Washington like the Hoover Institution and the Rand Corporation), at least two-thirds of them established since 1970.[31] Some, including the Economic Policy Institute and the Center for National Policy Studies, have a liberal outlook. The old-line Brookings Institution, once thought liberal, still is in vigorous operation but has moved to the center. The newer American Enterprise Institute has become an influential conservative group. The greatest growth in recent times has come in "advocacy tanks" that are "unabashedly partisan and ideological." The most influential of these, and one of the largest of all, is the far-right Heritage Foundation.[32] But even some of the older think tanks, including Brookings, have taken to marketing their ideas, blurring the lines between research and advocacy.[33]

The Causes of the National Advocacy Explosion

Foremost among the causes of the national advocacy explosion were the same societal changes—the nationalization of American society—detailed in Chapter 2, especially those involving rapid mass communications, higher levels of education and standard of living, new social protest movements, and the expansion of the national government. For example, a 1986 Congressional Research Service study of the growth in the numbers and activities of American interest groups found that America's "communications revolution," by shortening distances and compressing time, had facilitated the "rapid transfer of information between group leaders and followers" and made "rapid mobilization" for lobbying possible. Television especially, the study said, had spawned new interest groups, not only because television immediately and directly informs people about events—about the fighting and turmoil of the Vietnam War, for example—but also because it "intensifies the feelings people have" about such events.[34]

The Congressional Research Service study also noted that rising levels of education and standard of living had given an increasing number of Americans a kind of "luxury of choice about ideas and values," had stimulated them to create and join politically oriented groups, and had "enlarged the numbers of potential group organizers."[35] The study quoted approvingly from an earlier Cigler–Loomis book that pointed to the 1960s as the beginning of a special "period of opportunity" for policy organizers, or policy "entrepreneurs" as the book called them, at a time when "idea-orientation" was resulting from skyrocketing college enrollments, as well as from new and "powerful forces such as civil rights and the antiwar movement."[36]

The types of interest groups became much more diverse; many new issue-oriented, ideological, consumer, and citizen-activist groups formed.[37] The numbers of these organizations grew much faster than those of the more traditional economic and occupation-based pressure groups, like trade and professional associations, business and industry groups, and labor unions. Still, business interest groups continued to be the "best represented sector in American society," bolstered, as interest-group expert Jeffrey M. Berry has noted, by "ample resources and a fierce determination to maintain its advantages in Washington," while the strength of "its traditional rival, organized labor, is on the decline."[38]

Barbara Sinclair has detailed the growth in social protest movements, starting with the civil rights movement of the late 1950s and early 1960s, which, she has written, not only was instrumental in "thrusting black civil rights to the center of the agenda," but "also served as an exemplar for the other social movements of the 1960s." According to Sinclair, these other movements, including the Chicano movement, the women's movement, the handicapped rights movement, the gay rights movement, and the movements against the Vietnam War and for the protection of the environment and consumer interests, owed a great deal, in some cases their very existence, to the civil rights movement and the "climate of political mobilization" that it had done so much to create. Many of these social movements "depended also upon the post–World War II growth of a large affluent middle class receptive to quality of life issues and other noneconomic appeals," Sinclair has concluded.[39]

The role of the federal government in the economic and social life of the country expanded greatly, and this contributed to the advocacy explosion (and partly resulted from it as well, as we noted in Chapter 2). Protest movements became organized pressure groups to push for new programs—as militant handicapped people banded together, for example, to lobby for the enactment of the Architectural Barriers Act of 1968 and the Education of the Handicapped Act of 1970—and then continued in existence to protect and enlarge these programs.[40]

New government programs spawned their own supportive new lobby groups and thus generated increased lobbying. The Community Action Program of the federal government's antipoverty efforts, for example, gave rise to the National Association of Community Action Directors. The government's B-1 bomber program stimulated increased lobbying activity in favor of it by Rockwell International and its subcontractors on the project.[41]

Often the increased actions of an expanded federal government in turn stimulated increased counteractivity, especially by business corporations and groups. As one major-company lobbyist has said:

> More and more groups and companies have recognized the increasing size of government and have therefore stepped up their involvement. Economically, the government is much more important these days than it was ten years ago. Great Society legislation and the environmental and consumer laws have all combined to make the companies feel the need to be more active in Washington.[42]

The federal government began actually to subsidize participation in policymaking in some cases—as for example when the Federal Trade Commission and the Environmental Protection Agency started to reimburse groups for taking part in their rule-making proceedings.[43] Government agencies, as patrons, actually began to sponsor interest groups and even to help found some in the "mixed, nonprofit, and citizen sectors" that were "built around professional specialties in areas like health care, education, welfare administration, mass transportation, scientific research, and other program areas that depend heavily on federal funds"; private charitable foundations did this, too.[44]

Interest-group expert Jack L. Walker summed up the various causes of the advocacy explosion in America, then, when he wrote:

> Long-term improvements in educational levels provided a large pool of potential recruits for citizen movements; the development of cheap, sophisticated methods of communication . . . allowed leaders in Washington to reach members in all parts of the country; a period of social protest that began with the civil rights demonstrations of the early 1960s called many established practices into question . . . and provided a powerful impetus for change. Once these mutually reinforcing factors led to the creation of massive new government programs, . . . newly created government agencies and foundations began to foster voluntary associations among the service providers and consumers of the new programs. During this period, the new regulatory legislation in civil rights, consumer protection, environmental preservation, pollution control, and occupational health and safety prompted business groups to organize in self-defense. . . .[45]

Not just greatly increased interest-group numbers, but also an enormous growth in the range, intensity, and volume of their activities have been a part of the advocacy explosion. In the modern high-tech age, lobbyists, and the specialized consultants they hire, take advantage of phone banks and direct-mail techniques to find and stir up grassroots sentiment in their favor, using computers to target the phone-bank calls and the direct-mail letters. "Instead of springing up spontaneously in the heartland," the *National Journal* has said, "many of today's grassroots communications are synthesized by Washington-based consultants."[46]

Lobbying consultants for some groups file the names of grassroots supporters in a Western Union computer "bank" which can spew them out almost instantaneously with messages for members of Congress when key votes come up. The National Education Association gets its members to agree in advance to annual "proxies" that are stored in computers. When an issue in which this organization is interested comes up in Congress, the proxy names are retrieved and affixed to about two-hundred thousand postcards, which are then flooded into congressional offices. It took eighteen months for the office of Democratic Senator Alan J. Dixon of Illinois to answer 669,000 postcards he received as a result of a similar, and successful, 1983 grassroots campaign by the banking industry for the repeal of tax withholding on interest and dividends.[47]

Some lobbying firms specialize in manufacturing, as well as mobilizing, grassroots organizations. Bonner and Associates is one of these. A Bonner advertisement offered the following services to prospective clients:

- Personally educating and recruiting the *precise number of on-record* constituent supporters (10–10,000 per target legislator) you feel you need to win.
- Enlisting constituents who have *no direct vested interest* but who strongly and actively support your position.
- *Organizing local coalitions* of groups which politically count with their legislators to show your position has potent broad-based support.
- Arranging *face-to-face meetings* between constituent supporters and legislators which *prove support is deeply felt* and goes far beyond any slick mail campaign.
- Generating on *very short notice (24 hours)* effective grassroots contacts for critical amendments and floor votes.[48]

Increasing numbers of lobbying firms, some of which call themselves "business-political consultants" or "communications and public affairs" groups, have broadened into "full-service" lobbying organizations. They organize grassroots efforts themselves, rather than going through specialty subcontractors (like Bonner and Associates). In doing this, they sometimes set up new umbrella fronts, like the Alliance for Energy Security, a front for natural gas producers, or Consumers United for Rail Equity, a front for a group of shippers that sought limits on freight charges.[49]

Senatorial Effects of the Advocacy Explosion

The issues and questions pressing policymakers, including senators, for attention and action—the national public agenda—grew enormously as a result of the advocacy explosion that took place in America following the 1950s. Starting with the 1960s, there were many more big national issues and groups looking for senators to serve as their national spokespersons and advocates than there had ever been before—many more causes and groups than there were senators, in fact.

"More issues and more groups result in more views striving for representation and consequently more leadership slots," Barbara Sinclair has written.[50] There was an enormous increase in the number of opportunities that an individual senator, not just a committee chair or a party leader, might have to star. The new opportunities were attractive to senators, no matter what their particular mix of personal goals or motivations—an orientation toward policymaking, influence in the Senate, or reelection to office; the desire for national prominence; or aspirations to be president.[51]

Individual senators began to find that strict conformity to some of the old Senate norms was often less rewarding than *not* conforming to them, especially those of apprenticeship (being seen and not heard for a time, deferring to those more senior); a very narrow specialization on issues, legislative work to the extent that it meant staying away from publicity; and reciprocity, in the sense that it meant not putting one's whole heart into a cause. Senators began to press for more real participation in Senate operations and decision-making.

The advocacy explosion amounted to a kind of a nationalization of American issues and interest groups. This, together with, and as a part of, the great societal and electoral changes detailed in Chapter 2, themselves constituting a nationalization of American society and of Senate elections, worked together in a major way to help cause the changes and "reforms" that took place in the Senate in the 1960s and 1970s. The Senate became a more nationally oriented and more open, democratic, and individualistic place.

Coming full circle, interestingly, these Senate changes and reforms themselves gave added impetus to the advocacy explosion. Before the changes in Senate rules and norms, Senate power was more concentrated in the committee chairs and party leaders. Afterward, power became more decentralized and fragmented, in subcommittee chairs and individual senators. Staff members became both more numerous and more important. Previously closed committee and subcommittee "mark-up" sessions, where important decisions were made about legislative measures, were opened up to the public—and to the lobbyists.

America's constitutional system, and within it Congress, was always a fragmented and decentralized system, of course, with many points at which pressure could be exerted effectively. Still, before the Senate changes and reforms of the 1960s and 1970s, one lobbyist has said, "If you knew the chairmen, that's all you had to know."[52] Afterward there were many more access points in the Senate (and to a somewhat lesser extent in the House, too). There were more bases to cover. Interest groups had to "escalate the range and volume of their activities,"[53] and they did.

The Explosion in Campaign Costs

The influence of national interest groups and lobbyists increased greatly, not only because of their greater numbers and the enlarged scope of their activities, but also because of the enhanced intensity of what they did. This was especially true about their involvement in the financing of Senate campaigns, one way in which money is translated into political power in America. The enormous sums spent in the 1988 Kerrey-Karnes Nebraska Senate contest—$3.46 million by Kerrey, $3.41 million by Karnes[54]—show the exponential increase in campaign spending that had taken place over the years preceding this contest.

In American elections through 1964, campaign costs stayed pretty close to the rate of inflation.[55] Then came the intertwined effects of the advocacy explosion, the communications revolution (particularly the advent of television), and the nationalization of Senate elections. With each two-year election cycle, Senate campaign costs shot upward. Total (primary and general election) expenditures for all U.S. Senate candidates who made it to the general election went from $28.8 million in 1974 to $189.7 million in 1986 (not counting independent and party-committee expenditures on their behalf).[56] This was an increase of nearly 660 percent, virtually three times the rate of inflation during that period.[57]

Winning Senate campaigns in 1988 spent 18.5 percent more money than winning Senate campaigns in 1986.[58] The average Senate victor in 1988 spent $3.9 million. As one reporter put it, this "comes to $1800 a day, seven days a week, for the entire six years of a senator's term."[59] The average Senate victor in 1990 spent only slightly less—$3.7 million.[60] Appointed in 1991 to fill a Senate vacancy when Republican Senator Pete Wilson was elected California's governor, Anaheim businessman John F. Seymour immediately launched a drive to raise at least $15 million in seventeen months in order to try—unsuccessfully, as it turned out—to hold on to the job in the 1992 elections.[61]

Fund-raising has become almost a year-round activity and concern for incumbent senators. Speaking on the eve of his 1986 retirement after eighteen years in the Senate, Senator Tom Eagleton declared that the cost of a Missouri Senate race had gone from around $500,000 to about $4 million in the years since 1968, when he had first run, and then added:

> Back when I came, fund-raising was basically a one-year process. Now, had I been running for reelection, I would have been fund-raising, at the very minimum, for a three-year period, maybe a four-year period. And I know some illustrations where it's six years, where senators have started fund-raising within a few months of their being sworn in.[62]

The increased time and effort required for fund-raising by incumbent senators "detracts from legislative work" and "cuts into your time to actually meet with constituents or tour your state," one of the Senate's younger Democratic members has said. Other senators say much the same thing.[63]

The increasing time and effort required to raise money also add to senators' growing feelings of political vulnerability and electoral insecurity (more evident in the Senate than in the House). Senate campaigns are, on average, eight or nine times as expensive as House campaigns.[64] Senate candidates usually have to reach much larger constituencies, in both population and area, than do House candidates. Television being an efficient way to do this, Senate candidates usually spend a greater percentage of their campaign budgets on this costly medium, especially since candidates for the House represent districts that may be only a small part of a television market.[65]

Fund-raising for Senate candidates, then, is much more "arduous" and "burdensome" than for House members, and Senate candidates, compared with House candidates, must "manage a much vaster constituency of contributors."[66] Senate candidates are more likely to face "independent" negative spending against them (such as advertising paid for by a group like the National Conservative Political Action Committee [NCPAC] without a candidate's knowledge or cooperation). Incumbent senators can usually expect to have better-financed challengers than House members; Senate challengers have a better chance of raising PAC and other campaign funds than House challengers "simply because they stand a better chance of victory against Senate incumbents."[67] Senators feel more vulnerable and politically insecure than do House members because they *are*.

The Nationalization of Senate Campaign Financing

Where do the great sums of campaign money come from? As we saw with the 1988 Senate race of Robert Kerrey, some of it is raised in small amounts in the home state. But, as with that campaign too, more than half of Senate campaign funds now are raised in large chunks and from out of state.[68] "The increasing reliance on outside money shows the growing nationalization of Senate races," a *New York Times* reporter has stated.[69]

A big part of Senate campaign money is "axe to grind" money. That, and the fact that so much of it comes from out of state, means, as political scientist David Magleby has put it, that many senators have divided loyalties: "The senators end up having two constituencies—the constituents who vote for them and the constituents who give to their campaigns."[70]

A large part of campaign funds of Senate incumbents come from PACs—over 27 percent in 1988, 23 percent ($33 million) in 1990.[71] "Alarming. Outrageous. Downright dangerous." Those are the words Fred Wertheimer, president of Common Cause, the "citizen lobby," has used to characterize what he calls "the threat posed by the torrents of special interest campaign cash being offered up to our Representatives and Senators by the special interest political action committees."[72]

The growth in the number of PACs and the amount of their campaign contributions is recent, but their existence goes back to as early as 1943.[73] It is a federal crime for labor unions or corporations to use their "treasury," or general, funds for political campaigns. Labor organizations thus set up the first PACs to gather political campaign war chests from the individual contributions of their members—the old Congress of Industrial Organizations in 1943, and the American Federation of Labor in 1947. When the AFL and the CIO merged in 1955, so did their PACs, into the powerful Committee on Political Education (COPE). By the 1960s, a few trade and professional groups had followed labor's lead and had set up their own PACs, for example, AMPAC of the American

Medical Association, and BIPAC (Business–Industry PAC) of the National Association of Manufacturers.

Congress took a step toward reforming federal campaign financing in 1971, with its passage of the Federal Election Campaign Act (FECA). The Act tightened disclosure requirements in regard to contributions and expenditures. It also, incidentally, legitimized PACs by expressly permitting unions and corporations to create and raise money for them. Right after that came the Watergate scandal of the Nixon administration. It revealed, among other things, that a good many of America's best known business corporations had, in violation of federal law, contributed corporate treasury funds to President Richard Nixon's 1972 reelection campaign. Strong public backing developed in favor of much more stringent campaign financing reform. Common Cause, which had been formed in 1970, helped marshall public opinion toward this end and pushed Congress for action.

In 1974, comprehensive campaign-reform amendments were added to the basic FECA law. Strict limits on contributions and expenditures in federal elections were established. Public financing was provided for presidential campaigns—though not for congressional campaigns. (The Senate had adopted a public-financing provision for congressional races, but the House had killed it.) In the 1974 FECA amendments, with both labor and business support, PACs were again, and more specifically, approved. Labor did not foresee how greatly it would be outstripped by business in the raising and contributing of campaign funds.

Common Cause did clearly foresee the consequences of approval of public financing for presidential campaigns alone, without also providing financing for congressional campaigns.[74] We were headed toward a vast leap in the number of PACs and the sums of money that they would spend in House and Senate races. Contributions that formerly went to presidential campaigns, and now were strictly limited in conjunction with public financing, were diverted to congressional candidates.[75]

There were only 608 PACs when the 1974 reform law went into effect. By 1977 that number had more than doubled, to 1,360; by 1982, it had more than quintupled, to 3,371. By 1990 there were nearly 4,200 PACs.[76]

Like the burgeoning PAC numbers, their campaign contributions rose spectacularly also. For the 1972 elections, PACs contributed about $8 million to congressional candidates; for 1974, $12.5 million; for 1976, $22.5 million; for 1978, $35 million; for 1980, over $55 million; for 1982, $83.6 million; for 1984, over $105 million; for 1986, $132.6 million; and for 1988, $172.4 million.[77] In current dollars, adjusted for inflation, that is an increase of more than sixfold in sixteen years (from $8 million in 1972 to $50.6 million, adjusted for inflation, in 1988). PAC spending for federal candidates in 1990 was the same as in 1988.[78]

The importance of PACs can be seen clearly by comparing two of Kansas Republican Senator Robert Dole's campaigns, twelve years apart. In his 1974 race Dole spent a total of $1.1 million, with only $82,555 of it coming from PACs. For his 1986 campaign, his PAC contributions alone amounted to $1.3 million—and Dole raised another $1.2 million for that race in individual donations of $500 or more.[79]

PACs spent around $50 million on Senate candidates alone in the 1988 elections, with the total divided about evenly between Republican and Democratic candidates.[80] Senate incumbents got about three-fifths of the $50 million total, a little over $31 million ($14.9 million for Republican incumbents, $16.1 million for Democrats). Challengers against sitting senators got about $8.3 million of the total (about $3.6 million for Republicans, $4.6 million for Democrats). Candidates for open seats, with no incumbent running, received about $10 million of the total contributed by PACs (Republicans $5.4 million, Democrats $4.6 million). In 1990, each of fourteen winning Senate campaigns and two losing ones were the beneficiaries of $1 million or more PAC dollars.[81]

Most of this PAC money came in direct contributions to the candidates. About a tenth of it was indirect, spent on *behalf* of the candidate—as when AUTOPAC, the Auto Dealers and Drivers for Free Trade PAC, associated with foreign car makers, independently paid out over a half million dollars for television and other advertising and activity in an unsuccessful effort to reelect Republican Senator Chic Hecht of Nevada in 1988.

Many PACs have no qualms about switching horses after their preferred candidate loses. For example, all seven Democrats who beat incumbent Republican senators in 1986 received after-election contributions before the end of the year from the same PACs that had earlier backed their opponents.[82] In 1988, $2.5 million of the PAC money spent in all congressional races was paid out to candidates *after* the election.[83] In seven Senate races studied by Common Cause that year, for example, sixty-two PACs were found to have switched sides after the election to give $1.7 million to the winners, including fifty-eight PACs that switched over and gave a total of over $300,000 to winner Robert Kerrey in Nebraska.[84]

Conservative PACs Strongest

There is a conservative bias in the PAC system, as labor and liberal groups have discovered to their dismay. "If the labor movement has suffered a worse self-inflicted political wound, it does not readily come to mind," a pro-labor, liberal observer has said.[85] He was referring to labor's failure to foresee, when it pushed Congress to legitimize PACs in 1974, how enormously aggressive labor's public-policy opponents— business and related conservative interests—would be in taking advantage of this new political-influence tool. Right away, after the passage of the 1974 law, the Sun Oil Company got a ruling from the Federal Election Commission that it could use general, or "treasury," funds to administer its PAC and to solicit funds for it. This ruling greatly facilitated the creation and expansion of corporate PACs. Trade associations and unions are allowed to do the same thing.[86]

The fastest growing and most numerous PACs have been those formed by individual business corporations. These represent nearly 43 percent of all PACs (accounting for over 50 percent of all congressional campaign spending by PACs in 1988), compared with labor's little over 8.25 percent of the total number of PACs (and about 32 percent of 1988 PAC congressional campaign spending).[87] The comparison of business corporation and labor union PAC numbers and expenditures does not fully indicate, though, the probusiness, conservative cast to the PAC world. There are also trade group PACs.

Considering spending in Senate elections only, the amounts laid out by PACs in the corporation and trade categories have been increasing. These PACs accounted for 60 percent of total PAC spending in Senate races in 1982 and 67.7 percent in 1990.[88] Meanwhile, as Professor Alan I. Abramowitz has pointed out, "Organized labor has not only been declining in its share of workers in the U.S. labor force but also in its share of contributions to congressional candidates."[89] The labor PAC share of total PAC spending in Senate races went down from 22 percent in 1982 to 14.5 percent in 1990.[90]

Raising and Giving Special-Interest Money

How do PACs raise and spend their money? The methods differ somewhat by the type of PAC involved. A labor union PAC can freely solicit the parent union's members without restriction. Rallies, special events, and face-to-face solicitations are favored fund-raising activities and methods. A trade association PAC can solicit the executive and professional employees of its member corporations once a year, with each corporation's prior approval; a trade PAC's fund-raising from other sources is fairly unrestricted by law. Nonconnected PACs may freely solicit anyone, and as often as they please; direct mail is the favored method.

The law permits a business corporation PAC to solicit, as often as it wants to, the stockholders and the administrative, executive, and professional employees of the parent corporation, and the PAC can mail solicitations twice a year to the homes of all other employees of the corporation.[91] Many corporate and labor PACs get their money automatically through payroll deductions, authorized in writing by the employee involved and binding until revoked.[92] In 1989, the major telephone companies inaugurated a new 900-number system that permits organizations to raise money through contributors' calls to an assigned number and have the contribution collected with the phone bill. The National Rifle Association was one of the first groups to use the new service.[93]

PAC giving is supposed to be voluntary, of course, and a lot of PACs use special promotions to raise their money each year. But much of PAC giving is actually "pressured" giving. The head of NBC, for example, once wrote in a memo to company officials that "employees who elect not to participate in a giving program of this type should question their own dedication to the company and their expectations."[94] Often the pressure is more subtle, though just as real. One bank company asks each of its executives to contribute one-half of 1 percent of salary to the corporation's PAC, and in their annual fund-raising drive, as a bank officer put it, "The captains perceive that to be involved in the political action committee . . . is good for their careers."[95]

Trying to track all the ways that contributions can legally be made under present federal campaign-financing law and how the law's ceilings can be evaded is like stumbling around in a maze blindfolded. The categories are confusing.

Individual Contributions. An individual is allowed to contribute $1,000 per election to each federal candidate—that is, $1,000 in the primary (and an additional $1,000 if the state has a runoff), plus another $1,000 in the general election. Individual contributions (not counting individual contributions through PACs or party committees) amounted to 65 percent of Senate campaign spending in 1986 and 59 percent in 1988.[96] But these contributions are often just as much "special-interest" money as PAC money is. For example, Democratic Senator David Boren of Oklahoma, an advocate of campaign-finance reform who refuses to accept PAC contributions, got fully a fourth of the million dollars he raised for his 1984 reelection in the form of individual contributions from executives of oil companies and banks (and probably more, since many of his contributors who listed their occupations as "homemaker" likely were spouses of the company executives).[97] Wall Street PACs gave New York Republican Senator Alfonse D'Amato $78,200 in his 1986 campaign, but he got nearly five times that amount—$360,780—from Wall Street executives, individually, that year.[98] It should be noted, further, that an individual can also separately give $5,000 a year to a PAC and $20,000 to a national party committee, although there is an overall limit of $25,000 for an individual's contributions to all federal candidates, PACs, and committees.

PACs. A PAC itself (that is, a multicandidate PAC, one that receives contributions from at least fifty individuals and contributes to at least five federal candidates; this includes the vast majority of PACs), can contribute $5,000 per candidate per election, with no overall limit.[99]

PACs maximize their muscle by "unofficially" coordinating their activities and giving; BIPAC, for example, the PAC of the National Association of Manufacturers, holds monthly meetings during election years with other PACs oriented toward business and industry, so that they can all review candidates and campaigns together.[100] The national committees of the two political parties also help with this kind of coordination.

Bundling. Some PACs use a practice called "bundling" to get around the limits on what they can give a single candidate. For example, when Republican Senator Bob Packwood of Oregon was up for reelection in 1986, AlignPAC, an insurance PAC that was grateful for Packwood's support of insurance interests in the Senate Finance Committee that he chaired, collected checks totaling $168,000 from insurance agents around the country and passed them on to the Packwood campaign in a bundle. Though this amount was more than thirty times what AlignPAC could itself have legally given to Packwood, it was a legal contribution because it was composed of many checks made out to the Packwood campaign directly, not to the PAC.[101]

Leadership PACs. It gets more confusing. One PAC can give $5,000 to another PAC. In the Senate (and House, too), leaders have set up their own individual PACs, so-called "leadership PACs," to channel money to their party's candidates for election and reelection to the Senate, in order to help build support for their parties in the Senate, and for themselves.[102] Leadership PACs solicit money from other PACs, as well as from individual donors.

Party Committees. A PAC can give $15,000 to a national party committee. The two Senate campaign committees—the Democratic Senatorial Campaign Committee (DSCC) and the National Republican Senatorial Committee (NRSC)—solicit money from both PACs and individuals. Each of these committees (alone or in combination with the Democratic National Committee and the Republican National Committee) can directly contribute $17,500 to each Senate candidate for each election cycle. And each can make "coordinated expenditures" (for polls, advertising, and even direct-mail fund-raising) for each of its Senate candidates, in an amount based on state population, ranging from $100,560 in Wyoming in 1990 to $2.1 million in California.[103] In the 1990 elections, the NRSC contributed $859,140 to its Republican candidates for the Senate and spent another $7.7 million in their behalf, while the DSCC typically lagged somewhat behind, contributing $510,133 to its candidates that year and spending $5.1 million for them.[104]

The central committee of each party at the *state* level, if it has the money—the Democratic State Central Committee of Iowa, for example, or the Republican State Central Committee of Nebraska—can also contribute $5,000 per election (primary election and general election, and runoff, if there is one) for a local Senate candidate.

Nationally, the Republicans are more successful at fund-raising than the Democrats. The Republican House and Senate campaign committees and the Republican National Committee, combined, raised a total of $80.5 million in 1989, an increase from the previous year ($35.5 million of the total being raised by the National Republican Senatorial Committee), compared with only $19 million, a decrease from the previous year, raised by the corresponding three Democratic committees (a little over $8 million of it by the Democratic Senatorial Campaign Committee).[105]

The Republican National Committee has acted as the agent for raising and spending this money on behalf of any of its state central committees unable to do so. The Supreme Court has upheld this practice, over Democratic objections.[106] Likewise, the NRSC (and the DSCC, too, with what money it has left over) sends money to state party organizations in states where there are contested Senate races. Republican state party committees received $1.1 million in this way from the NRSC for the 1988 elections and $2.4 million for those in 1990 (while Democratic state parties got $222,285 from the DSCC in 1988 and $430,274 in 1990).[107] Forty-two percent of the $1.5 million spent to help Montana Republican Conrad Burns beat incumbent Senate Democrat John Melcher in 1988 came from national Republican coffers, with a little thrown in from the Colorado party for good measure.[108]

Soft Money. Then there is the problem of so-called "soft money." In most states, contributions, including contributions from labor union and corporate treasury funds, can be made to state political parties for "party building" activities, such as registration and get-out-the-vote drives, outside federal campaign limits. So long as this "soft money" is not tied directly to the campaign of a federal candidate, even though the candidate may quite obviously benefit from it, the practice does not violate any federal law.

One study showed that $3.3 million in soft money was spent in just five of the states that had Senate races in 1986—California, Colorado, Florida, Missouri, and Washington.[109] In the presidential election year of 1988, $28 million in soft money was spent in nine states with Senate races.[110]

Independent Expenditures. There are also problems with so-called "independent expenditures" in federal campaigns. Individuals and PACs (and the great bulk of these expenditures are made by PACs) can spend freely, and without limit, to aid in the election or defeat of a Senate (or other federal) candidate, so long as there is no cooperation or communication between the spender and the candidate or campaign.[111] When Congress passed the 1974 federal campaign reform, it intended to limit independent expenditures like all others. But the Supreme Court held in *Buckley* v. *Valeo* that to do so would violate a person's First Amendment right to free speech and that limits on independent expenditures were therefore unconstitutional.[112] In 1990, independent expenditures in Senate races totaled $3 million ($1.7 million for Republican candidates or against Democratic candidates; $1.36 million for Democrats or against Republicans).[113]

Candidate's Own Funds. Another loophole in federal campaign laws opened by the Supreme Court—again, in *Buckley* v. *Valeo,* when the Court struck down certain provisions of the 1974 campaign-reform law—was one that allows any Senate candidate (or House candidate) to loan or give to his or her own campaign any amount of *personal* funds the candidate wants to, without limit. The Court did hold that when candidates accept federal campaign funds, as in presidential campaigns, they can be made to abide by limits on the spending of personal funds, but not otherwise (and, of course, there is no federal funding for Senate and House campaigns). This personal spending loophole makes it easier, naturally, for rich people to get elected; not surprisingly, a lot of the members of the Senate in recent years have been, and are, millionaires. Many Senate candidates "lend" their campaigns money, often borrowing personally to do so; 15 percent of candidates in the 1984 election gave or loaned their campaigns more than $100,000.[114] Candidate personal contributions or loans of more than $1 million are increasingly common.[115] Naturally, after a person is elected or reelected to the Senate, his or her personal campaign loans or debts are fairly easy to get covered by PACs or other special interests.

Fund-raising, Lobbying, and Influence

Even if one were to concede the truth of the dubious old adage, in all likelihood conceived as a consolation to poor people, that "money cannot buy happiness," it would still be folly to believe that political money does not buy influence.

Whether money is raised or given through PACs, party committees, individuals, or otherwise, fund-raising has "become an integral part of lobbying," a lobbyist for Beneficial Management Corporation of America has stated.[116] Big-time Washington lawyer-lobbyist J. D. Williams has said that, in 1969, when he opened his own office, "There were very few lobbyists. There were lobbyists around from previous administrations, including the Franklin D. Roosevelt and Harry S. Truman administrations. Trade associations were passive." Now, Williams notes, "there are tens of thousands of lobbyists," and what they do has changed, particularly in regard to raising money. "In the old days, a little money, especially early money, went a long way—a few hundred dollars or a thousand." But no more.[117] Now, a typical Washington lobbyist receives around ten fund-raising-event invitations a week when Congress is in session, and each invitation seeks a contribution of $250 or more; receptions and dinners that cost $50 or $250 several years ago now frequently charge four times that.[118]

What does all this money buy? There are those who feel that it cannot be shown or proven that PAC money, or campaign money generally, has a corrupting influence on the Senate and on American politics,[119] that in fact, as conservative Republican Senator Phil Gramm of Texas has said, the rise in the PACs specifically is a good sign of healthy democratic participation.[120] This rise has merely, as another conservative Republican senator put it, "diluted organized labor greatly and . . . resulted in somewhat more balance."[121]

Those are not the prevailing views. The colorful former Democratic senator from Louisiana, Russell Long, used to tell a humorous story on himself that contained more than a kernel of truth about the effect of money in Senate campaigns. He said that in one of his reelection races, a union group brought him a cash contribution of $5,000. "When he handed me the money," Long related, "the leader of the group said, 'All we want, Senator, is good government.' And I told him, I said, 'For a thousand dollars, you could have had good government; for five thousand dollars, you can have any kind of government you want!'"[122]

The fact is that money does buy influence, and now and again a PAC representative may say so, as former liberal Republican Senator Charles McC. Mathias of Maryland once reported:

> I did one of those TV debates for the U.S. Chamber of Commerce. I was shocked because one of the Chamber's people, in the course of this discussion, said, "You can't do away with campaign contributions because no public official would do anything for you. How would you ever get anything done?" In some circles that is the naked truth—that is, this is an outright lever that you can *buy* to get government action. And this person said so right on the air.[123]

It is true that a provable quid pro quo connection between a campaign contribution and a particular senator's vote on the Senate floor cannot be demonstrated. But PACs and other campaign contributors give far more to those who support them than to those who oppose them. And they tend to concentrate their giving to those who serve on the committees that handle the issues the particular PACs are interested in.[124] If they are not getting something substantial in return for their money, one would have to agree with political scientist Gary Jacobson that "Irrationality on this scale would be difficult to explain."[125]

No senator, of course, would openly admit to, or charge another senator with, trading a vote for campaign money. But here are the revealing comments of three of them in off-the-record interviews:[126]

- It is difficult to maintain a sense of integrity and self-worth when asking for money, and then trying to separate that from your decisions.
- In some cases, you feel one way, and vote the other.

- Congress will listen to big contributors. They have a direct influence. It is demeaning and wasteful and the money makes us ripe for corruption.

The influence sought by a contributor may be intervention with a government agency, rather than legislative action. Take the 1991 Senate ethics hearings on the so-called Keating Five, Republican John McCain of Arizona and Democrats Alan Cranston of California, Dennis DeConcini of Arizona, John Glenn of Ohio, and Donald Riegle, Jr., of Michigan. This investigation resulted from the intervention of the senators with the Federal Savings and Loan Insurance Corporation on behalf of Charles Keating and his then rapidly sinking Lincoln Savings and Loan Association, after Keating had made major contributions to, and to help, their reelection campaigns. The senators contended they had done nothing more than any senator would have done for a constituent. Smoking gun? What did Keating say? He said, "One question, among many others raised in recent weeks, had to do with whether financial support in any way influenced several political figures to take up my cause. I want to say in the most forceful way I can: I certainly hope so."[127]

A prior quid pro quo arrangement is not necessary for the influence of money to be exerted. If you already know how senators or Senate candidates are going to vote before you help elect or reelect them, you do not have to put pressure on them each time there is a vote. What one lobbyist told Senate observer Donald Matthews in the late 1950s is even truer today: "Ninety percent of what goes on here during a session is decided on the previous election day. The main drift of legislation is decided then; it is out of our control. There is simply no substitute for electing the right folks and defeating the wrong folks."[128]

And the need for campaign money can sometimes subtly begin to change a candidate's thinking while the campaign is under way. During the unsuccessful 1982 Connecticut Senate campaign of Toby Moffett, then a member of Congress from that state, Moffett at first rejected out of hand the suggestion of his hired fund-raiser that he solicit money from a dairy-industry PAC. But, as he became increasingly pinched for money, Moffett changed his mind, and he related the result like this:

> By the time I got to the last month of the campaign, I was telling my wife and my close friends that here I was, somebody who took less PAC money, I think, than anybody running that year, but I felt strongly that I wasn't going to be the kind of senator that I planned on being when I started out. I felt that they were taking a piece out of me and a piece of my propensity to be progressive and aggressive on the issues. I felt like, little by little, the process was eating away at me. One day it was insurance money, the next day it was dairy money.[129]

Sometimes a direct connection between campaign money and senatorial action is not provable because of the fact that what the PACs and other contributors want is *inaction*, rather than action, often in committee rather than on the floor. Inaction on a tax measure was, for example, what the banking industry wanted when it contributed $3.5 million to the 1986 campaigns of congressional banking committee members.[130]

Frequently, for incumbents, the power of money works in a negative way: a member of Congress may act, or fail to act, so as not to anger certain special interests enough to cause them to recruit and finance a serious opponent in the next election. Of course, this produces no evidentiary record, but the effect is there nevertheless.[131] As one member of Congress put it in a kind of muddled conversational way:

> Take anything. Take housing. Take anything you want. If you are spending all your time, calling up different people that you're involved with, that are friends of yours, that you have to raise $50,000, . . . you're in effect saying, "I'm not going to go out and develop this new housing bill that may get the Realtors or may get the builders or may get the unions upset. . . . I've got to raise the $50,000."[132]

PACs and other campaign contributors certainly buy "access," as everyone benignly puts it. There is no argument about that. And access is power, power that skews the representation equation. As Republican leader Robert Dole of Kansas has himself said, "There aren't any poor PACs or Food Stamp PACs or Nutrition PACs or Medicare PACs."[133]

The alarming increase in the costs of campaigns and the problems inherent in the way campaigns are financed "threaten the very essence of political equality," according to U.S. Court of Appeals Judge J. Skelley Wright, who has written that if corrections are not made, "the principle of one person, one vote could become nothing more than a pious fraud."[134]

Further, the money and campaign skills now marshalled by the PACs and the national party committees have increased senators' sense of electoral insecurity because they know that more money can be raised against them and, with PACs and parties coordinating their efforts, it can be raised quickly.[135]

The parties and the PACs have also helped to nationalize Senate campaign issues and the outlook of senators—helping to make senators more national and outward-looking and less local and inward-looking. Gary Jacobson has made these points:

> National funding sources force national issues onto the local agenda, and candidates find it necessary to campaign on issues that may be of marginal interest to their constituents but of deep concern to groups outside the state or district. . . . More time and attention is devoted to politics outside the institution, and outside influences on internal politics become stronger and more pervasive.[136]

What can be done to clean up the system of campaign financing that threatens our democratic system of one person–one vote, as Judge Wright put it—and the independence of the Senate?

The Possibilities of Campaign-Finance Reform

Unfortunately, there are separate Democratic and Republican answers to the questions about the rising costs of Senate campaigns and the evils that attend that growth. Eight times during the 1987 and 1988 sessions, Senate Democrats tried to get a vote on their campaign-finance reform plan, principally authored by Senator David Boren of Oklahoma. Eight times, Senate Republicans filibustered.

But sentiment for reform continued to grow, particularly as a result of the Senate Ethics Committee investigation of the Charles Keating–Lincoln Savings and Loan Association case.[137] The Ethics Committee eventually recommended that a bipartisan task force be created to develop a reform of campaign financing.

The struggle to pass an overall campaign reform bill began again in earnest during the 1989 session and lasted into 1992. The Senate had to consider three major plans: the basic Democratic proposal, a Senate Republican alternative, and a third put forward by President Bush.[138]

The Republicans had been angered by the fact that business and trade PACs gave more money to Democrats than to Republicans in the 1988 elections (primarily because all PACs tend to favor incumbents, and there were more Democratic incumbents running than Republican ones). Senate Republican leader Robert Dole of Kansas "dressed down" seventy leaders of business PACs for this apostasy in a special Washington meeting to which he summoned them immediately following the 1988 election.[139] A prominent Republican consultant was quoted as saying that "I have to conclude that Republican candidates would be better off if PACs didn't exist and candidates had to raise their funds from individual contributors."[140]

Little wonder, then, that the Republican plan proposed to cut PAC contributions drastically, to $1,000 per candidate per election.[141] Democrats were quick to point out, though, that the Republican proposal sought to raise what an individual could give (up from $1,000 to $2,000 per election). President Bush went even further on PACs. He called for the total elimination of all business, trade, and union PACs—precisely those, Democrats pointed out, that had favored them in the last election—while allowing nonconnected PACs to continue in operation, though with a new contribution limit of $2,500. Democrats also pointed out that both the Bush plan and the Senate Republican plan sought to raise the limits on coordinated expenditures by state and national parties—where Republicans enjoyed a fund-raising advantage—by 150 percent.

The guts of the Democratic plan (and the basis for the most determined Republican opposition to it) had to do with overall spending limits and public financing. When, in late 1991, Senate Democrats finally despaired of ever being able to work out an agreement with the Republicans, they passed their own plan in a near party-line vote and sent it to the House. There, also, House members split almost exactly along Democrat–Republican lines and, spurred on by a need to accomplish some reform in the wake of the House bank and post office scandals, adopted a House Democratic version of the campaign finance bill, sending it to conference. In party-line votes, again, both houses then adopted the 1992 compromise conference report.[142]

What made such a House–Senate agreement possible—the first on campaign-finance reform in more than a decade and a half—was that the new rules agreed to were different for each house. The bill provided for Senate spending limits: between $950,000 and $5.5 million in the general election, depending on state population (with a total cap of $600,000 for House campaigns). Participating Senate candidates who accepted the spending limit and also raised a substantial portion in contributions of $250 or less, half of it in-state, would receive public-fund vouchers worth up to 20 percent of the limit to buy television advertising time, would be able to purchase television time at half the regular rate, and could make campaign mailings at a discount rate. (Participating House members would be eligible for direct public funds to match a portion of the money they raised privately and could also make discount-rate mailings.) Republican opponents of the public funding provisions called it "food stamps for politicians."[143]

Candidates accepting public funds could contribute out of their personal funds only 10 percent of the general-election spending limit, to a maximum of $250,000. And the spending limits would be lifted, and additional public funds provided, if a candidate faced a nonparticipating opponent who refused to abide by the spending limit or an independent campaign that spent more than $10,000. Too, restrictions were placed on soft-money spending by state and national party committees.

In regard to PACs, the 1992 bill was sadly deficient. Senate conferees had backed away from their original position in favor of banning PACs altogether, fearing this might be held unconstitutional. Instead, the final bill reduced the amount a PAC could give a Senate candidate to $2,500 and set a cap on PAC contributions to a Senate campaign of no more than 20 percent of the general-election spending limit. (The bill still allowed a PAC to give $5,000 to a House campaign and put an aggregate limit on total PAC contributions a House candidate could accept of one-third of the overall spending limit.) Still, the leader of Common Cause and Democratic backers of the bill felt it was a major step forward.

President Bush had vowed to veto any bill that treated the two houses differently, provided for public financing, or set overall spending limits. The 1992 bill did all three, plus something else Bush opposed: it restricted party spending. The presidential veto came down like a guillotine. Without any Republican votes to speak of, House and Senate Democrats had no chance for an override.

There is still a fundamental need for campaign-finance reform. The passage again of the 1992 measure, this time with a ban or at least tighter restrictions on PACs, would go a long way toward interrupting the process in America by which money is translated into political power.

The Problem of Senate Salaries

The question of money and its influence on senators is bound up with the matter of their pay. Whether or not Capitol Hill votes are influenced by campaign contributions, it is a fact that most senators (and House members) found it impossible until recently to make ends meet on their Senate salaries.

True, members of Congress have provided generous retirement systems for themselves; they enjoy a $3,000 tax credit to offset Washington living costs; they take advantage of large expense allowances for trips home; and they draw salaries that are, to say the least, several levels above the national average. But the $3,000 tax credit is the same now as when it was enacted in 1952; there is no allowance for the travel and other expenses of spouses; and no perquisite comes close to making up for the great expense of maintaining two houses, one back home, one in Washington, as good politics mandates.

One member of Congress illustrated the extremely high cost of Washington real estate when he made the following observation about a type of house that is too small for most Senate families: "You can't explain that a two-bedroom row house here costs $150,000, and back home in Mississippi, that gets five bedrooms and five acres."[144] A more suitable Virginia suburban house that one senator bought in 1966 for $55,000 and sold ten years later for three times that amount was by 1991 worth a half million dollars—nearly a tenfold increase during a period when a senator's salary increased a little over threefold.[145] "Money in Capitol Hill is not being put to evil purposes," a Democratic campaign consultant has said. "In most of these cases, it's being put to survive. These members have got families, kids in school."[146]

Being forced to vote on increases in their own salaries has been a political embarrassment, and worse, for senators and representatives ever since the republic was founded. The Constitution (in Article I, Section 6) declares that members of Congress "shall receive a compensation for their services, to be ascertained by law"—that is, by Congress itself. When the very earliest Congress set salaries for the first time, American newspaper editors were heavily critical of the action. Salary increases in 1816 and 1873 resulted in a virtual electoral slaughter of the members of Congress who had voted for the measures. In fact, the congressional pay-raise problem has been a "two-hundred-year dilemma."[147]

Judges and cabinet secretaries cannot be charged with self-serving actions when their salaries are raised from time to time: Congress sets their salaries. But Congress cannot escape, try as it might, the charge of self-serving motives when congressional salaries are raised: Congress controls the purse strings of the country, and this includes, of course, the strings to the purse out of which members of Congress themselves are paid. When members of Congress vote to increase their *own* salaries, this is self-serving action by definition, as James Madison clearly foresaw it would be. He said back in 1789 that "there is a seeming impropriety in leaving any set of men, without control, to put their hands into the public coffers, to take money to put in their pockets."[148]

A modern device that worked part of the time to get Congress off the hook on the pay-raise issue was the delegation of the power to set congressional salaries (as well as executive and judicial salaries, all linked together) to an independent commission, with the provision that the raises decided on, once endorsed by the president, would go into effect automatically unless both houses of Congress acted to block them.

In December 1988, this quadrennial commission, as it was called (officially, the President's Commission on Executive, Legislative, and Judicial Salaries), recommended a 51-percent pay hike for members of Congress (and executive and judicial officials), stipulating that this recommendation should go into effect only if Congress prohibited its members from receiving outside honoraria—fees for speeches and appearances.[149] The recommended increase would have brought House and Senate pay to $135,000 annually (with greater amounts for the leaders), restoring congressional salaries, the Commission said, to their 1969 buying power. President Ronald Reagan endorsed the report before leaving office,[150] as did his successor, George Bush, soon after being inaugurated.

The next move was up to Congress, although a nonmove would suffice. Under the law creating the quadrennial commission, if either house *failed* to act on the recommended raise, members would receive it. A vote on the raise could not be prevented in the Senate, of course, since Senate rules allow amendments, even nongermane amendments, to be freely offered to almost any measure under consideration. And many senators assumed that they could vote down the increase (which they in fact then did) with reasonable assurance that they would still get it because then Speaker of the House Jim Wright, under that body's more restrictive rules, could, and surely would, see to it that no vote would be allowed in the House.

Citizen lobby Common Cause supported the salary increase as the best way to secure the elimination of the honoraria system. Activist Ralph Nader, on the other hand, opposed both the raise (which he said was particularly inappropriate in view of the fact that Congress had failed to act for eight years to increase the minimum wage for American workers) and the system that allowed members of the House to duck a vote on it.

Public opinion was against the raise, and there was special outrage over what many felt was a cynical attempt by members to get the increased money while evading a vote on it. Congressional offices soon had clogged phone lines and a glut of protest mail. Speaker Wright reversed himself and allowed a vote on the raise, and the House killed it like a poisonous snake.[151]

The commission-recommended pay raise was dead. And, in effect, so was the independent-commission device for setting congressional salaries. And the honoraria system, and its potential for corruption, remained very much alive. The inadequacy of congressional salaries persisted as an ever-sharper reality too. As public opposition began to cool some, new proposals were made, including one backed by President Bush. Finally, in late 1989, as noted earlier, Congress acted, this time right out in the open, without the commission blind. The House first passed a bipartisan bill that, among other things, gave senators and representatives (and judges and senior executive-branch officials, too) immediate cost-of-living salary increases, a permanent salary increase to take effect after the 1990 elections, regular cost-of-living increases thereafter to keep up with inflation, and a ban on personal acceptance of honoraria. But Senate backers, who included the leadership, were unable to construct a majority in that body in favor of the measure. Some senators who otherwise probably could have been counted on to vote for the raise and honoraria ban were facing 1990 reelection challenges and feared voter disapproval. The final bill agreed to by both houses thus had different provisions for the House and Senate: lower salaries for senators and no prohibition against honoraria for them.

Clearly, something still needed to be done about Senate salaries and honoraria. In 1991, something was. Senate salaries were raised to the same level as those for the House—coupled, of course, with an honoraria ban. As one senator put it during debate on honoraria, "The issue really before us is whether or not our salaries should be paid in total by the taxpayers of the United States or whether or not they should be paid in part by the taxpayers of the United States and in part by the special interests."[152]

The national advocacy explosion that America has experienced, then— an enormous increase in the number of interest groups and in their diversity, as well as in the range, volume, and intensity of their activities—has resulted in a nationalization of issues and interest groups. It has provided more opportunities, and demand, for senators to become national spokespersons, and it has been a major cause of increased insistence by senators on fuller individual participation in Senate operations and processes. The advocacy explosion; the communications revolution, particularly the advent of television; and the nationalization of Senate elections have caused senatorial campaign costs to skyrocket and senators to feel increasingly vulnerable, electorally.

Senate campaign financing has become increasingly nationalized, too, adding to the pressure for senatorial issues and outlook to become more national and external and less local and internal. Growing fund-raising pressures have begun to interfere with the work of senators and to affect their behavior. The need to raise great sums of campaign money has become an ever more worrisome source of potentially corrupting influence.

Reform is urgently needed: all Senate campaign costs and contributions should be strictly and enforceably limited, with provisions for free television and other subsidies in Senate campaigns, as well as the possibility of additional public funding, in order to curb rich-candidate self-financing of Senate races and to circumscribe and offset independent expenditures. The power of money could thus be reduced and senators made more free to pursue the public interest.

III

THE SENATE
AS A NATIONAL
INSTITUTION

4

National, More Individualistic Senators

"He was a young white man from the right side of the tracks, but he dreamed of becoming a basketball player. His neighbors laughed and ridiculed him, but he persevered and broke the chains of race-conscious behavior."

The Reverend Jesse Jackson was teasing Bill Bradley, Democratic senator from New Jersey.[1] The occasion was a 1988 Washington, D.C., fund-raising "roast" of the kind so commonly seen in the entertainment industry, but rare in national politics. This particular event, sponsored by Independent Action, a liberal political action committee, attracted the attendance of all the major 1988 Democratic presidential candidates as "roasters."

The six-foot, five-inch Bradley had, indeed, been a star forward in basketball, initially in high school at Crystal City, Missouri, then at Princeton University, and finally, for ten years, with a professional team, the New York Knicks. It was his U.S. Senate stardom, though, rather than his basketball stardom, that had drawn the presidential heavyweights, as well as the great crowd of contributors, to the Bradley roast.

On that June 1988 night, the presidential hopefuls joked about how glad they were that Bradley was not himself a candidate for the Democratic nomination, and there was an undertone of seriousness in their quips. Bradley had earlier declined to run for president that year, saying that his "internal clock" had told him that he was not ready to do so. Alluding to this at the roast, Senator Biden said, "We hope you keep listening to your internal clock, and that your batteries hold out until 1996."[2]

That evening's humor also focused on Bradley's celebrated less-than-scintillating speaking style (a problem he shared with a number of the others, including Dukakis, and something he later worked to improve), as well as on his equally celebrated reputation for hard work on important but somewhat unexciting issues. Bradley himself joined in these jokes when he rose to speak at the end of the program. "Why not use this opportunity," he asked, "to talk about what people care about—third-world debt, floating exchange rates, Soviet economic reform, North–South dialogue and, my favorite, the Strategic Petroleum Reserve?"[3]

Without doubt, Bill Bradley was—and is—a person of influence and respect in the U.S. Senate. His colleagues say so.[4] When in 1991 Bradley took the lead in attacking President Bush's civil rights record, and separately in opposing the president's nomination of Robert Gates to head the Central Intelligence Agency, the Senate paid attention. "Bill Bradley is someone who people take seriously around here," Republican Senator Warren Rudman of New Hampshire said. "People might not agree with Bill, but he brings a lot of credibility."[5] Senate administrative assistants and other senior staff members also give Bradley a high ranking: a 1987 scholarly survey of forty administrative assistants in the Senate placed Bradley fourth among all senators in "senatorial respect," behind only Republican Minority Leader Robert Dole of Kansas and Democrats Sam Nunn of Georgia and Daniel Inouye of Hawaii.[6] A 1988 popular survey of senior congressional staffers found Bradley tied for third place with Democratic Majority Leader George Mitchell of Maine for "most respected" senator (after only Dole and Nunn).[7]

Administrative assistants surveyed by *Washington Magazine* in 1988 picked Bradley for two of the magazine's annual designations, "Mensa Material" and "Going Places" (while also, for example, selecting Chic Hecht, Republican of Nevada, later defeated for reelection, as "Densa Material," Majority Leader Mitchell as a "Credit to the Institution," and Jesse Helms, Republican of North Carolina, as a "Debit to the Institution").[8]

In 1989, the authors of *The Almanac of American Politics 1990*, careful observers of the Senate, wrote:

> If Democratic nominees for President were chosen by secret ballot of their peers, the winner would almost certainly be Senator Bill Bradley of New Jersey. A Senator for just a decade, Bradley has already produced a major reform of our tax system and attracts respectful attention whenever he weighs in on a major issue.[9]

What is there about Bill Bradley that has made him a respected and influential senator in a relatively short time, even though he neither chairs a major committee nor holds a Senate or party leadership position? To put the question more broadly, What is it that gains respect in today's Senate? In other words, what are the norms of behavior expected of a senator? And how have these norms changed in recent times?

1950s Norms and Senatorial Behavior

The first systematic study of Senate norms, or folkways, was Donald Matthews's *U.S. Senators and Their World*, published in 1960.[10] Matthews interviewed twenty-five senators and former senators who had served between 1947 and 1957, as well as eighty-four staff members, lobbyists, and journalists. He also discovered a great deal of information about senators from published primary and secondary sources. His conclusion was that:

> The Senate of the United States, just as any other group of human beings, has its unwritten rules of the game, its norms of conduct, its approved manner of behavior. Some things are just not done; others are met with widespread approval. "There is great pressure for conformity in the Senate," one of its influential members said. "It's just like living in a small town."[11]

Senators told Matthews that influence and effectiveness in the Senate depended on the degree to which a senator was respected by colleagues and that "the safest way to obtain this respect is to conform to the folkways, to become a 'real Senate man.'"[12] Testing this assertion, Matthews looked at what percentage of their legislative proposals various senators were able to get passed, and developed from this an index of "legislative effectiveness." He concluded that "Conformity to the Senate folkways does . . . seem to 'pay off' in concrete legislative results," and "Nonconformity is met with moral condemnation, while senators who conform to the folkways are rewarded with high esteem by their colleagues. Partly because of this fact, they tend to be the most influential and effective members of the Senate."[13]

Donald Matthews found that the principal operative norms, or folkways, of the U.S. Senate of the 1950s were, as noted earlier: apprenticeship (deferring to those more senior, being seen and not heard for a time); legislative work (attending to detailed legislative business, being a work horse, not a show horse); specialization (focusing on a few subjects, connected with committee assignments or constituency); courtesy (being able to disagree without being disagreeable, and avoiding excessive partisanship); reciprocity (keeping one's word, helping other members and expecting to be helped in return, and refraining from using one's powers to the fullest); and institutional patriotism (upholding the Senate and its reputation).[14]

There is reason to believe that these norms were actually operative in the U.S. Senate as far back as the nineteenth century.[15] In fact, some legislative norms may have originated even earlier and may have been fairly universal among American legislative bodies. George Washington, for example, served as a member of the Virginia House of Burgesses long before he became president of the United States.[16] In that legislative service he conformed to certain norms that he later urged upon his nephew, Bushrod Washington, when the younger man was elected to the Virginia legislature in 1787:

> Should the new legislator wish to be heard, the way to command the attention of the House is to speak seldom, but to important subjects [specialization norm], except such as relate to your constituents and, in the former case, make yourself perfectly master of the subject [legislative work]. Never exceed a decent warmth, and submit your sentiments with diffidence [courtesy norm]. A dictatorial style, though it may carry conviction, is always accompanied with disgust.[17]

The widespread conformity of 1950s senators to the norms of that body had, according to Matthews, great impact on Senate operations—facilitating decision-making, making the Senate more responsible, more efficient as a legislative body, better able to act:

> These folkways . . . are highly functional to the Senate social system since they provide motivation for the performance of vital duties and essential modes of behavior which, otherwise, would go unrewarded. They discourage frequent and lengthy speech-making in a chamber without any other effective limitation on debate, encourage the development of expertness and a division of labor, in a group of overworked laymen facing unbelievably complex problems, soften the inevitable personal conflicts of a problem-solving body, and encourage bargaining and the cautious use of awesome formal powers. Without these folkways, the Senate could hardly operate with its present organization and rules.[18]

The viewpoint of senators of that earlier time was more local and inward-looking. And the senatorial norms they observed served to expedite action and make the Senate more efficient.

The Senate's Environment Is Nationalized

As discussed in Chapter 2, American society became nationalized—with more rapid and national communications and transportation systems, a rising standard of living and level of education, and a more urban and mobile populace. America became less and less a collection of isolated and markedly different states and localities.

With this nationalization came a tremendous growth in the size and activities of the federal government and the work of the Senate. And as activity increased, the numbers of news correspondents in Washington grew as well, and coverage of senators became more detailed and aggressive. Voters became much more attentive to public and senatorial affairs. Highly assertive national civil rights and derivative protest groups mushroomed.

Senate campaigns and elections were nationalized also, converted into little presidential campaigns. Senate races became increasingly partisan, and the Senate Republican and Democratic campaign committees became quite importantly involved in each of them. All Senate seats became party-competitive. Senators learned that they had to campaign almost year-round.

The financing of senatorial races became increasingly nationalized, and the power of money in the Senate grew. There was a nationalization of issues and interest groups, too, a national "advocacy explosion." Senators were lobbied much harder. The national agenda was greatly expanded, with many more big national issues; there were far more national groups looking for Senate advocates, pressing senators to take up their special causes. There was a consequent and parallel expansion in the number of opportunities for individual senators to star as national spokespersons on particular issues.

With all these changes, senators became much more accessible and more exposed to public scrutiny and personal contact. They became more politically insecure, more electorally vulnerable, and subject to increased interest-group pressure.

Norms or folkways in any organization are followed only so long as most members feel that conformity by all, or nearly all, helps them achieve their goals; norms change when most come to feel that there is more reward and less penalty for nonconformity than for conformity.[19] This began to be the feeling in the Senate from the end of the 1950s on, as senators increasingly needed and wanted to participate more, individually, in Senate decisions—to have more opportunities for Senate influence and limelight. They became less willing, less able, to "go along, in order to get along," to quietly follow Senate leaders and committee chairs. Senators became more individualistic, senatorial issues more national, and the Senate outlook more external.

Political scientist Nelson Polsby describes how the Senate evolved into "a predominantly nationally oriented body":

> The senatorial generations of the 1960s and thereafter have pursued a style of senatorial service that in their search for national constituencies and public visibility [has] little in common with the old Senate type. . . . The more common pattern today is for senators to seek to become national politicians, something the mass media have made increasingly possible. . . . The Senate is now a less insular body than it was in former times, and the fortunes of senators are correspondingly less tied to the smiles and frowns of their elders within the institution.[20]

The "Churn Factor"

One former Senate staff member used the words "churn factor" to describe the type of turnover after the 1950s that, in his view, impaired the Senate's "institutional memory."[21] Political scientist Randall Ripley has asserted that membership turnover can cause a change in norms—

and, in fact, is the "single most important" cause of such change. "New people are rarely socialized perfectly in an institution in the sense that they understand and accept its norms totally. . . . The greater the number of newcomers to an institution the more likely that the norms will change."[22]

But the "churn factor" that helped to change the Senate of the 1950s was not simply a matter of turnover in numbers. It also involved a turnover in ideology and geography, particularly at what might be called "critical-mass" times when relatively large freshman classes, especially within the controlling majority party in the Senate, arrived in Washington to take their oath of office and then, almost at once, began to challenge the old ways of doing things.

The Senate of the 1950s was a very traditional and conservative place. A substantial majority of senators were ideological conservatives, and most of these could be classified as extreme conservatives.[23] During that decade the Democrats were in the majority during all but one Congress (1953–1954), although their margin was always fairly tight. The seniority system (under which the majority-party member who has served on the committee longest becomes its chair) placed "a premium on regular re-election to the Senate once a seat is won," as Donald Matthews put it back then, which meant "that senators from small, rural, one-party states are over-represented among the chairmen."[24] Southern Democrats made up nearly half (46 percent) of all Senate Democrats in 1955, for example,[25] and predominated as chairs during that decade, controlling the major committees.[26]

"The sustained electoral success of many conservative members over the 1940s and 1950s produced a conservative elite in the committee system and thereby in the Senate as a whole," one observer has noted.[27] Each committee chair was a Senate baron. Each had the power, pretty much alone, to decide which committee members would be allowed to chair their own subcommittees or serve as conferees on committee bills, and each virtually controlled the committee scheduling of, or refusal to schedule, bills for hearings and mark-up.[28]

These Democratic committee chairs, together with the enormously skillful majority leader of the time, Lyndon B. Johnson of Texas (perhaps the most powerful Senate majority leader ever, who both served and led the chairs), constituted an entrenched "Senate establishment." Their control over both policy matters and Senate operations went largely unquestioned and undisputed by most senators. In those days, in fact, most bills recommended by a standing committee were not even contested on the Senate floor and were passed without a roll-call vote.[29]

This concentration of power in the Senate committees and party leaders is called "decentralized" power by political scientists (compared with "centralized" authority—power located in the party leaders—or "individualized" power—that located in the individual senators and *sub*committee chairs).[30] The Senate norms that sustained such concentration and muffled challenges to it developed in a time of relatively stable Senate membership.[31]

Beginning with the close of the 1950s, however, Senate membership became less stable. An average of twelve new senators were sworn in every two years during the thirty-year period from 1959 through 1989.[32] In each decade there was a steady reduction in their average age—down from an average of about 57.5 years in the 1950s to an average of 54 years in the 1980s.[33]

There was some change in the background and experience of Senate newcomers, too. For example, the percentage who had previously served in the U.S. House of Representatives went up a little, from an average of 28 percent in the 1950s to around 30 percent in the 1980s (36 percent in 1988).[34] But there was a steady decline in the percentage of senators who were former governors, down from an average of 22 percent in the 1950s to 11 percent in 1982, and to 8 percent in 1987. The lengthening of governors' terms, as well as the number of terms allowed, seemed to have reduced the interest of these state officials in running for the Senate.[35] There was a corresponding increase in the percentage of Senate members with only local political experience before they came to Washington (from an average of 4.67 percent of Senate membership during the period 1914 to 1956 to 12 percent by 1986), as well as an increase in the percentage of senators with no previous political experience at all (from an average of 8.22 percent of the Senate during the period from 1914 to 1956 to 15 percent in 1986).[36]

Senate turnover had the most effect on senatorial norms during one of two critical-mass times (periods when there was a relatively large influx of freshman members within the controlling Senate majority party). The first such time began in 1958, when the Senate majority was Democratic.[37] President Dwight Eisenhower was finishing his second term and could not run again. The country was in a severe economic recession, the Russians had beaten us in the space race with Sputnik, the president's chief of staff had been forced to resign because of corruption, and the civil rights pot was about to boil over.

There was something to a humorous saying, current back then about senators, that "few die, and none resign." But in 1958 five incumbent Republicans did decide to retire. Other GOP senators, as it turned out, would have been well advised to follow the lead of their partisan colleagues and quit too, because, of the sixteen Republicans who did run again that year, only five won.

The Democrats gained eleven of the Republican seats and also picked up three of the four new seats from the recently admitted states of Alaska and Hawaii. Moreover, they won enough contests for open seats, when added to eleven their own incumbents rewon, to widen their margin in the Senate overnight from inches to yards—from a scant 49 to 47 majority over the Republicans to a grand thirty-vote, 64 to 34, advantage. Additional Democratic gains in succeeding elections, capped by Lyndon Johnson's landslide presidential victory in 1964, increased the Democratic margin to 68 to 32 in the Senate of the Eighty-ninth Congress (1965–1966).[38]

The main story of this critical-mass time was not that of the growing Democratic-over-Republican margin. It was the growing percentage inside the controlling-majority Democratic party of new members with a geographical base and an ideological bent different from those of most of the powerful old senior members. The Senate majority had become more liberal and more northern.[39] Democrat Robert C. Byrd of West Virginia, himself first elected in 1958, said of that year's Senate class that it "increased the ranks of Northern and Western Democrats over Southern Democrats, and [gave] Senate Democrats their most liberal complexion in years."[40]

The class of 1958 included liberal Democrats Philip Hart of Michigan, Vance Hartke of Indiana, Eugene McCarthy of Minnesota, Frank Moss of Utah, Edmund Muskie of Maine, Harrison Williams of New Jersey, and Stephen Young of Ohio. Additional liberal Democrats—Edward Kennedy of Massachusetts, Birch Bayh of Indiana, Abraham Ribicoff of Connecticut, George McGovern of South Dakota, Gaylord Nelson of Wisconsin, and Daniel Brewster of Maryland—were elected in 1962. Liberal Democratic ranks were augmented further with the 1964 election of Joseph Tydings of Maryland, Walter Mondale of Minnesota, Robert Kennedy of Massachusetts, and Fred Harris of Oklahoma. Senate Democrats began to wrest loose from the conservative hold of the South.[41]

The result, political scientist Michael Foley has found, was that in the early 1960s there was a "conscious reaction against the folkways" that "effectively changed the form and significance of the Senate's culture."[42] As "poverty, civil rights, housing, environmental protection, urban renewal, crime, and consumer rights" became "fully national issues" that were "apparent in most centers of population," according to Foley, many of the new liberal senators became "national senator[s], who developed a national base outside the chamber" and were "less concerned with cloakroom politics and internal status and more with publicizing issues, making direct appeals to national audiences, and mobilizing forces to pressure the government (including the Senate) into action."[43]

Another, more recent critical-mass time in the Senate began with the years 1978 and 1980. This period brought a relatively large influx of freshman members within what then became the controlling Senate majority party, the Republican party.[44] Twenty new members, eleven of them Republicans, entered the Senate as a result of the 1978 elections, and just two years later, as Ronald Reagan was winning a first presidential term in 1980, another eighteen new members were elected, sixteen of them Republicans.[45] In the Ninety-seventh Congress (1981–1982), more than half the members of the Senate—55 percent—were in their first terms.[46]

With their 1980 victories, Republicans gained the majority in the Senate. They were a relatively young group. Most were between the ages of forty-six and fifty-five, compared with Senators in 1953, the year the Republicans had last controlled the Senate, when most were between fifty-six and sixty-five.[47] A large number of the 1980s Senate Republicans (fifteen in 1981) had held no previous elective office before coming to Washington—compared with only three politically inexperienced Senate Republicans in 1953.[48] Senator Nancy Landon Kassebaum, Republican of Kansas, who had served on the staff of a senator before being elected in her own right in 1978, came to believe that the large influx of Republican senators without prior political experience contributed to the weakening of old Senate norms. She pointed to an earlier time when the Senate functioned better because senators "had great respect for the Senate as an institution," and she said that "when so many Republicans came in with no experience in 1980, [this] broke down."[49] Former Indiana Republican Senator Dan Quayle, who came to the Senate in 1980 and was later elected vice president, agreed. "So many of us were new," he said. "We didn't think that much about the institution of the Senate."[50]

As with the critical-mass group of new Senate Democrats of the early sixties, the critical-mass group of new Republicans of the early eighties changed the ideological and geographical makeup of their party in the Senate. In 1953 when the Republicans controlled the Senate, 50 percent of Republican senators were from the Midwest, Mid-Atlantic, and New England states; by contrast, in 1981 the percentage of Republican senators from those states had declined to only 26 percent of all Senate Republicans (this went down further, to a little over 25 percent, in 1983), while the percentage of Republican senators from the South, Rocky Mountain, and Pacific Coast states climbed to 51 percent (and rose to nearly 53 percent in 1983).[51] Dominance within the Senate Republican Party had shifted to the South and West. There was an ideological shift, too. Many of these new Republican senators were from the hard right and much less willing to compromise their views than were their more moderate-right fellow Republicans in the Senate.[52]

The critical-mass time in the Senate of the early 1980s further affected Senate norms. As a Senate aide put it then:

> We've gone from older people to younger people, from more experienced types to less experienced types, from guys who know you don't always get your way to those who are bound and determined to get their way, no matter what. There's a lack of responsibility in governing. They don't think they have to be a part of compromise.[53]

Another Senate staff member agreed and, after saying that formerly "newcomers were absorbed by the comity, the tradition, the club," added that "now they set the rules of the body."[54]

The radical changes in the Senate's external environment, then, produced a different kind of senator, with different pressures, needs, and aspirations. These new senators steadily replaced the old-style senators, with extra replacement surges coming during the critical-mass times of the early sixties and the early eighties. The new senators forced changes in both the formal rules and the informal rules of the Senate.

Current Senate Norms

To some Senate observers it began to seem that all Senate norms had "died." That was not true. Every organization has unwritten rules of behavior, and today's Senate is no exception. How do we know that there are still norms of behavior in the Senate? We know because some senators are respected more, and some less, than other senators. We know from talking with senators themselves,[55] and as a result of a number of surveys of the congressional community, the best, most pointed, and most recent of these being a 1988 survey of Senate administrative assistants by political scientists John R. Hibbing and Sue Thomas.[56]

What difference does it make whether senators are respected or not? It makes a great deal of difference in the degree to which they are effective, or ineffective, in the Senate. When questioned, senators today, as in former years, rapidly and readily identify their fellow members who have little "respect and influence," as well as those who are greatly respected. And members of the larger congressional community, particularly senatorial staff members, can do so too, and just as rapidly and readily.

There has always been what might be called an "easy no" in the Senate. Some senators are accorded so little respect that, when they offer an amendment or present a bill on the Senate floor, a goodly number of other senators consider their vote on the measure an "easy no," requiring little thought (so long, of course, as the vote does not involve some issue of principle or the interests of constituents). That was certainly true in the 1960s, for example. Take the case of one senator, Jack Miller, a Republican from Iowa, who served from 1960 until he was defeated (by Democrat Dick Clark) in 1972.

Senator Miller was a tax lawyer, and he had a tax man's penchant for the picayune.[57] He seemed unable to resist the urge to try to improve a little, even in minor details, many of the bills that came through the Senate. This was true whether or not the bill under consideration dealt with a subject that could be said to be within his special competence. In the Oklahoma legislature such behavior used to be called "dancing every set," and other members looked on it with great disdain and irritation. This was true in the U.S. Senate too. Jack Miller's colleagues, when called away from their committee meetings for any Senate vote, would typically ask each other, in the halls and in the subway cars on the way over to the floor, what the vote was to be on (political scientists call this "taking cues"). If someone said, "It's a Miller amendment," many would say, "Easy no," without waiting for further explanation.

Often when the Senate was winding up a long, late-session consideration of some bill and senators were restless to leave for the night, and as the presiding officer started to move toward "third reading," a final vote on the measure, announcing, "If there are no further amendments . . . ," Senator Miller would burst through the doors of the Republican cloakroom waving a hurriedly typed amendment and yell, "Mr. President, Mr. President, I have an amendment." The chamber would resound with exasperated groans and a cacophony of senatorial voices shouting at once, "Let's vote! Third reading! Final passage!"

Undeterred, Senator Miller would insist on offering his amendment, and the presiding officer would order it read. But as soon as Miller would begin to explain the amendment, he would be cut off by another general cry for an immediate vote on it without debate: "Let's vote! Let's vote!" Like as not, Miller would compound his offense by demanding a roll call on the amendment, rather than allowing it to be disposed of more rapidly by voice vote. Recorded vote or not, the amendment would be summarily rejected, and senators would, at last, answer to the final roll call and rush away. Upset by Miller's last-minute amendment practice, Senator Everett Dirksen of Illinois, the Republican minority leader of the time, ordered that the only typewriter in the Republican cloakroom be carried away someplace where it would not be so readily available to the insensitive Iowan.

Are there similar "easy no's" in today's Senate? Senators at once say yes when asked that question.[58] They say there are several, and even when told that they need mention no names, many do—for example, the name of Jesse Helms, the hard-right Republican from North Carolina. Senators who are today accorded little respect by their colleagues are individuals who are identified, among other things, by the extreme nature of their ideological views and the overly persistent and obstructionist tactics they use in advocating them.

What about the reverse side of the question? Are there senators whose names immediately leap to the minds of their colleagues as the most respected, most influential, and "best" senators? There certainly are. Why is it that a senator like Bill Bradley is so respected? A person who talks with members of the Senate hears a number of such respected names over and over—in addition to Bradley's, names like Robert Dole and Sam Nunn.

These senatorial comments track well with the 1987 Hibbing–Thomas survey of Senate aides, which found that the Senate's twenty-five most respected members were then (in order from first to twenty-fifth):

Robert Dole (R., Kan.)   Lawton Chiles (D., Fla.)
Sam Nunn (D., Ga.)   Edward Kennedy (D., Mass.)
Daniel Inouye (D., Hawaii)   Warren Rudman (R., N.H.)
Bill Bradley (D., N.J.)   Dale Bumpers (D., Ark.)
Lloyd Bentsen (D., Tex.)   John Danforth (R., Mo.)
Mark Hatfield (R., Ore.)   Howell Heflin (D., Ala.)
Pete Domenici (R., N.M.)   David Boren (D., Okla.)
Alan Simpson (R., Wyo.)   Robert Byrd (D., W. Va.)
Richard Lugar (R., Ind.)   J. Bennett Johnston (D., La.)
George Mitchell (D., Maine)   Bob Packwood (R., Ore.)
Nancy Kassebaum (R., Kan.)   William Cohen (R., Maine)
Ernest Hollings (D., S.C.)   Carl Levin (D., Mich.)
John Stennis (D., Miss.)

What was it in their behavior that made these senators so respected? And what were their own views about right conduct in the Senate?

Some of the old norms of the Senate have changed; there is less conformity to others, now, than there used to be; and some norms have been replaced. But the Senate still has norms, unwritten rules of behavior.

Courtesy Norm Still Significant

Describing the norm of courtesy in the 1950s Senate, Donald Matthews wrote that a "cardinal rule of Senate behavior is that political disagreements should not influence personal feelings":

> The Senate of the United States exists to solve problems, to grapple with conflicts. Sooner or later, the hot, emotion-laden issues of our time come before it. Senators as a group are ambitious and egocentric men, chosen through an electoral battle in which a talent for invective, righteous indignation, "mud-slinging," and "engaging in personalities" are often assets. Under these circumstances, one might reasonably expect a great deal of manifest conflict and competition in the Senate. Such conflict does exist, but its sharp edges are blunted by the felt need—expressed in the Senate folkways—for courtesy.[59]

Today, during debate, senators still address the presiding officer rather than another senator directly: "Mr. President, I would like to explain to the distinguished senior senator from Kansas that . . . " They refer to each other in the third person, and with showy courtesy, saying, "the distinguished senator" or "my able friend," whether or not such a salutation comports with the facts. They are expected to refrain from personal attacks on each other—to avoid what are called ad hominem arguments, which the Greeks held to be the lowest form of debate because they distracted attention from the real issues. And, though these and other forms of Senate courtesy may sometimes sound odd or forced, they can be very important to the operation of the Senate because, as Barbara Sinclair has written, they facilitate "productive bargaining among senators and, consequently, Senate decision making."[60]

There are some experts (and Barbara Sinclair is one) who appear to feel that the courtesy norm has been seriously weakened in the Senate. Without doubt, in recent years there have been a number of instances of intemperate, objectionable words by one senator about another on the Senate floor. Sinclair, for example, cites a 1981 incident when, vexed by a statement by Republican Senator John Heinz of Pennsylvania, Lowell Weicker, then a senator from Connecticut, declared on the Senate floor that "Anyone who would make such a statement is either devious or an idiot. The gentleman from Pennsylvania qualifies on both counts."[61] Weicker later had these words stricken before they could be printed in the *Congressional Record*. During the 1991 debate on the confirmation of Clarence Thomas as an associate justice of the Supreme Court, Utah Republican Senator Orrin Hatch, disagreeing with Massachusetts Senator Edward Kennedy's criticism of the tactics of another Republican senator, said "Anybody who believes that, I know a bridge up in Massachusetts that I'll be happy to sell to them." This was an apparent reference to Kennedy's accident at Chappaquiddick Island in 1969 in which a woman passenger was killed. Senator Hatch later apologized and said that he had meant to say Brooklyn, rather than Massachusetts.[62]

But there has probably never been a period in Senate history when such breaches of senatorial courtesy did not take place, and in some cases, the threat of violence attended them. For example, during the classical Senate era of the Great Triumvirate—Clay, Calhoun, and Webster—Henry Clay made such offensive remarks about Senator William King in an 1841 Senate debate that King challenged Clay to a duel, Clay hotly accepted the challenge, and the duel was only prevented by a judge's action in putting both men under peace bond.[63]

During an 1850 Senate debate on slavery, one of the great senators of all time, Thomas Hart Benton of Missouri, was so offended by the mean and bitter words of Senator Henry S. Foote of Mississippi that he leaped from his seat and charged toward the Mississippian with violent intent. Foote immediately whipped out a pistol and pointed it at his would-be attacker. A tumult ensued. Senators attempted to disarm Foote and hold back Benton, who declared heroically, "I have no pistols! Let him fire! Stand out of the way! I have no pistols! I disdain to carry arms! Stand out of the way, and let the assassin fire!"[64] Order was finally restored, and the debate continued.

During the Civil War, in 1863, Senator Willard Saulsbury of Delaware, in a speech primarily attacking President Lincoln, referred to earlier remarks by other senators as the "blackguardism that can be uttered on this floor." The vice president, presiding, ordered Saulsbury to take his seat, and when he refused ordered him removed from the chamber by the sergeant at arms. While the written record mildly reports that "It was understood that Mr. Saulsbury refused to retire, but at a subsequent period he left the chamber," the real facts were that the offending senator had pulled a gun on the sergeant at arms and had threatened to shoot him.[65]

Senatorial discourtesy was sometimes disingenuously couched in words of indirection, but the meaning was just as sharp. In 1879, for example, senators Roscoe Conklin of New York and Lucius Q. C. Lamar of Mississippi got into a nasty confrontation on the Senate floor. Conklin, angered by Lamar's having appeared to call a statement of his a falsehood, declared, "I have only to say that if the Senator—the member [this title meant as an insult] from Mississippi—did impute or intended to impute to me a falsehood, nothing except the fact that this is the Senate would prevent me denouncing him as a blackguard and a coward." The record shows that this statement was followed by applause. Then Lamar arose again and this time said, "Mr. President, I have only to say that the Senator from New York understood me correctly. I did mean to say just precisely the words, and all they imported. I beg pardon of the Senate for the unparliamentary language. It was very harsh; it was very severe; it was such as no good man would deserve and no brave man would wear." Applause again.[66]

During the 1930s, no senator was the subject of more discourteous attacks from his fellows than Huey Long of Louisiana. Long provoked this behavior because of his too-frequent use of the filibuster to try to get his way, once blocking an extension of the National Recovery Administration program for fifteen hours by reading Southern recipes at length, championing the qualities of "pot likker" and explaining how to fry oysters properly. This caused Vice President John Nance Garner to pronounce from the chair that it was "cruel and unusual punishment" for senators to have to sit and listen to Long.[67] On another occasion, Senator Kenneth McKellar of Tennessee, speaking in the Senate, said of Long, "I do not believe the Senator could even get the Lord's Prayer endorsed in this body if he undertook to do so," adding that "the Senator from Louisiana has an idea that he is a candidate for President. For Heavens sake!"[68] And Senator Alben Barkley of Kentucky, after the Senate's presiding officer had admonished the people in the galleries against laughter during a Long speech, declared on the Senate floor, "I appeal to the Chair not to be too harsh with the occupants of the galleries. When people go to the circus they ought to be allowed to laugh at the monkey."[69]

But Huey Long could give as good as he got. Against his own Democratic leader in the Senate, Joe Robinson of Arkansas, Long once brought a copy of the *Martindale-Hubbell Law Directory* to the Senate floor and read from it a lengthy list of corporate clients represented by Robinson's firm, thereafter adding this comment: "Now I do not say that just because a man is being paid by the special interests, that would influence his activities as a U.S. Senator. Oh no." Long was called to order under Senate rules and made to take his seat.[70]

In the Senate of the 1950s that Donald Matthews studied, Democratic Senator Robert S. Kerr of Oklahoma was notorious for his sharp and mean words in debate. He once took part in a heated exchange with the heavy-bodied—and, some said, heavy-witted—Republican Senator Homer Capehart of Indiana over President Eisenhower's economic policies. After charging that Eisenhower had "no brains" (a declaration he later changed, for the *Record*, to "no fiscal brains") Kerr, with little indirection, said that Capehart was "a tub of ignorance." When called to account for this, he corrected himself by saying that he had meant to say that the senator from Indiana was "a *rancid* tub of ignorance" (neither statement about Capehart survived in the printed record).[71] Liberal Democratic Senator Paul Douglas of Illinois was more of an intellectual match for Kerr, but the Oklahoman was so "ugly and personal" in debates between the two, as one senator put it, that Douglas sometimes wound up with tears in his eyes.[72]

For a time after the late Hubert H. Humphrey, Democrat of Minnesota, entered the Senate in 1949, he occasionally drove home crying at the end of the day because of the way he was treated by the powerful old Southern bulls of the Senate, whom he had angered by his successful fight for a strong civil rights plank at the Democratic National Convention the preceding year.[73] Once, for example, as Humphrey was walking down the Senate's center aisle past the desk of the enormously powerful Senator Richard Russell of Georgia (for whom one of the Senate's office buildings is now named), the Georgian said to a nearby colleague in a voice that Humphrey would hear, "Why would anyone send a son of a bitch like that to the Senate?"[74]

There have always been lapses, then, in the observance of the Senate's norm of courtesy. And it may be that there has been a certain amount of slippage in recent years. Some senators feel that the 1981 shift of control of the Senate to the Republicans, led after the 1984 elections by a highly partisan majority leader, Robert Dole of Kansas, brought a greater degree of Senate partisanship than had previously been seen and that this reduced somewhat the level of compliance with the Senate norm of courtesy.[75] The influx at that time of inexperienced and hard-right senators may have played a part, too. Democrat Bill Bradley has stated:

> There is civility. . . . But there are certain senators that evoke a kind of visceral reaction—five or six [very conservative] senators that poison the atmosphere, who choose amendments on a substantive question solely for the specific, planned use in a coming campaign; they already know in advance how they will frame the thirty-second TV ad. Otherwise, there is less confrontation than I expected.[76]

But, weakened or not, there is no question that courtesy is still a norm that senators think is important and a significant aspect of right conduct in the upper chamber. As expert Senate observer Ross Baker has put it, "Almost all senators profess distaste for highly dogmatic and inflexible colleagues, and there are always a few senators in every era pointed to by their fellow members as troublesome and obstructionist, but widespread and enduring incivility is not characteristic of the Senate."[77] Baker interviewed fourteen present and former senators in 1986 and 1987 and found that all of them felt that "interpersonal comity was still a defining feature of the Senate."[78]

A 1988 series of interviews with senators confirmed Baker's conclusion about Senate courtesy.[79] When asked how important they felt this norm currently was, senators said:

> Still important—very, very important. There is a real contrast between the House and Senate. Personal relationships are very important in the Senate. (David Boren [D., Okla.])

> There is less [courtesy] than when I arrived, but probably still more than in any legislative body in the world. (Sam Nunn [D., Ga.])

> Important. And there is still a lot of it. (Dale Bumpers [D., Ark.])

> There has been some deterioration, I suppose. There is an increasing number of senators who feel that. But [Robert] Byrd's [Senate] history will show that the Senate was not always a courteous place in the past. (Richard Lugar [R., Ind.])

> It works, has to work, with some exceptions. You can disagree, but you should keep it as a clash of issues, rather than of personalities. A senator should not become vicious; those who do pay a price. It is necessary to do this, to work shoulder to shoulder, in order to work together on another issue. (Warren Rudman [R., N.H.])

> Courtesy—decency, comity—is still important. The Senate is the epitome of civility. (Pete Domenici [R., N.M.])

> Courtesy is still important. There's not been much change since [Edmund] Muskie's day. (George Mitchell [D., Maine], who served on Muskie's staff in the 1960s)

Collegiality and Courtesy

Increased Senate partisanship has affected an aspect of the norm of courtesy known as "collegiality"—the idea that the Senate is a kind of club, "the world's most exclusive club," as some like to put it. Will, for example, one club member try to unseat another? There was a time when the answer was no, although there were always exceptions—Huey Long campaigning against Joe Robinson in Arkansas in the 1930s, for example, or Joseph McCarthy, the red-baiting Republican senator from Wisconsin, campaigning against Senator Millard Tydings of Maryland in the 1950s. "The old rule was that a senator did not campaign against another senator at all," Democratic Senator David Boren of Oklahoma has said.[80] And still today, no senator would think of helping a challenger unseat a senator of his or her *own* party, no matter how much disagreement or dislike there might be between them. But times have changed in regard to helping a challenger of one's own party against a senator of the opposite party. According to Senator Boren: "That has broken down. Some do. Sometimes bitterness lingers, and that's bad."[81]

The breakdown, senators feel, has come as a result of the increased partisan activity of Senate Republicans in the late 1970s and early 1980s. As Senator James Sasser, Democrat of Tennessee, has stated, "They [senators] do that now [get involved in campaigns against a Senate incumbent], at least *for* the opponent. The NRSC [National Republican Senatorial Committee] is responsible for this."[82]

Some of the more senior members of the Senate still will not take part in *any* way in a campaign against a sitting senator. Republican Senator Strom Thurmond of South Carolina is one of these.[83] For most senators, though, this stricture has been relaxed somewhat:[84]

> I seldom at first campaigned for a challenger against a sitting senator, and I will not campaign against a senator personally. (Sam Nunn [D., Ga.])

> The campaign committees now recruit challengers, but people still won't campaign against a senator personally. (Mitch McConnell [R., Ky.])

> I don't campaign against an incumbent personally, and it has only been in the last two or three elections that I have gone out to campaign for a challenger at all. (J. Bennett Johnston [D., La.])

> My preference is for positive campaigning. But I would have no hesitation in pointing out disagreements on issues. (Bill Bradley [D., N.J.])

In fact, senators can think of only three examples in recent years when a senator has campaigned against another sitting senator, criticizing him or her by name. One occurred when Nebraska Democratic Senator J. James Exon made a public statement against the incumbent Republican senator from Nebraska, David Karnes (who had been appointed to fill a vacancy), during the 1988 campaign.[85] Senator Exon has admitted this lapse but in defense claims that he himself was the target of personal criticism by a senator: "It happened to me when I ran for reelection; [Senator Richard] Lugar [Republican of Indiana] criticized me by name. He was the [campaign finance] man for the Republicans at the time."[86]

The 1990 campaign saw another of the rare instances when one senator has campaigned against another by name. Democrat Wendell Ford of Kentucky raised eyebrows when, campaigning in South Dakota for the Democratic nominee there against incumbent Republican Senator Larry Pressler, he was quoted as saying that Pressler "doesn't work," "doesn't mingle," and "doesn't prepare for anything."[87]

More typical senatorial behavior is illustrated by what happened in New Mexico Democratic Senator Jeff Bingaman's reelection campaign in 1988. His Republican opponent, Bill Valentine, ran a television ad charging that Bingaman's Senate votes had differed "over 700 times" from those of the state's popular Republican senator, Pete Domenici, whose picture was used in the ad. Valentine's efforts were nullified when

Domenici immediately announced to the local press that he had neither authorized the advertisement nor the use of his picture in it, further volunteering that "On many issues that pertain to New Mexico . . . we [Bingaman and Domenici] vote similarly most of the time."[88]

There is still a sense of collegiality in the Senate, although some feel there is not as much close personal friendship and socializing among senators as there once might have been, perhaps a result of busier schedules and increased staff. According to Republican Senator William Cohen of Maine, for example, "the Senate is not quite so clubby as it once was. There is less personal contact now. Contact is mostly through mail and through staff."[89] Similarly, Democratic Senator David Boren of Oklahoma has said, "Senators know each other less than we used to. There is less socializing. Traveling together is the best time to talk personally. The size of our staffs [is a problem]. When you talk to a colleague about an amendment, the answer is 'Who's your guy on that? Have them call my guy.'"[90]

Collegiality is not the same as friendship. According to Senate observer Ross Baker:

> Friendship is a habitually misunderstood feature of the upper chamber. Few senators are buddies, although there was more social interaction outside the walls of the Senate in the past than there is now. What senators call friendship, and what is vitally important to the smooth operation of the Senate, is a kind of stable business relationship based on trust and reciprocity that grows up between people who may have never exchanged a confidence, or traded volleys on a tennis court, or drunk bourbon together.[91]

Among senators the general view is that, while there should be more of it, there is still a goodly amount of that type of business relationship in the Senate. Republican Senator Bob Packwood of Oregon, while pointing out that there is less outside senatorial social contact than before, because "now, senators are more inclined to go home for a couple of days, and demands on evenings are infinitely greater—fundraisers, charity events, openings, and so forth," has stated, "But there is just as much business intercourse."[92]

Former House members who have come to the Senate feel that, whereas they may have had a few tighter friendships in the House, there is more collegiality in the Senate. "In the House, you get to know guys on your committee well," Democratic Senator Wyche Fowler, Jr., of Georgia said in 1988. "Here [in the Senate], you can really get to know the Democrats and whatever Republicans you want to. It is more like a club. There are twice as many [senators] I feel close to."[93] Another former House member, Democratic Senator Barbara Mikulski of Maryland, has said that, in comparison with the House, the fact that the upper chamber is a "smaller body allows more contact; committee work is more intimate."[94] And Democratic Senator Timothy Wirth of Colorado stated:

One of the things that's so impressive about this place is the personal relationships that exist over here. Because of the filibuster and the fact that so much gets done by unanimous consent and the comity that's needed to support that kind of system, what grows up is a great deal of personal rapport among members of the Senate that is real.

People talk about this place being a club. It has to be a club in order to get unanimous consent. But underneath that grow up very strong, very real relationships so that people know your style and approach enough to trust you and take you seriously.[95]

Today there is actually at least as much, and perhaps more, personal contact among senators during business hours, albeit along party lines, as there was in the 1960s. In former times, senators saw other senators in meetings of the committees on which they happened to serve together, when they traveled together, when they milled around on the floor together after a roll-call vote (and they still do all these things). But party conferences in the Senate of the 1950s, for example—that is, private, separate conclaves of Senate Democrats and Senate Republicans—met quite infrequently. For the Democrats especially, party conferences were few and far between during the 1960s and into the 1970s. But after the Republicans gained control of the Senate in 1980, the Democrats, under then Majority Leader Robert C. Byrd, began to meet regularly and even to hold party retreats outside Washington. That practice continued, so that now both parties hold a separate party-conference luncheon every week, as well as other conference meetings. These are well attended. Thus, Republican Senator Don Nickles of Oklahoma, when asked whether there is today sufficient personal contact among senators, has said, "Within parties, yes. Each meet once a week. Across party lines, no."[96]

As was always true, especially in the old days for Southern Democrats, many senators use the ultra-exclusive senators' private dining room in the Capitol—actually segregated by party into two separate rooms—for socializing with other members, and they assiduously take advantage of the opportunity to see senators at the party luncheons. Senator Mikulski has said, "One of the things I most enjoy is the Tuesday luncheon of the Democrats. You can't do that in the House. It is the only thing on my schedule that is fixed."[97]

Specialization Norm Becomes Expertise Norm

In the Senate of the 1950s, Donald Matthews found that widespread observance of the specialization norm was important to the efficient operation of that body: "If many more senators took full advantage of their opportunities for debate and discussion, the tempo of the action would be further slowed. The specialization folkway helps make it possible for the Senate to devote less time to talking and more to action."[98]

To conform to this norm, Matthews said, senators were expected to restrict their activities to two kinds of issues:

> According to the folkways of the Senate, a senator should not try to know something about every bill that comes before the chamber nor try to be active on a wide variety of measures. Rather, he ought to specialize, to focus his energy and attention on the relatively few matters that come before his committees or that directly and immediately affect his state.[99]

But today, restricting a senator's energy and attention to those matters that "come before his committees" or that "directly and immediately affect his state" would, in reality, amount to much less of a limitation than it did in the 1950s. For one thing, as we have seen, American society has been nationalized, and nearly every major national issue now affects every state. Michael Foley, for example, has shown how, through the 1960s and 1970s, the new problems of "poverty, civil rights, housing, environmental protection, urban renewal, crime, and consumer rights" became national problems that were apparent and pressing in most population centers throughout the country and how, as a result of this, "all senators to some extent became generalists. In order to serve their increasingly large and complex constituencies, senators had to master a wide range of issues and to spread themselves over the full gamut of their state's problems."[100]

In addition, senators now serve on more committees and subcommittees than they did in the 1950s. Back then the average senator served on two standing committees and fewer than five subcommittees; by 1990 the average senator served on three or more standing committees and seven subcommittees.[101]

Take Bill Bradley, for example. On the eve of his reelection to a third term in 1990, Bradley served on four Senate committees—Finance; Energy and Natural Resources; Intelligence; and Aging. He also served on seven subcommittees, chairing two of them (the International Debt Subcommittee of the Finance Committee and the Water and Power Subcommittee of the Energy and Natural Resources Committee). If Bradley were to confine his energy and attention to just those matters that fall within the general purview of the committees and subcommittees on which he serves or that could be said to directly and immediately affect his home state of New Jersey, very few of the major subjects that come before the Senate today would be off limits for him. Little wonder, then, that the 1988 Index of the *New York Times* lists forty news items in which Bradley figured prominently that year and that the subjects involved ranged from arms control to credit to homeless persons to the Soviet Union to taxation to water pollution.

A more informed and attentive public, the advent of television, year-round Senate campaigns, a proliferation of national interest groups and issues—all these have also played a part in pressing senators away from the narrow focus of earlier times. As Senator Richard Lugar of Indiana has said, "Even among senators who want to specialize, their constituents expect that they will be able to speak on all subjects."[102] Similarly, according to Senator David Boren of Oklahoma, "There is some erosion [of the specialization norm] because of the pressure for amendments. Television has changed this. Everybody is pushed into introducing some suggested, high-intensity amendment."[103]

In the Senate, every measure is freely open to amendment, and amendments do not usually have to be germane to the measure. On a single day, then, as various amendments are offered to a pending bill, issues that senators have to consider on the floor can gyrate wildly across a wide spectrum. This, combined with the fact that more and more Senate decision-making has moved to the Senate floor in recent times, as well as new budget laws and rules that require the Senate annually to adopt an overall, integrated federal budget, offer much greater opportunities for individual senators to have a say on a broad range of policy questions.

Considering all these developments, it should be no surprise to learn that there has been a decline in the degree of specialization by senators. Barbara Sinclair has come up with two different ways to measure the number of "generalists" in the Senate at different times in its history, both of which show a significant increase in generalists over time. By one measure, Sinclair has calculated that the number of senators who offered floor amendments to bills that came out of four or more different Senate committees increased from four senators in the Eighty-fourth Congress (1955–1956) and nine in the Eighty-sixth (1959–1960) to twenty-nine in the Ninety-ninth Congress (1985–1986). By the other measure, she has shown that the number of senators who offered floor amendments dealing with three or more (out of six) broad issue categories increased from six senators in the Eighty-fourth Congress and fifteen in the Eighty-sixth to thirty-seven in the Ninety-ninth.[104] (Significantly, both measures show that there was a greater number of Senate generalists in the 1970s than there was in the 1980s.)

Still, it should be noted that if, by these measures, twenty-nine or thirty-seven senators can be considered generalists, then a substantial majority of senators are *not* generalists. Sinclair herself concluded as a result of interviews of Senate aides that committee assignments give senators special issue-oriented platforms and focus and "it is easier to exercise influence on a matter that comes before a committee on which he serves"; thus, as a practical matter, it continues to be true that "a senator is better off choosing a committee issue [that is, specializing] over a non-committee issue if the payoffs of involvement are roughly equal."[105]

The specialization norm has not disappeared. It has broadened into what can be called a norm of *expertise*. David Rohde and his co-authors, who interviewed senators in the early 1970s, expressed the norm change in this way: "All of the senators with whom we discussed the matter said there was no bar to members being active in areas outside their committees if (as many offered the caveat) they know what they're talking about."[106] Senators interviewed in 1988 endorsed the same view, whether or not they felt, as many did, that the old specialization norm had been weakened or changed somewhat:[107]

> People [in the Senate] who are looked to are the ones with expertise; people who speak on every subject don't have the kind of respect here needed to get things done. (Sam Nunn [D., Ga.])

> The most important decision senators make is how they spend their time; it requires planning and discipline, but you must choose a few issues. (Bill Bradley [D., N.J.])

> Show me a jack of all trades, master of none, and I'll show you an ineffective senator. (Warren Rudman [R., N.H.])

When a former West Virginia governor with a famous name, John D. (Jay) Rockefeller IV, came to the Senate in 1984, he decided to follow "the Bradley model," as he called it. "I used to talk about it with Bill Bradley," Rockefeller has said. "Don't say much, work hard, learn stuff, and when you have, then go get it."[108] By 1990, Rockefeller was beginning to be listened to in the Senate on health policy, one of his fields of concentration (and one within the jurisdiction of the Finance Committee, of which he, like Bradley, is a member). His colleague from West Virginia, senior Democrat and Senate President Pro Tempore Robert C. Byrd, encouraged Rockefeller in this expertise-norm approach. "The way to become influential in this body is to know your subject well, and the other people will start to listen to you."[109]

Political scientist Steven Smith would call the expertise norm the "Sam Nunn model":

> Developing a reputation for work in a particular field can establish a claim to a leadership role on a wider range of issues than committee jurisdictions would allow. . . . The model to whom many senators refer is Sen. Sam Nunn, D-Ga. Nunn chairs the Committee on Armed Services and has been involved in defense issues thus far, but he has extended the reach of his personal involvement to many foreign policy and economic issues that belong to the much larger field of national security policy.[110]

Legislative Work Norm Metamorphoses

In describing the legislative work norm as he found it to exist in the 1950s Senate, Donald Matthews quoted a Senate baron of the time, the late Senator Carl Hayden. Hayden told Matthews that he had been admonished when he came to the Senate that there were two kinds of senators, "show horses" and "work horses," and that "If you want to get your name in the papers, be a show horse. If you want to gain the respect of your colleagues, keep quiet and be a work horse."[111] Declaring this was still true in the 1950s, Matthews wrote that senators who devote a major share of their time, energy, and thought to the "highly detailed, dull, and politically unrewarding" work of the Senate "are the senators most respected by their colleagues," and he added that those "who do not carry their share of the legislative burden or who appear to subordinate this responsibility to a quest for publicity and personal advancement are held in disdain."[112]

Not all senatorial publicity was bad, though, Matthews wrote. Publicity necessary to get elected and reelected was all right, and so was publicity calculated to further a legislative cause, as well as publicity that flowed naturally from a senator's position or performance. The main thing, according to Matthews, was that publicity not "interfere with the performance of legislative duties," that "a senator give first priority to being a legislator."[113]

These sentiments from the 1950s are still generally operative in today's Senate. But times have changed, and so has the application of these senatorial understandings. The 1950s choice between being a show horse or a work horse does not make much sense now. In fact, it is almost certain that there is no empirical way today to divide senators into one or the other such category reliably.[114] In large measure, they are expected to be both—a result of the modern pervasiveness of television and media attention generally, the changed nature and almost constant necessity of campaigning, greatly heightened constituency contact, and enormously enlarged opportunities and pressures for national advocacy. Two prominent Republican members of the Senate made this point in 1988 interviews. Senator Warren Rudman of New Hampshire said, "Work horses get things done and get respect here, and then wind up getting more shows than show horses." Senator Orrin Hatch of Utah added, "Because of television, it's possible to be both (and it's not bad to be)."[115]

James Miller in his excellent micro report on one week in the life of the U.S. Senate, during April 1983, pinpointed how important television had become for senators. He noted that the old-timers who had scoffed at television were no longer there, and he called the new Senate "a centipede that dances to a television tune." Television offers members, Miller wrote, both an opportunity "to prove themselves to the folks back home—thus alleviating their insecurities about losing their jobs in the next election" and a "vehicle to national political stardom, if they know how to use it."[116]

Primarily for that same year of 1983, Stephen Hess did an analysis of how the principal national television networks, newspapers, and weekly news magazines covered senators, and concluded, among other things:

> The distinction between show horses and work horses (that is, between those who get publicity and do little legislative work and those who do not get publicity but do most of the legislative work), while once valid, is no

longer a distinction the news system makes. Today those who do the work get most of the publicity.[117]

That is as true for home-state publicity as it is for national publicity. Political scientist Steven Smith has written that changes in local media coverage, including those made possible by the reach of satellite communications, "have undermined the traditional distinction between work horses and show horses," so that "Even the most serious legislators now head to the Senate television studio or the Capitol steps to talk to local reporters after debating an issue of either national or local importance."[118]

Along the same lines, *New York Times* reporter Steven Roberts has spoken clearly about two developments that have permanently and significantly changed the relationship between senators (and representatives) and the news media:

> Congress itself has altered its rules, to make its inner workings much more open and visible; and the pervasiveness of television has broadcast those inner workings to the country. More importantly, television has given individual lawmakers, even very junior ones, a chance to make a public mark without the approval, or even the knowledge, of their leaders. . . . As Senator John Danforth, a Missouri Republican, has put it: "For politicians, the key to success is to be in the news."[119]

And senators certainly take advantage of the increased opportunities they have to be in the news. Stephen Hess showed that the national media he studied* generally concentrated their Senate attention on the committee and party leaders of that body.[120] In a later analysis, Barbara Sinclair came to the same conclusion, but she also discovered that national coverage of senators was somewhat wider: in 1985 all but seven senators were mentioned at least once on the television network evening news programs; the typical senator made the nightly news 8.6 times and and was mentioned in an average of 7.7 stories in the *New York Times*. Sinclair went on to point out that a majority of senators, fifty-three, appeared on the Sunday television interview programs ("Meet the Press," "Face the Nation," and "This Week with David Brinkley") an average of 1.6 times during 1985 and 1986; through 1985, at least eighty-two senators had appeared on public television's MacNeil/Lehrer Report at least once, and many several times; and C-SPAN, Cable News Network, and National Public Radio provided additional multiple opportunities for senators to make national news.[121]

Sinclair reported that the national news media gave heaviest coverage to committee chairs and to senators who were "legislative activists."[122] Again, though in a somewhat broader sense, this forces the conclusion that in today's Senate the show horses *are* the work horses.

---

* These were principally three television network news programs, three Sunday television interview programs, and five large metropolitan newspapers, all for the year 1983.

The changed nature of Senate campaigns and the almost year-round necessity for campaigning have also been factors in the demise of the show-horse–work-horse dichotomy. James Miller has written that the Senate has become more like the House in that, among others things, it is "more centered around perpetual campaigning than it used to be."[123] Interviews in 1987 with nine senators who had formerly served in the House showed that every one of them felt that senators had become more media oriented because of their heightened reelection concerns, agreeing, as one of them concluded, that:

> Senators are as worried about re-election as House members, in some sense even more worried. Sure there are six-year terms, but a vote at the beginning of a six-year term means exactly as much, has the same impact as a vote at the end of that six-year term because your opponent could take that vote six years ago, do some blow-up TV ad, and they could distort the issue. . . . There's . . . greater insecurity here than there is in the House, by and large; emotional insecurity; that is, reelection insecurity in this body.[124]

The disappearance of the show-horse–work-horse choice in the Senate has also resulted from the fact that senators must and do pay tremendously more attention to their constituents than they did in the 1950s. They must put much more emphasis on what Richard Fenno has called "home style."[125] James Miller noted in observing the Senate that a "job that never ceases" for a senator's offices, in Washington and at home, is "the dissemination of mail, newsletters, and other miscellaneous missives to the all-important . . . Folks Back Home," adding that senators "have to be public relations minded now, and treat their offices like small businesses."[126]

Senators now go back to their home states an average of twice a month during Senate sessions.[127] Days spent by senators in the home state have greatly increased too:[128]

1959–1960 sessions—average 7 or 8 days annually
1965–1966 sessions—average 18 days annually
1969–1970 sessions—average 35 days annually
1973–1974 sessions—average 75 days annually
1979–1980 sessions—average 84 days annually

Though the most respected senators do not go overboard,[129] increased "home style" has come to be encouraged by the Senate through expanded staff, broadcast studios, and increased allowances and a more relaxed Senate schedule to facilitate home travel.

Finally, the show-horse–work-horse choice is no longer a valid one because of the vastly increased pressures and opportunities for senators to become national advocates or spokespersons. Political scientist Nelson Polsby has written that the prototype of today's national senator was Democrat Hubert H. Humphrey of Minnesota, who in the late 1940s and early 1950s sensed (much earlier than most of his colleagues) the great possibilities there were for a senator to take the lead on national issues—using Senate speeches and activity, as well as outside media, public appearances, and writings, to publicize the issues, get them on the public agenda, and build support for them. Humphrey became an early advocate of national issues involving civil rights, Medicare, housing, aid to farm workers, food stamps, job corps, area redevelopment, disarmament, and others. Polsby notes that "A little over a decade later most of them were law, and Humphrey had in the meantime become a political leader of national consequence. The force of his example was not lost on younger senators."[130]

There was a 1970s proliferation of this type of "national senator"—Democrats Edmund Muskie of Maine on air and water pollution, George McGovern of South Dakota on cutting military expenditures, Edward Kennedy of Massachusetts on health. This new type of senator saw himself, Michael Foley has put it, "as having a roving commission that permitted him to expand the area of his activism in response to personal interests and public demands," and "he was less concerned with cloakroom politics and internal status and more with publicizing issues, making direct appeals to national audiences, and mobilizing forces to pressure the government (including the Senate) into action."[131]

Over the last three decades, then, Polsby has written, the Senate changed from a states'-rights-thinking "inner club" to a "predominantly nationally oriented body"; it also became the main institutional source of presidential candidates, but "even senators with small realistic hopes of advancement to the presidency now frequently seek national recognition for their substantive legislative work, and not merely the approval of interest groups and citizens in their home states."[132] Even New York Republican Senator Alfonse D'Amato, who at first took pride in his local focus and called himself a "pothole senator," gradually broadened his interests and activities—from the problem of drugs in New York City to U.S. drug policy generally and from the effect of acid rain in the Adirondacks to concern about national environmental policy.[133]

Today's senators feel that their constituents expect them, as contrasted with House members, to be involved in national issues.[134] The fact that so much of their campaign money comes from national out-of-state sources also increases the pressure on them to take on these issues.[135] Senators are "inundated with requests" to become the advocates and spokespersons for national causes.[136]

Today the old legislative work norm has been replaced by two new norms. One can be called a *national advocacy norm*, an expectation that a senator will be a national activist and spokesperson for a cause, or causes. The other can be called a *diligence norm*, the requirement that a senator work hard at the job. This is the basic conclusion to be drawn from the study by John Hibbing and Sue Thomas concerning which senators are most respected and why. Hibbing and Thomas found that, for the most part, respect is accorded "to senators who are involved in substantive legislative activities" and that, "rather than being viewed with suspicion, senators who are media-savvy and interested in seeking the presidency are accorded additional respect in the modern Senate."[137]

In regard to the diligence norm, one senator has been quoted as saying that Senate influence is the result of "just year after year of patience—willingness to carry at least your fair share of the work."[138] Other respected senators agree, including Republican William Cohen of Maine, who has said that "The key people [in the Senate] are the people who spend long hours on legislative work."[139] And Barbara Sinclair has concluded that activism has replaced institutional position as the most important factor in Senate influence, adding that this "requires long hours and very hard work."[140]

National advocates in the Senate use the media "to publicize problems, to shape debate on an issue, and to build pressure for action."[141] They often use the media, too, to communicate with each other.[142]

Democrat Bill Bradley lives by the Senate norms of expertise, diligence, and national advocacy—and the respect he is accorded by his colleagues is the result. Bradley has explained his approach this way:

> I kind of work through issues. I take a subject and essentially learn it, think about how it works, what needs to be done and come up with ideas and fight for those ideas. I did that with energy [issues before his Energy and Natural Resources Committee], then with tax reform—that was a four year effort. It's not simply studying ideas, ideas separated from action. The legislative process is a matter of combining the two, of making ideas happen.[143]

Whatever the issue, the Bradley model is, first, to become an expert on it, then diligently work on it in the Senate and concurrently become a national advocate for it outside, as well as inside, the Senate. That is the course Bradley followed successfully in regard to tax reform, for example.

In 1982, when Bradley was ready with a plan to lower individual rates, simplify the structure, and close loopholes, he went public to build support for it and to get it on the public agenda. He used various methods to gain this national support, including a book; numerous op-ed page and other newspaper articles; speeches everywhere, including the 1982 off-year Democratic National Convention; television talk shows; news interviews of all kinds; and a private session with the 1984 Democratic presidential candidate, Walter Mondale, as well as with numerous convention delegates. In the Senate, and especially in the Senate Finance Committee on which he serves and which has jurisdiction over taxes, Bradley stayed constantly and persistently on the job, talking with members, lobbying them, ironing out disagreements. It all eventually worked. The basics of his plan were embodied in the Tax Reform Act of 1986.[144]

Reciprocity Norm Weakened

The old Senate norm of reciprocity has almost changed into a norm of individualism. For Donald Matthews, the norm of reciprocity that he found functioning in the Senate of the 1950s had three parts: (1) a senator should be slow to make a commitment to another senator, but, when such commitment was made, the senator should "live up to his end of the bargain, no matter how implicit the bargain may have been"; (2) when able to help out a colleague, a "senator should provide this assistance and . . . be repaid in kind"; and (3) a senator should not "push his formal powers to the limit."[145] In regard to the last, Matthews wrote:

> A single senator, for example, can slow the Senate almost to a halt by systematically objecting to all unanimous consent requests. A few, by exercising their right to filibuster, can block the passage of all bills. Or a single senator could sneak almost any piece of legislation through the chamber by acting when floor attendance is sparse and by taking advantage of the looseness of the chamber rules. While these and other similar powers always exist as a potential threat, the amazing thing is that they are rarely utilized. The spirit of reciprocity results in much, if not most, of the senators' actual power not being exercised.[146]

Most senators still try to work together in some harmony, keep their word (though at least a couple think this is not as rigid and as universal a norm as it once was), and help each other when they can.[147] It is in regard to refraining from use of the rules to the fullest that the reciprocity norm seems to have been weakened most. This weakening is evidenced particularly in two respects: by senators offering floor amendments to other senators' bills; and by filibusters and other obstructionist tactics.

Fully sixty senators offered three or more amendments to bills on the floor in the Ninety-ninth Congress (1985–1986), compared with only fifteen who did thirty years earlier.[148] In the Ninety-ninth Congress, eighty-six senators offered floor amendments to bills reported by committees on which they did not serve, whereas thirty years earlier only one-third of the senators did that.[149] Amending activity in the Senate generally peaked in the 1970s and then decreased some, but it is still greater than in the 1950s.

Senate rules have always made individual senators potentially quite powerful. Now, more senators are willing to use those powers. And senators have always claimed the right to speak at length on the floor of the Senate in order to delay or prevent a majority from acting—activity that came to be called a filibuster.[150] Present rules provide that a three-fifths vote of all senators—sixty votes—is required for cloture (cutting off debate). Southern conservatives once were the most-practiced wielders of the filibuster, to block civil rights legislation. But more recently, senators of all viewpoints and from all parts of the country have come to use the filibuster, even to block bringing a bill up for debate, and other obstructionist tactics, seemingly at the drop of a gavel.

In the 1950s, Senate filibusters averaged fewer than one per Congress. The average was four per Congress in the first half of the 1960s, five per Congress in the last half of that decade, eleven and a half per Congress in the 1970s, and a little over twelve per Congress during the 1980s.[151] Senator David Pryor of Arkansas was quoted in 1987 as saying that there had been more filibusters (eighty) in the 19 years from 1968 to 1987 than there had been in the 127 years before that period.[152]

Starting when their six-year control of the Senate ended after the 1986 elections, Senate Republicans have regularly used a party-backed filibuster against Democratic proposals that they oppose, such as public financing of congressional campaigns.

Some senators have made a career out of obstructionism. Democrat Howard Metzenbaum of Ohio sees himself as a kind of watchdog of the Senate, particularly against special-interest, pork-barrel, and anticonsumer legislation, and he uses the considerable powers of an individual senator, which are especially great in the waning days of a session, in furtherance of that self-assumed role.[153] Metzenbaum's willingness to risk the displeasure of his colleagues to protect the people's interests, as he sees it, has made him a power to be reckoned with. As fellow Democrat David Pryor of Arkansas (who has been a leader in efforts to make the Senate more efficient and improve senators' "quality of life") has said: "When legislation is being prepared by staff, they're preparing it anticipating problems with Sen. Metzenbaum. He's like the security guard at the airport: You know he's going to X-ray your bag, so you have to be clean."[154]

But a senator on the hard right, Republican Jesse Helms, is a more persistent, and more exasperating, practitioner of the obstructionist's arts, frequently pushing his senatorial powers to the last centimeter of their limit for his ultra-conservative causes. Helms kept the Senate in session almost all night in September 1988, for example, trying to stop federal funding for what he termed "obscene and indecent" art, was tromped finally by a vote of 62 to 35, and then came right back the next morning with the same issue and effort, finally losing again by a vote of 74 to 22.[155] Helms's critics call him "Senator No," and the overuse of his dilatory powers makes him quite unpopular with his colleagues. As one senator reported in 1988: "After Helms kept us all here all day yesterday, somebody said to David Pryor in the cloakroom: 'You don't have to have all those studies to determine how to improve the quality of life around here; just get rid of Jesse Helms, and that'll improve the quality of life more than anything!'"[156]

But the increase in recent years in the use of obstructionist tactics in the Senate has been general, although they may now be decreasing some (only one cloture vote was taken during the first eight months of 1989, compared with fifteen during each of the preceding two years[157]). The reciprocity norm has been weakened in the Senate. But it is not dead. Senators can still go too far with obstructionism, and those who do lose respect. Some of the Senate's most respected senators had this to say on the subject of reciprocity in 1988:[158]

> It [reciprocity] has broken down some. Now, the rules are used to the fullest. We now have constituents and constituent groups that push filibusters as a tactic, and it is not limited to major matters. (Sam Nunn [D., Ga.])

> It [reciprocity] has broken down some. Every senator is prepared to use the rules to the fullest, including the filibuster. There is not much antagonism about it, and it's not necessarily bad. (Bob Packwood [R., Ore.])

> The cause [of the breakdown in reciprocity] is single-minded devotion by a small group—so many things are of vital importance to them—and their duty to constituents or overall ideas causes them to feel that all the rules of the Senate must be mobilized to serve them. The filibuster is over-used. . . . So many are using the rules so much and with increasing frequency that there is an unhappy feeling around here. Often, they get their way, and this encourages more of this. Long delays, holds, use of the rules, threats of filibuster, deter the leaders. (Richard Lugar [R., Ind.])

> This [reciprocity] has been reduced. There are too many amendments on the floor, not enough deference to committees. There is an abuse of the filibuster, now, longer debate. (William Cohen [R., Maine])

> Oh yes, this [reciprocity] has broken down. Filibustering is overdone—and is used on motions to proceed. We should give more power to the chairs of committees, re-establish their authority. (Nancy Landon Kassebaum [R., Kan.])

> Yes [there are senators who cause an almost sure "no" vote] because they are more polarizing, politically and personally—that is, both ideologically and because of their Senate behavior. Reciprocity has broken down some. (George Mitchell [D., Maine])

Apprenticeship Norm Reduced to Seniority Norm

Donald Matthews wrote that in the 1950s the norm of apprenticeship was the "first rule of the Senate." The new member of the Senate in those days, he said, was "a temporary but very real second-class senator" who received the committee assignments, office suite, and committee and chamber seats that other senators did not want. The new senator was required to share in the "thankless tasks of the Senate," like presiding over sessions, and was expected to "keep his mouth shut, not to take the lead in floor fights, to listen and learn." Any senator who failed to comply with these strictures of the apprenticeship norm would be "met with thinly veiled hostility."[159]

Already, though, in Matthews's time, the norm of apprenticeship was changing. Veteran senators told him "rather wistfully" that the practice was "on the way out," and, he wrote, "they are undoubtedly correct." Matthews quoted one such senior senator as saying, "A new senator today represents millions of people. He feels that he has to *do* something to make a record from the start."[160] That is certainly true in today's Senate.

Changes in the Senate's external environment caused the apprenticeship norm to weaken, as was true with the other norms we have discussed. In addition, three internal developments played a part: Lyndon Johnson, as Democratic majority leader, changed the pattern for making Democratic assignments to committees in 1953 and decreed that, thereafter, every new Democrat would receive at least one good committee assignment; the Democratic class of 1958, large in number, was more willing to challenge Senate traditions; and Mike Mansfield of Montana, who followed Johnson as majority leader, operated on the principle, often expressed, that every senator was as important as any other.[161]

By the late 1960s, new senators could and did speak on the Senate floor as soon as they felt they had something worth saying—although even then a senator's first major speech, or "maiden speech" as it was called, was still looked upon as a special occasion.[162] Other senators would be notified of this impending rite of passage. They would come to the Senate floor and, when the new senator had finished, would rise and make effusive, complimentary remarks about the new senator's speech.

By 1973, just two senators out of forty interviewed thought new senators still needed to serve any kind of an apprenticeship.[163] Although freshman senators of the majority party were (and are) still expected to take their regular turn at the pledge chore of presiding over the dull and routine sessions of the Senate, the apprenticeship norm was, for all intents and purposes, gone.

New senators not only could speak out whenever they wanted to; they could also, and did, freely offer amendments to bills on the Senate floor. For example, in the Eighty-fourth Congress (1955–1956), only about one-fourth of the freshman members (with two or fewer years service in the Senate) had offered any floor amendments at all, and the per capita sponsorship of such amendments by this freshman group had been only two and a half amendments; by contrast, in the Ninety-second Congress (1971–1972), every freshman member offered at least one floor amendment, and per capita sponsorship of these amendments had risen to a little over sixteen amendments.[164] In the Ninety-ninth Congress (1985–1986), the per capita floor amendment sponsorship by freshman senators went down some, to about thirteen, but this was still five times the rate of the 1950s.[165]

Further, new senators no longer had to wait years to get a good committee assignment. In the 1950s and 1960s it was especially unlikely that a freshman senator would be appointed to one of the four most coveted and powerful Senate committees—Appropriations, Finance, Armed Services, and Foreign Relations. Even during most of the 1970s the number of freshmen who won such prestigious appointments was still small (only four Democratic freshmen from 1971 to 1976, and only two Republican freshmen from 1971 to 1978). For the next four or five Congresses (1977 through 1986 for the Democrats and 1979 through 1986 for the Republicans), however, 60 percent of Senate freshmen were assigned right away to one of the top four committees.[166]

In 1989 two of the five freshman Democrats and four of the six new Republicans gained immediate appointment to one of these committees. Three of the Democrats were named to the Democratic Policy Committee, which schedules legislation and reviews and recommends concerning substantive policy positions.[167] In 1990, twenty of the eighty-seven total positions on the four most powerful standing committees were held by senators still in their first six-year term.[168]

The seniority system in the Senate (the other side of the Senate apprenticeship coin), though, did not disappear when the apprenticeship side went blank. In today's Senate it is still an unbroken tradition that the most senior member of the majority party (in 1993 Democrat Robert C. Byrd of West Virginia) will be elected president pro tempore of the Senate. The one hundred personal suites in the three Senate office buildings are still unfailingly assigned on the basis of seniority: when a suite becomes vacant, it is offered first to the most senior of the one hundred members, and then to the next down, then the next, and so on. Only after all those more senior have turned down the suite is a more junior member, still in order of seniority, allowed to choose it. (Senators may prefer one office suite over another because of its size, configuration, or location—that is, for example, because it is closer to a senator's committee rooms, or closer to the subway to the Capitol, making it easier for the senator to get to the floor for votes.)

It is still seniority, too, seniority within the party (the parties being separated by the middle aisle), that determines the order in which a senator may choose a desk on the Senate floor. (Some senators prefer one desk over another because they want to sit next to a particular colleague, because the desk is closer to an aisle or to a door and therefore easier to get to, because it is more visible to the press gallery, or because it is a desk that was once occupied by a particularly famous senator, such as Daniel Webster or John C. Calhoun, all the names of senators who ever sat at each desk being listed inside the desk drawer.)

To the left and right in committee and subcommittee hearings and meetings, separated according to party affiliation, senators sit near the chair or far away purely on the basis of party seniority within the committee or subcommittee. The chair, while alternating back and forth between the parties, recognizes senators for questions or comments in the order of their seniority.

Leaders of committees and subcommittees are chosen on the basis of seniority. The majority-party senator who has served on the committee or subcommittee longest becomes the chairperson. The minority-party senator who has served on the committee or subcommittee longest is chosen as the ranking member. In 1971, the recommendations of a Democratic reform committee, named by Senate Majority Leader Mike Mansfield of Montana (and composed of Fred Harris of Oklahoma, as chair, and Hubert Humphrey of Minnesota and Herman Talmadge of Georgia, as members), made it possible for the Senate Democratic Conference to vote, by secret ballot, at the beginning of each Congress on approving or disapproving all committee chairs (and committee assignments).[169] Two years later, Senate Republicans adopted a rule under which the Republican members of each committee could elect the ranking member of the committee, subject to a vote of the Senate Republican Conference. In 1975, Senate Democrats adopted a formal resolution providing that nominations of committee chairs by the Senate Democratic Steering Committee would be put to a secret ballot in the Senate Democratic Conference when one-fifth of its members demanded such a vote.[170]

These new procedures almost surely affected the attitudes and behavior of committee chairs and ranking members, making them less autocratic. Quite significantly, however, the changes still have never caused any Senate chair or ranking member to be deposed, nor have they ever caused a senator other than the most senior party member on a committee to be chosen for committee leadership.

Senator Helms precipitated a challenge to this system in 1987. Helms had come to the Senate four years ahead of fellow Republican Richard Lugar of Indiana, so he was listed as senior to Lugar on the Foreign Relations Committee on which they both served (although both had gone on that committee the same year). In 1984, while the Republicans were in control of the Senate, the then chair of the Senate Foreign Relations Committee was defeated for reelection to the Senate. Helms was the senior Republican on both the Foreign Relations Committee and the Senate Agriculture and Forestry Committee. In line with a reelection campaign promise, Helms stepped aside on Foreign Relations, allowing Lugar to chair that committee, and Helms became chairman of Agriculture. Later, in 1989, after the Democrats had regained power in the Senate, Helms decided that he wanted to be ranking member of Foreign Relations rather than Agriculture. When instead the other Republicans on the Foreign Relations Committee voted unanimously to choose Lugar as their ranking member, Helms appealed the decision to the Republican Conference on the basis of his seniority. The Conference, by 24 to 17, voted for Helms—and seniority.[171]

The system as it relates to chairs and ranking members remains intact, and there is something to be said in favor of such a system. It elevates to committee leadership senators with experience in the matters within the committee's jurisdiction. It avoids the likelihood of bitter and time-consuming pecking-order fights at the beginning of each Congress. Former Democratic Senator Eugene McCarthy of Minnesota once said, approvingly but wryly, that "The Senate seniority system is like the old rule of primogeniture; you got some bad kings that way, but it saved a lot of trouble."[172]

Nobody would find it particularly surprising that in any organization the members who have been around the longest are likely to have more power and enjoy more respect than more junior members. That is the way it is, generally, in the U.S. Senate: the more senior the member the more respect received.[173] The more senior the senator, the more likely it is, too, that he or she is a committee leader—and being a committee leader is "clearly correlated with respect" in the Senate.[174] As we have seen, committee leaders get more media attention, and as we have also seen, "media exposure in recent Senates is more likely to attract respect than it is to repel it."[175] Thus, whereas the apprenticeship norm has largely vanished, or perhaps more accurately has been reduced to a norm of seniority, the seniority norm continues to be important in the U.S. Senate.

Institutional Patriotism Norm Changed

Senators of the 1950s were expected to have a kind of "school spirit" about their institution, to have an emotional commitment or investment in the Senate. They were, according to Donald Matthews, "expected to believe that they belong to the greatest legislative and deliberative body in the world. They are expected to be a bit suspicious of the President and the bureaucrats and just a little disdainful of the House. They are expected to revere the Senate's personnel, organization, and folkways and to champion them to the outside world. Most of them do." Senators were not to bring the Senate or senators as a class into disrepute or use their position for "self-advertising or advancement," Matthews found.[176]

Members of the Senate who today run for president or who go outside the body to publicize and build support for their causes, and incidentally themselves, do not lose respect in the Senate, as noted before. They gain more of it, even though their emotional commitment to the Senate is thus shown to be less than "total."

Senators may—and do—criticize some of the old ways of the Senate, the unpredictability of its schedule, and the obstructionist tactics of some of their colleagues, for example. But these are mostly constructive criticisms. Senators generally love being in the Senate and are loyal to it as an institution. Senators do not "run against" their body or criticize it as an institution the way members of the House do.[177] Even leaving the Senate because he had lost his zest for the job, Democrat Lawton Chiles of Florida, for example, went out of his way to emphasize to an interviewer that "this is a great place, and I wouldn't take anything for the experience."[178]

The same can be said for two senators who worked hard in the 1980s to improve the procedures and "quality of life" in the Senate, David Pryor (D., Ark.) and Daniel Evans (R., Wash.). Despite his attempts to reform it, Pryor has made clear that he feels that serving in the Senate "is probably one of the greatest jobs and challenges," that the Senate is a "great place," and that even its foibles are "sort of the sinew, the bone marrow that makes this place what it is."[179] Similarly, when Evans announced in 1988 that he would not run again for a second term, he wrote an article about reforms needed to make the Senate more efficient and its floor debate more meaningful. (Separately, he told an interviewer that his real reason for leaving the Senate was his feeling that he had come to the Senate too late and was too old [sixty-three] to build up much future seniority.)[180] He was careful to point out in the article that "I have spoken out not in disgust or discouragement but out of admiration—for my colleagues, but most of all for the Senate."[181]

Senators will quit the Senate to become president or vice president, but that is nothing new. Powerful and respected senators of the 1950s, such as Republican Robert A. Taft or Democrats Lyndon Johnson and Richard Russell, had their eyes on the presidency. But few senators will leave the Senate to run for governor (except, infrequently, in a large state like California), and seldom does a senator quit to take a cabinet position (although Democrat Edmund Muskie resigned to become secretary of state during the administration of President Jimmy Carter). Senators do not leave that body to run for the House, but the reverse happens all the time. And when, in 1987, Ross Baker interviewed fourteen senators who had first served in the House, he found all of them strong in the belief that they had made a good move, that the Senate was a great place to be.[182]

The institutional patriotism norm of the Senate has changed some with the times. Senators are not now expected to restrict themselves to a monk's cell of isolated, unquestioning devotion. But institutional patriotism still exists as a Senate norm and senatorial conviction.

The 1950s norms of the U.S. Senate were weakened or changed, then, as a result of the nationalization of the Senate's external environment, as well as by a continuing turnover in Senate membership, particularly with the influx of large numbers of majority-party senators during the critical-mass times of the early 1960s and the early 1980s. But the Senate still has norms of behavior, as we know from the fact that some senators are highly respected and others are not.

The courtesy norm is still important, although it has suffered some from heightened partisanship and hardened ideology. The specialization norm has been changed by the nationalization of issues and the increase in the size and number of committees and subcommittees. Limiting senators to subjects that concern their states or that come within the jurisdictions of their committees is much less of a limit now than it once was; in any event, senators are not today expected to focus only on such subjects so long as they conform to a norm of expertise—knowing what they are talking about—a norm that has taken the place of the old specialization norm.

The norm of legislative work—paying attention to legislative details and being a work horse, not a show horse—has changed into two other norms, the norm of national advocacy and the norm of diligence. Today the work horses and the show horses are largely the same senators, the "show" flowing justifiably from the work. All senators have greatly increased their "home style" attention back home, too.

The reciprocity norm has weakened perhaps more than any other. Senators now push their powers close to the limit. There are more filibusters and more obstructionism. Senators still like serving in the Senate, however, and while it may not be the total focus of their lives, they respect the norm of institutional patriotism.

Thus today's senators are more national and individualistic, and this has produced a nationalized Senate.

5

A National, Individualistic Senate

The United States Senate marked the two-hundredth anniversary of the first day a quorum was present in that body by meeting in extraordinary session in Washington on April 6, 1989. This bicentennial convocation brought together present and former senators in the capitol building's carefully restored Old Senate Chamber where the Senate had met from 1800 to 1859 and where the voices of Clay, Calhoun, and Webster seemed to resonate still. Following a brief ceremony and commemorative remarks in that historic place, the contemporary and former senators proceeded double-file down a hallway to the Senate's present chamber for a continuation of the special program.

There the Senate clerk called the roll of the fifty American states in the order of their admission to the Union. The senior senator from each state arose in turn to answer the roll call and to announce the date on which his or her state had been admitted to the Union (Delaware was the first to ratify the Constitution on December 7, 1787), as well as to present the names and dates of service of the state's first two senators.*

Democratic Senator Wendell Ford of Kentucky then discoursed on the first Senate, "The Senate of 1789." He was followed by Republican Senator Mark Hatfield of Oregon, who spoke on the hundred-year-old Senate, "The Senate of 1889." Hatfield mentioned during the course of

---

* At his turn, Senator Strom Thurmond (R., S.C.) provoked both laughter and gasps when, after making the standard announcement, he added: "South Carolina was not the first state to join the Union, but it was the first to secede."

his otherwise serious remarks that the centennial Senate still had a standing Committee on Revolutionary War Claims "long after there were any living Revolutionary War veterans to make claims." The reason this committee had continued in existence, Hatfield said, to the great amusement of his audience of senators and former senators, was "only to provide its chairman with a committee room and a clerk."[1] (Back in those days, most senators had virtually no personal staff and no offices other than their individual desks on the Senate floor or the parlors of their Washington boarding houses.)

All the past and present senators who were listening to Hatfield's history could have readily testified to the insatiable senatorial appetite for ever more space and staff. Indeed, the expansion of Senate offices and staff through the years, and especially since the 1950s, can serve quite well as a metaphor for the change the Senate has undergone.

When the Senate was a hundred years old, the total number of its staff members had not yet reached eighty—fewer than forty personal staff members for the eighty-eight senators and about forty other staff members for all the Senate's standing committees together.[2] Even so, the Senate wing of the Capitol did not provide nearly enough room for senators and staff. So in 1890 the Senate decided to buy a nearby building, Maltby House, for use as an office building. What forced this decision, finally, was the unusually prolonged and bitter-cold winter of 1889, when for many weeks senators had to travel back and forth each day on Washington's icy streets from their private boarding houses, in the poorly heated parlors of which they saw visitors and handled correspondence and other office business, to the Capitol, where committee meetings were held and the Senate met.

Maltby House, which the Senate bought for $138,000, was a five-story apartment building located a few steps away from the Senate side of the Capitol.[3] Senators and their standing committees moved into these new quarters in 1890, but they were disappointed to discover that the upper rooms were ovens in the summer and that the whole edifice was a perilous firetrap. Too, the old floors quivered and shook from the foot traffic, and the superintendent of the Capitol grounds, making a belated inspection, soon found out why: Maltby House had been constructed on unstable "made ground," the former site of a livery barn. "When its foundations were exposed," this official stated in declaring Maltby House unsafe, "they were found to rest on what was practically an old bed of manure, over two carts of which were taken out."[4] With admirable restraint, the superintendent refrained from playing this situation for laughs and soberly recommended that the Senate find other office space for itself.

Accordingly, in 1904 the Senate (then with ninety members) provided funds for the purchase of a site on the northeast corner of Delaware and B streets and authorized the construction of a new C-shaped, three-sided Senate office building. The cornerstone for this new edifice was laid in 1906, and senators were finally able to occupy the completed building in 1909.

Yet by 1931, just a little over twenty years later, this structure had already become too crowded. The number of personal staff members who worked for the now ninety-six senators had grown to nearly 300, the number of Senate committee staff members to nearly 170. So that year, the Senate authorized an addition to the building—a fourth side, enclosing an open courtyard in the middle. When the expenditures for this purpose and for other alterations, including a more decorative C Street facade and a subway to connect the building to the Capitol, were added to the original costs, the total price for the completed four-story building (plus a basement) came to $8.39 million.[5] First simply called the Senate Office Building, and later the Old Senate Office Building, it was named the Russell Senate Office Building in 1972.*

Senate staff numbers continued to grow inexorably. By 1948, senators' personal staffs totaled nearly 600 people, double their 1931 size; the number of Senate committee staff members had grown to nearly 250. The Senate, therefore, voted to purchase half of a city block just across C Street to the east of the Senate Office Building for a New Senate Office Building to comprise six floors of offices, plus a basement and a ground floor. The Korean War caused construction on this building to be delayed, however, and it was not completed and occupied until 1958. The price for this second building, today called the Dirksen Senate Office Building,† came to about $24.2 million. By the time the Dirksen Building was first occupied in 1958, the number of personal staff members for the ninety-six senators (Alaska and Hawaii not admitted to the Union until 1959) had grown to over 1,100, nearly double what they had been ten years earlier when the building was authorized, and committee staffs had already risen to the 400 mark.

From 1958 to 1972, Senate staff numbers doubled again—until there were 2,400 personal staff members for the one hundred senators, and committee staffs totaled nearly 850 people. In 1972 the Senate Public Works Committee reported that space provided for the Senate was no longer adequate and that "toilets are sometimes used for offices and meetings are held in hallways."[6] The Senate voted to purchase the rest of the large block on which the Dirksen Building sat and to build a connecting addition to it of nine floors of offices (plus a basement and a ground floor). This construction began in 1977 and was completed in 1982, at a total cost of nearly $138 million.[7] By the time the building was ready for occupancy, its lavishness had become a source for press criticism, and some senators eligible to do so delayed moving into it for

---

* In honor of the late Senator Richard Russell, Democrat of Georgia, who served in the Senate from 1933 to 1971.

† After the late Senator Everett McKinley Dirksen, Republican of Illinois, who served in the Senate from 1951 to 1969.

fear of being spotlighted by the adverse publicity. In 1982, as the first senators gradually began to move into the Hart Senate Office Building, as it was called,* there were over 4,000 people on senators' personal staffs, plus over 1,000 committee staff members. After that, staff growth slowed, but even so, by 1991 staff numbers (including personal staff, committee staff, and full Senate and leadership staff) totaled 7,200.[8]

The really dramatic surge of growth in Senate staff—a sixfold increase—came in the years between World War II and the present. Senators' personal staffs are today so large that Democratic Senator Wyche Fowler of Georgia laughingly commented in 1989, soon after coming to the Senate from the House, that "in the Senate, I have so many staff members that one of these days I'm going to walk into the reception room and introduce myself to someone as the Senator from Georgia and get the response, 'Why, senator, I've been working for you for three years!'"[9]

The completion of the Hart Building, and the move there by a number of Senate members, committees, and subcommittees, allowed the Senate offices that remained behind in the Russell and Dirksen buildings to spread out. Senator David Pryor of Arkansas was one of those who elected to stay on in the historic old Russell Building, with its high ceilings and ornate fireplaces. Pryor noted in a television interview in 1989 that the one big room he was then using solely for his private office had once constituted in the early 1940s the entire office space for Senator Harry S. Truman of Missouri and all his staff members. By contrast, Pryor's full senatorial offices, he said, encompassed *nine* such large rooms.

Senators' offices grew enormously in size because their staff grew enormously in numbers. With the great changes in the Senate's external environment after the 1950s—the nationalization of American society, issues, and interests, as well as of Senate elections and campaign financing—came growing waves of new, more individualistic and more nationally oriented senators. These newcomers were increasingly unwilling to sit quietly in their committees and on the floor of the Senate and follow the lead of the committee chairs and other, more senior members. They caused structural changes in the Senate—in greatly increased staff numbers, which allowed them simultaneously to become more national while shoring up their home bases, as well as in the proliferation of subcommittees and an increase in total membership on the most important standing committees. The new senators also caused

* When Senator Philip Hart (D., Mich.), who had served in the Senate since 1969, died in office in 1976, the Senate voted to name the new building then under construction after him. Ironically, the Senate took this action despite the fact that, in 1972, Hart had opposed naming the Senate's first two buildings after senators Russell and Dirksen so soon after their deaths, when senators were still moved by grief and when "history's verdict has yet to be received."

process changes—altering the Senate's norms and rules to make it a much more open and accessible place, but one in which greater individualism and democracy were unfortunately accompanied by seriously increased obstructionism. We want to consider each of these structural and process changes in turn.

Staff-Growth Causes and Effects

As senators became more individualistic and more national in their outlook, more and more of them began to demand an increase in staff resources.* One of the first modern pleas for Senate staff increases had come earlier, though, in 1946, from Senate reformer Robert M. La Follette, Jr., Republican of Wisconsin, who declared:

> Undoubtedly one of the great contributing factors to the shift of influence and power from the legislative to the executive branch in recent years is the fact that Congress has been generous in providing expert and technical personnel for executive agencies but niggardly in providing such personnel for itself.[10]

After the Monroney–La Follette Legislative Reorganization Act of 1946 was adopted, neither the Senate (nor Congress as a whole) was ever again to be called niggardly on the subject of its own staff assistance. Later general reforms and reorganizations,[11] as well as specific increases, resulted in new jumps in staff numbers and periodic doublings, as already noted.

The principal argument for enlarging congressional staff was that the congressional work load had increased enormously, primarily from the advent of a much more attentive and activist public. For example, a 1984 congressional management study reported:

> The amount of constituent correspondence entering the House and the Senate has skyrocketed over the last few years, from 15 million pieces of mail in 1970 to 300 million pieces today—a 2,000 percent increase!
>
> Casework requests have doubled over the last decade—some offices report as many as 5,000 to 10,000 requests per year.
>
> The number of bills introduced in 1956 was 7,611. By 1980, that number has risen to 14,594. While this trend dropped off slightly in the last Congress, members today are expected to know a great deal more about many more issues than their predecessors.
>
> The number of record votes is up by 1,250 percent since 1956. You are expected to know the details of these votes and to be able to defend your position on them as they are scrutinized by the press and the public.[12]

Growth in Senate personnel came in both committee staff and each senator's personal staff. But even into the 1970s, a Senate committee's principal, majority-party staff members were still chosen and controlled by the committee chair alone (and of course, the committee's minority-

---

* Under much the same outside pressures, House members won staff increases, too.

party staff members were chosen and controlled exclusively by the ranking minority member of the committee).[13]

Committee mark-up sessions, where the critical decisions about bills
and amendments are made, used to be closed to the press and the public.
Individual committee members were often prohibited from bringing a
personal staff member into these important meetings. In closed mark-
up sessions of the Senate Finance Committee, for example, dealing with
highly complicated bills involving tax, trade, social security, and welfare
matters, members had to rely on the committee staff for on-the-spot
advice. This, of course, gave extra power and influence to the chair and
the ranking minority member, who had chosen the committee staff. No
wonder that when a junior member of the Finance Committee made a
motion in one of its 1968 meetings to permit each member of the committee to bring a personal staff aide into mark-up sessions, the motion
was soundly rejected in the face of implacable opposition from the committee chair.[14]

No wonder, either, that most of the old Senate barons strongly
opposed Senate Resolution 60 when it came to a vote in 1975. This measure, a landmark in the development of the contemporary Senate (some
say the real beginning of the "staff revolution," as it has been called[15]),
provided that every senator would be allowed to hire an additional legislative assistant for each of his or her principal committee assignments.[16] Junior senators, emboldened by their growing strength and
pressured by a changing environment, were able to secure the adoption
of Senate Resolution 60 over the objection of the more senior members
of that body.[17]

What specific tasks do all these new staff members perform? The
answer is: a little of everything that senators' offices are expected to do
these days. Senate (and House) staffers take care of casework and constituency service. They deal with the press, constituents, and interest
group representatives. They do research and write senatorial testimony
and speeches. They draft bills and brief senators on pending legislation.[18]

In addition to expanding their own staffs, senators joined with
House members to create a number of new congressional support agencies. These included, particularly, the Congressional Research Service,
the Office of Technology Assessment, and the Congressional Budget
Office. Of course each of these agencies has since developed a large and
growing professional staff of its own.[19]

Some political scientists have called congressional staff members
"unelected representatives" who act as "policy entrepreneurs," increasing Senate work, and are barriers between senators, interfering with their
direct communications with each other.[20]

Some senators have made similar points. Herman Talmadge, then a
Democratic senator from Georgia, complained in 1975, for example,

that "when you get more staff. . . they spend most of their time thinking up bills, resolutions, amendments"; if half of them were fired, he declared, "we would complete our business and adjourn by July 4."[21] Democrat Ernest Hollings of South Carolina made a similar complaint about the increase in the number of Senate aides. "There are senators who feel that all they are doing is running around and responding to the staff," he once said, adding that, "Now it is how many nutty whiz kids you get on the staff to get you magazine articles and get you headlines and get all of the other things done."[22]

Other senators have also expressed concerns about there being too many staff members working for them, one declaring that "Staff begets staff. Instead of the staff taking work off us, it works the opposite way. Additional staff adds work for senators."[23]

Swollen staff numbers have helped cause Senate inefficiency, according to Democratic Senator David Boren of Oklahoma, who listed that as one of the problems to be addressed in the new effort for congressional reform that he launched in 1992. Staff members have their own agendas, he said, "increasing the glut of inconsequential matters that block the arteries of an already sluggish system."[24]

Still, most senators continue "hungering for the impact that extra staff person or two, or three, or even dozens can give to them and their political careers."[25] The great growth in Senate staff numbers came as a result of a kind of chain reaction that cannot be reversed. Senators are simply not ever going to go back to the modestly staffed days of the past.

The "staff-saturated" Senate, as one observer has called it,[26] is a different kind of Senate from the Senate of the 1950s. It is more individualistic. In committee, senators are better prepared to challenge committee leaders; on the Senate floor, senators not on a particular committee are more willing to oppose the recommendations of the committee.[27] Increased staff assistance is one of the important reasons for this. Steven Smith has written that "nearly all members gained the means to write amendments, publicize their proposals, prepare for floor debate, and attract supporters."[28]

The increase in their staff has also helped senators to become more national in their outlook. Democratic Senator Barbara Mikulski of Maryland, who served in the House before coming to the Senate, has said that senators are "more staff-dependent," less "hands-on," than House members, but that the trade-off is that senators are freed up by staff to spend more of their time on the important issues.[29] Political scientist Barbara Sinclair has made much the same point. Senate staff members, she has written, concentrate on issues and other matters that are tied to a senator's reelection, and this permits "senators to reserve their own time for issues that interest them personally, that they believe to be of national significance and that are also likely to contribute to the establishment or maintenance of a national reputation."[30] With their increased staff help, senators can now take on a much wider range of issues, not just those tied to their home states or to their committee assignments alone.[31]

A changed environment, then, produced a more individualistic, nationally oriented Senate, whose members demanded, and got, more staff assistance. Coming full circle, increased staff assistance for members allowed the Senate to become even more individualistic in its operations and even more national in its outlook. Increased staff assistance thus both fueled the process and was the product of it.

As Senator Mikulski indicated, senators have been able to use their increased staff assistance for greater attention to expanded communications and contacts with their constituents.

Greater Support for Home Style

It may at first seem somewhat paradoxical that today's Senate is more attentive to constituents than was true of the Senate of the 1950s. But this is very much tied in with the nationalization of the Senate and its environment. With the nationalization of American society and mass communications, senators became more politically exposed and electorally vulnerable back home. They needed to work harder at shoring up their home bases at the same time that they began to move more aggressively onto the national scene. Enlarged staffs helped them do both.[32] But differences between a senator's national orientation, or "Washington style," and greater communications and contacts with constituents, or "home style," became blurred. As issues became nationalized after the 1950s, the same issues concerned a senator's local constituents too. Moreover, working on national issues, being a national advocate, gaining prominence in the national media—all became good local politics and helped a senator back home (unless the senator was known for excessive contentiousness or divisiveness).[33]

By the 1980s it could correctly be said that "Senators [and especially their staffs] now perform most of the constituent services House members perform to keep themselves afloat; the political difference between the two jobs is disappearing."[34] Senators' staffs in the home state were beefed up. A large portion of constituent service and "casework" (helping individual constituents with complaints and problems, such as pressing the Social Security Administration to approve a claim previously denied) is now done in an office back home. In 1988, sixteen hundred members of senators' personal staffs, which represented over a third of the total, were based in home-state offices. This compared with only three hundred staff members working back home in 1972, 12.5 percent of the total.[35]

In addition to getting more staff and stationing more of them back home, senators pressed for, and got, increased allowances for home travel and related expenses. These changes facilitated and encouraged home style. So did changes in Senate sessions. As a result of the 1970 Reorganization Act, the Senate (as well as the House) now takes a month-long summer recess each year.[36] Congress usually adjourns sine die by the end of November in off-years, earlier in election years.

A more relaxed Senate schedule also furthered home style. Democrat George Mitchell of Maine was elected Senate majority leader in 1989. In campaigning for that office, he had promised to improve the "quality of life" for senators—among other things, to make the Senate's schedule more predictable, so that members could know in advance when their presence in the Senate would not be required, and they could thus count on more free time to spend with their families and for fence-mending trips back home. Once in office, Mitchell made good on this campaign pledge.

The new majority leader also announced that he would continue the "three weeks on–one week off" Senate schedule that had been instituted by his predecessor, Robert C. Byrd of West Virginia. This schedule confines official Senate meetings and votes to the first three weeks of each month that Congress is in session, with a one-week recess thereafter. Mitchell went further than Byrd. He decreed that during the three weeks of Senate meetings each month there would be no evening sessions after seven o'clock except on Thursdays, and that there would be no roll-call votes on Mondays, to allow long weekends back home.[37]

Home style was also facilitated by the Senate's provision of increased allowances for the mail and office expenses of senators. Each of the 378 senatorial home-state offices boasts its own personal computers. These computers are linked, in a state-of-the-art network, with the extensive and constantly upgraded Senate systems in Washington.[38] Senators use 80 percent of their office computer capacity for mail to constituents.[39] For fiscal 1990, they allowed themselves $29.4 million for mail, 90 percent of it for sending out unsolicited mail. This figure actually represented a 14-percent reduction from the high of the previous year, but still was enough to permit at least three "occupant" mass mailings a year by each senator.[40]

In addition, each Washington Senate office now has a highly advanced facsimile (FAX) machine. Senators can FAX a document to any number of recipients simultaneously. These FAX machines are perfect for targeted mail, as well as for the widespread dissemination of news releases back home.[41]

Senatorial allowances for mail and office expenses also pay for frequent newsletters to constituents. One tongue-in-cheek report called these newsletters "'advertisements for myself' that would make even a Norman Mailer blush," and broke down the typical newsletter stories into four categories:

Taking credit. Republican Senator Alfonse D'Amato of New York: "In 1981, we were in the grip of the worst crime wave in our Nation's history. . . . I pledged to you that I would work hard for tougher criminal laws. . . . Working with my Senate colleagues, I have kept [that] promise. . . . By last year the number of crimes had been reduced by 14 percent."

Sucking up to voters. Democratic Senator Paul Simon of Illinois: "I've traveled the length and width of Illinois."

Stoking good feelings. Former Republican Senator Mark Andrews of North Dakota: "It's hard to say no when someone asks so nicely. The third grade at Franklin School is getting the letter and picture they asked for, and I'm making plans to drop in for a visit. . . ."

Hanging tough on motherhood. Former Democratic Senator Lawton Chiles of Florida: "Frankly, I find crack cocaine a frightening drug."[42]

As noted in the last chapter, senators beam themselves back home by television now, with great facility—and facilities. As late as the 1970s, members of the Senate stood little chance of getting on home-state television newscasts while in Washington. Their offices could produce filmed or taped interviews in Senate studios, but, by the time the tape or film was delivered airmail, most stations would not use it because it was not timely. New technology, especially the availability of satellite uplinks, has changed all that.

When then Republican Senator Pete Wilson (later governor) of California spoke on the Senate floor on a drug bill one day in 1988, for example, taped excerpts of his remarks were carried in the newscasts of a number of California television stations that very night. Similarly, the day after former Minnesota Republican Senator Rudy Boschwitz made a speech at the United Nations on the so-called "greenhouse effect," Minnesota television stations used parts of the speech in their news programs. Both reports had been transmitted to the home states by television facilities and satellite hookups maintained in the Hart Senate Office Building, with taxpayer subsidization, by the National Republican Senatorial Committee (the campaign arm of Senate Republicans).

The Democratic Senatorial Campaign Committee provides similar facilities and assistance for Senate Democrats, and with similar subsidization.[43] For a time in the fall of 1988, Democratic Senator Howell Heflin of Alabama, for example, was sending satellite feeds twice a week to his home-state television stations, emphasizing a drought-relief bill that he had sponsored. On a typical night during the same period, Senator Wyche Fowler would sit in the Senate Democratic studios, linked up to the satellite, for live interviews, five minutes apart, with television anchors in three different Georgia cities. One former Senate Democratic staff member has said, "Now it is possible for a senator to answer the question every day, 'What are you doing in Washington?'"[44]

Thus, even as the Senate was becoming more national in outlook, it was also looking homeward more. Enlarged staffs were a cause and a result of this. But the Senate's loss in legislative efficiency because of increased home style would no doubt have been greater if staff numbers had not grown so.

The proliferation of Senate subcommittees during the 1970s also increased the number of staff members available to individual senators (in their new positions as chairs and ranking minority members of the subcommittees). Indeed, the desire for more staff was one of the main reasons why junior senators pushed for an increase in the number of sub-committees.

Importance of Committees and Subcommittees

Senate power has been individualized as a result of the growth in the number of Senate subcommittees and the increase in the number of members on most of the Senate committees. This is highly consequential because of the critical functions of committees in Senate operations.

Woodrow Wilson, in his 1885 treatise *Congressional Government*, declared that "Congress in session is Congress on public exhibition, whilst Congress in its committee-rooms is Congress at work." He wrote that the functions of the Senate (like those of the House) "are segregated in the prerogatives of numerous standing committees."[45] Committees are not as strong, by any measure, as they were then, but they are still the gears of the legislative machinery.

A 1984 Senate committee-reform panel headed by then Indiana Republican Senator Dan Quayle set forth the ideal functions of a Senate committee in this way:

> The . . . Senate committee system should serve as a legislative filter and refiner. Bills should be referred to the committee of appropriate jurisdiction so that the committee members, a group of experts in their area, may carefully analyze, critique and alter proposed legislation, then report it to their colleagues on the Senate floor after thorough consideration or not report it at all, as they deem best. In this way not only would poorly crafted or ill-considered legislation be filtered out and kept off the Senate Calendar, but a bill emerging from committee would be a refined product, technically sound, thoroughly understood by committee members and ready for consideration by the Senate. The committee report and minority and additional views would further crystallize the issues for floor consideration.[46]

Committees, then, are subject-matter experts that filter and refine proposed legislation before, as "gatekeepers" to the floor, they allow proposals to be considered by the full Senate. Committees also act as "agenda setters" in their fields; originate and incubate policy proposals; and exercise an oversight, or "watchdog," function in regard to particular agencies and programs of the executive department.

Senate committees also have great influence over policy decisions in their respective jurisdictions because, in effect, they get a "second look" at policy when measures in dispute between the two houses go to conference; the conferees for the Senate (like those from the House) have the job of working out a House–Senate compromise, and the senators appointed to a conference committee are usually the most senior members of the Senate committee that had jurisdiction over the measure originally.[47]

The jealously guarded jurisdictions of Senate committees are spelled out in detail in the rules of the Senate.[48] Political scientists sometimes categorize committees on the basis of whether they are standing or select committees. The standing committees generally are permanent, with legislative (including budgetary) jurisdiction; the select, or special, committees generally are temporary study panels without legislative jurisdiction.*

Another way to categorize Senate committees is to separate the authorizing committees from the Appropriations Committee. Quite confusingly, in both the U.S. House and Senate (unlike most state legislatures), authorization bills, which usually precede appropriations bills, establish programs and goals and set upward limits on the amounts of money that can be spent for the authorized purposes, and separate appropriations bills then grant the actual moneys within these guidelines. Authorizing committees—all those with legislative jurisdiction, Senate committees like Environment and Public Works, Commerce, and Armed Services, for example—handle authorization legislation, and the Appropriations Committee considers the actual spending bills. One authority on Congress has explained its authorization and appropriations process this way:

> The best way to look at the authorization and appropriations process is to view authorizations as creating the authority to open a checking account at the Treasury for a designated amount of money, and appropriations as putting money into the account. Once the appropriations legislation has deposited a specified sum with the Treasury, which generally is less than the authorization ceilings, the agencies may draw upon their accounts, making outlays, which is analogous to writing checks on their Treasury accounts.[49]

Does the separation of the authorization and appropriations functions mean—for goodness sake!—that money for a weapons system, say, must be dealt with in two different bills, be approved by two different Senate committees, and then be debated and voted on two different times in the Senate? Well, yes, it means that and more. In fact, with the budget law and rules adopted in 1974, *three* legislative measures, counting the budget, are required, as well as three committees, counting the Budget Committee, and at least three debates and Senate votes, count-

---

* The trouble with these categories is that, in fact, the Senate's Select Committee on Indian Affairs and Select Committee on Intelligence both have legislative jurisdiction, and both, together with the Special Committee on Aging and the Select Committee on Ethics, have taken on a permanent character—although the Intelligence Committee has a rotating membership.

ing a required budget resolution (all, of course, duplicated in the House, too). As they might say on the street: Is this cumbersome, redundant procedure calculated to bog the Senate down and make it inefficient—or what?

Increased Committee and Subcommittee Positions

However cumbersome the Senate committee system, there is no doubt that committee work is important work, and committee assignments are viewed by senators as vital to their influence in the Senate—especially vital since the 1950s to their increased national advocacy and, as a result of their greater visibility and exposure, to their increased need to be seen doing something. A structural change wrought in the nationalized Senate, then, was a more widespread dispersal of committee influence: an increase in the number of seats on the most important committees; multiple committee assignments for senators; and a great increase in the number of subcommittees.

The four most powerful committees in the Senate, those that might be called the power-prestige committees, are: Appropriations; Finance; Foreign Relations; and Armed Services. These are the committees to which senators most want to be assigned. During the 1950s, when power in the Senate was more concentrated, only half the members of the Senate served on one of these most important committees.[50] The ruling elite—the senators with most seniority, respect, and influence—often served on two, and even more, of the four power-prestige committees (and the chair of an authorizing committee was, as well, an ex-officio member of the Appropriations subcommittee that dealt with the money for the agencies and programs within the jurisdiction of his or her authorizing committee).

Since the 1950s, membership on the power-prestige committees has been expanded, and a rule has been adopted that prohibits a senator from serving on more than one of the four.* As a result of these moves to spread Senate power and prestige around more, eighty-seven of the one hundred members of the Senate were by 1990 assigned to one of the four power-prestige committees.[51] The membership of other committees was also expanded.

It is especially significant that the number of subcommittees also increased after the 1950s. There were only 61 Senate subcommittees in 1948.[52] The number was 85 in 1959, a figure that had increased to 116 by 1972[53] and had more than doubled to 174 by 1976;[54] in 1976, the average senator served on four committees and fourteen subcommittees.[55] As a result of the 1977 recommendations of the Stevenson Com-

---

* Senators who were already serving on more than one power-prestige committee were "grandfathered" in, and Democrat Robert C. Byrd of West Virginia, who serves on both Armed Services and as chair of Appropriations, is the last of these.

mittee (the Temporary Select Committee to Study the Senate Committee System, chaired by then Democratic Senator Adlai Stevenson III of Illinois), the number of subcommittees was cut to 110 and the average senator's assignments were reduced to three committees and 7.5 subcommittees.[56] But, as had happened in the past, senators were granted exceptions to assignment restrictions, and some committee and subcommittee proliferation began again.

The 1983 Pearson–Ribicoff Study Group (the Study Group on Senate Practice and Procedures, headed by former senators James Pearson, Republican of Kansas, and Abraham Ribicoff, Democrat of Connecticut) recommended some consolidation of Senate committees, but there was little Senate action on this recommendation.[57]

Similarly, in 1984 the Quayle Committee (the Temporary Select Committee to Study the Senate Committee System, co-chaired by senators Quayle and Wendell Ford, Democrat of Kentucky) recommended additional changes. These included a reduction in the number of committee slots, strict enforcement of committee-assignment restrictions previously agreed to in principle, the reduction of subcommittees from 110 to 80, a limit of nine on the total number of committees and subcommittees on which a senator could serve (with an exception for Appropriations Committee members), and a limit of two on the number of standing committees and subcommittees that a senator could chair.[58] While most of the Quayle Committee recommendations were not agreed to by the Senate, there were some modest subsequent reductions.

For purposes of committee assignments, the present Senate Rule XXV divides Senate committees into three main categories, in addition to a catchall grouping, "all other committees," for those of least importance. The twelve most important A committees are: Agriculture, Nutrition, and Forestry; Appropriations; Armed Services; Banking, Housing, and Urban Affairs; Commerce, Science, and Transportation; Energy and Natural Resources; Environment and Public Works; Finance; Foreign Relations; Governmental Affairs; Judiciary; and Labor and Human Resources. The seven B committees are: Budget; Rules and Administration; Veterans' Affairs; Small Business; Aging; Intelligence; and Joint Economic. Three C committees include: Ethics; Indian Affairs; and Joint Taxation.

The present assignment situation, then, as stated by the Senate Rules and Administration Committee, is as follows:

> Senators *shall* serve on two A committees, *may* serve on one B committee, and may serve on as many C committees as needed. They may serve on three subcommittees of each A committee and two of each B committee. Senators also are limited to one full committee chairmanship among all class A and B panels and one subcommittee chairmanship per committee. . . . Both [Senate parties] designate "exclusive" committees and limit a Senator's service to only one such unit (Appropriations, Armed Services, Finance, and

Foreign Relations for both parties; Budget for the Republicans, and Commerce for the Democrats). . . .[59]

By 1990 there were fifteen standing committees of various types in the Senate, eighty-six subcommittees (with 785 places), four select or special committees, and four joint congressional committees that, together, had eight subcommittees.[60] While in the 1950s the average senator was a member of only two standing committees and fewer than five subcommittees, by 1990 the average senator served on three or more standing committees, seven subcommittees of standing committees, and one other committee or subcommittee, for a total of eleven official units.[61]

In addition, most senators now hold committee or subcommittee leadership positions (chair or ranking minority member). In 1990, ninety senators* served as either chair or ranking minority member of at least one standing committee or standing-committee subcommittee, eighty-one held two or more such leadership positions, and thirty-five held three or more such positions (counting, in this case, the Select Committee on Indian Affairs and the Select Committee on Intelligence).[62] Seventy-two senators in 1990 served as chair or ranking minority member of one of the four power-prestige committees or one of the subcommittees of those committees.[63]

Committee Influence Diffused

If you take a look at the assignments and positions of a couple of typical senators who were in their second terms in 1990—Republican Don Nickles of Oklahoma, first elected in 1980, and Democrat Frank Lautenberg of New Jersey, first elected in 1982—you get an idea of how stretched senators are in their committee responsibilities. In 1990, Nickles served on three standing committees—Appropriations; Budget; and Energy and Natural Resources—and ten subcommittees (a total of thirteen official units). In the Appropriations Committee he was the ranking minority member of the Subcommittee on the Legislative Branch, and in the Energy and Natural Resources Committee he was the ranking minority member of the Subcommittee on Energy Regulation and Conservation.[64] In addition to the Senate's official committees, there are also party committees. Nickles served on a party leadership committee, the Senate Republican Policy Committee, which advises on party action and policy, and was chair of the party campaign committee, the National Republican Senatorial Campaign Committee, which raises money to elect, and reelect, Republicans to the Senate.[65] (Another party

---

* Not including the majority leader, who traditionally holds no committee or subcommittee leadership position.

leadership committee, the Senate Republican Committee on Committees, fills party vacancies on official Senate committees.)

Lautenberg was spread just about as thinly. In 1990 he served on three standing committees—Appropriations; Budget; and Environment and Public Works—and nine subcommittees (a total of twelve official units). In the Appropriations Committee he chaired the Subcommittee on Transportation, and in the Environment and Public Works Committee he was chair of the Subcommittee on Superfund, Ocean and Water Protection.[66] In addition, Lautenberg was a member of one of two party leadership committees, the Senate Democratic Policy Committee, which schedules legislation, reviews legislative proposals, and makes recommendations.[67] (Another party leadership committee, the Senate Democratic Steering Committee, fills party vacancies on official Senate committees; the Senate Democratic Campaign Committee is the campaign-finance arm of the Democratic party in the Senate.)

The average standing committee and its subcommittees together hold about 150 public hearings a year.[68] In addition, there are mark-up sessions. No wonder, given the multiplicity of assignments, that so few senators are found on the Senate floor except when roll-call votes are being taken. As one senator has said, "Sometimes at lunch in the Senate dining room I see more senators than I do on the floor of the Senate."[69]

No wonder, either, that it is a rare senator who does not have many days when two or more of his or her committees or subcommittees are meeting at the same time. In a 1987 survey of Senate staff, more than 90 percent said that committee scheduling conflicts for the senators they worked for were frequent.[70] One senator has said: "Yesterday I had four subcommittees meeting at the same time. We are spread so thin in the Senate. We could do a better job if we concentrated on fewer committees and subcommittees. It is frustrating to me, trying to do my job."[71]

These scheduling conflicts cause low attendance at committee meetings. According to one senator, "When you have hearings of a subcommittee on appropriations in the Senate, generally only the chairman is there."[72]

Senators and their staffs have to set priorities and make choices among conflicting meetings. Staff members cover those that the senators cannot attend. A staff member for a senior senator has been quoted as saying: "At any given time, he will have three to five simultaneous committee meetings on his schedule. So we must choose. With hearings, the rule is: is his attendance essential—is he chairing or must he be there to represent some viewpoint? With the floor it's the same test: is one of his issues up?"[73]

It is often very difficult for a committee or subcommittee mark-up session to obtain a quorum. Senators frequently give their proxies to the chair or ranking minority member so that their panels can act. According to a staff member for a junior senator: "He goes if there's something specific that he's got to participate in. If he's comfortable with it, he gives his proxy to the chairman. He just doesn't have time to attend them all."[74]

Committee Reform

Senators support reform of the Senate's committee system. A 1987 study found that nearly 60 percent of those surveyed favored reducing the number of Senate subcommittees, and over 70 percent supported more stringent limits on the number of committee and subcommittee assignments each senator could have. In addition, 80 percent of the senators interviewed said that they thought that Senate committees with overlapping jurisdictions should hold more combined hearings (rather than each holding its own separate hearings), and a little over 55 percent supported the idea of having more joint committee hearings with the House of Representatives.[75]

The best and most comprehensive committee-reform proposal of recent times is one first put forward jointly in 1987 by Senator Nancy Landon Kassebaum (R., Kan.) and Senator Daniel Inouye (D., Hawaii).[76] Senate changes since the 1950s, Senator Kassebaum has said, particularly the proliferation of Senate subcommittees, were adopted with the "avowed intent of making everyone chiefs and eliminating all the Indians." The "result is verging on chaos," and the Senate "is losing its ability to make policy." Her solution is to reconcentrate, recentralize power in the Senate "to increase the efficiency of the deliberative process."[77]

Not only has power in the Senate been too individualized and diffused, according to Senator Kassebaum, but the presently duplicative and "haphazard approach to fiscal policy" is a "legislative nightmare":

> Today, effective action by Congress requires a level of consensus that is painfully difficult to achieve. Before we can appropriate a single dollar for a B-1 bomber or a sewage treatment plant, the matter frequently must be debated three times on the Senate floor alone. We debate programmatic funding levels in annual budget resolutions. We then debate the same issues all over again in authorization bills. Then we refight the same battles in annual appropriations bills.[78]

To deal with the various committee problems, the Kassebaum–Inouye measure would, first of all, combine the authorization and appropriations functions; the same committees would handle both. Second, the proposal would reduce and restructure Senate committees. There would be ten legislative policy committees, of seventeen members each: Agricultural Policy; Commercial Policy; Defense Policy; Economic Policy; Energy Policy; Environmental Policy; Foreign Policy; Governmental Policy; Judicial Policy; and Social Policy. These would essentially replace the present twelve A committees, and no senator would be allowed to serve on more than two of them. A senator could chair or serve as ranking minority member of only one of them and, in that case, could serve on no other policy committee.

There would be four legislative program committees: Native American Programs; Veteran American Programs; Senior American Programs; and Entrepreneurial American Programs. These would generally be substituted for the present seven B committees.

There would be three administrative committees: Rules; Ethics; and Intelligence.

The third provision of the Kassebaum–Inouye proposal would create a Committee on National Priorities, a leadership committee that would set policy priorities and replace the Budget Committee. The Committee on National Priorities would be made up of the chair and ranking minority member of the new policy and program committees and up to five other appointees, presumably including the party leaders.

There is no doubt that the Kassebaum–Inouye proposal is long overdue and should be adopted. The Senate Rules and Administration Committee, not yet prepared to endorse the measure outright, has called it "a thoughtful and innovative approach to revamping the Senate's committee system and redistributing its work," saying that it offers the Senate "a vision of a more coherent and coordinated system."[79]

Until the Senate is ready to move on such a comprehensive committee reform, that body should at least enforce its present rules limiting committee memberships, sizes, and assignments, as the Committee on Rules and Administration has recommended.[80] It should also encourage more combined hearings by Senate committees with overlapping jurisdictions and more joint hearings by counterpart House and Senate committees. These reforms would go a long way toward aiding the Senate in regaining some of its lost efficiency in law-making.

In addition to the structural changes—staff increases and the proliferation in committee positions assignments and the numbers of subcommittees—the great changes in the Senate's external environment also caused changes in Senate processes. Today's Senate is both more interested in the public and more accessible to it. It is less insulated from the outside world—less walled off, now, from the folks back home and from the nation.

Increased Senate Openness

For most of its history, the Senate of the United States could not have been described as a particularly public-regarding institution. During its first six years, until 1795, the Senate's sessions were held behind closed doors. And even after the press and the public were admitted to the Senate galleries, these visitors were frequently unable to make out what the speakers on the floor were saying. In fact, it was not until 1969—many, many years, of course, after electronic amplification was invented—that the Senate finally got around to installing microphones and a sound system in the chamber. Before then, in the Senate of the 1950s, for example, leaders such as Democratic Majority Leader Lyndon Johnson and Republican Minority Leader Everett Dirksen were notorious for purposely half-mumbling their procedural motions and unanimous consent requests, sotto voce, in order to secure time limits on debate or fixed times for votes, without objection.*

It took six years after 1789 for the Senate to decide to open its floor sessions to the public. It took 186 years for senators finally to agree to open their most important *committee* meetings to the public.

Public hearings, of course—the "show and tell" sessions of Senate committees and subcommittees—had always been open meetings. But until 1975, as incredible as this now seems, when those same committees and subcommittees met in "mark-up" sessions to make decisions about legislation under consideration—on proposed amendments and on whether to report a measure to the Senate floor at all—these sessions were routinely closed to the public and the press.

In the late 1960s a junior member of the Senate Finance Committee moved that its mark-up sessions be opened to the public. The motion got nowhere. The chair of the committee argued that accessible meetings would increase the power of lobbyists and that, further, the tax, trade, and other matters within the jurisdiction of the Finance Committee were so complex that open consideration of them would "confuse the public."[81] In fact, one result of closed meetings was an enhancement of the power of the chair and of those lobbyists who had special influence with him.

During the 1970s a new concept of "government in the sunshine" began to radiate throughout the country, and the Senate could not, at last, deflect its shine. The state of Florida was a leader in this open-government development, and one of its senators, then a junior Democrat, Lawton Chiles, joined by Republican William Roth, Jr., of Delaware and backed by the citizen lobby Common Cause, was able to push through the Senate a resolution in 1975 to open committee and subcommittee mark-ups, as well as conference committee sessions, to the public.[82] The Senate ordered that these meetings would henceforth be open meetings unless members of the committee or subcommittee voted publicly, for national security or other important reasons, to close them.

The opening up of committee and subcommittee mark-ups spread lobbyist influence around more, improving the position of interest groups that had not previously been close to committee chairs; it made committee operations more democratic, further individualizing Senate power; and it made the Senate a more "public enterprise," more exposed

---

* Johnson's and Dirksen's words on these occasions were inaudible not only to those seated in the press and public galleries, but to the senators themselves, who as a result frequently had to crowd around the leaders in the well of the Senate to try to find out what was going on.

to public view and outside pressure and more respectful of national public opinion (perhaps even encouraging senatorial "grandstanding," in preference to legislating, some thought).[83]

Opening the Senate to Television

Since 1948, beginning with the televised hearings of the Senate Armed Services Committee on the issue of universal military training,[84] television cameras had been allowed to cover the public hearings of Senate committees. Opening Senate committee mark-up sessions to the public meant that television cameras were permitted into these more important sessions as well. But it was not until 1986 that the Senate finally permitted its floor sessions to be televised.

A very long debate, almost as old as broadcasting itself, preceded that decision. America's first radio station (KDKA in Pittsburgh) went on the air in 1920; within just two years the first resolution was introduced in Congress to provide for live radio broadcasting of House and Senate proceedings. The measure failed. Two years later a Senate resolution was passed to require a *study* of session broadcasting, but little came of the resultant report.

Year after year thereafter, individual senators continued to press the issue, and year after year they lost to the more traditionalist Senate majority. Senator Claude Pepper of Florida took up the broadcast cause in 1944. But Pepper's plan got no better reception from the majority of his fellow senators than earlier proposals had. And, not too long after Pepper's failed try, the debate about Senate-session broadcasting switched to television, because television by then had become America's principal broadcast medium.

The first senator to become a "household" name because of television was Democrat Estes Kefauver of Tennessee. As head of the Senate's Organized Crime Committee,* Kefauver chaired Senate hearings that, beginning in 1948, captured, and captivated, a huge national television audience. Partly because of this "show horse" use of television, Kefauver was never considered a member of the Senate's influential "inner circle," but the national fame that television brought him outside the Senate made Kefauver into a serious Democratic presidential contender in 1952 and won for him his party's vice presidential nomination four years later.[85]

Observers expressed the opinion that the televising of committee hearings caused changes in senatorial behavior and altered the nature of the very proceedings being broadcast. One critic wrote:

---

* The Special Senate Committee to Investigate Organized Crime in Interstate Commerce.

By laying undue emphasis on publicizing its investigations, as seems currently in vogue, the congressional investigating committee can all too easily lose sight of its legitimate objective of searching for needed information and thus degenerate the whole proceeding into nothing more than a three-ring circus constituting a mere entertainment spectacle for the public, or a propaganda extravaganza for ambitious politicians.[86]

Television created its first senatorial monster in the person of Joseph McCarthy—and then destroyed him. In the early 1950s, McCarthy (R., Wis.) chaired the Senate's Committee on Government Operations and its Permanent Investigating Subcommittee. His demagogic and persecutorial hearings on supposed communist infiltration of the federal government became a national television sensation and McCarthy himself a mighty and feared political force. But McCarthy finally went too far when he questioned the loyalty of leaders of the U.S. Army, and in the famous "Army–McCarthy Hearings" of 1954, televised live to a massive and spellbound national audience, he was at last exposed as a mountebank and a bully. The Senate soon censured and thus ruined him.[87]

Perhaps it was the McCarthy experience in the Senate that caused Speaker Sam Rayburn of Texas to block television coverage of House committees—both mark-up sessions and public hearings—for so long. The Senate, though, continued to allow televised coverage of its committee hearings, and by the 1970s the filmed or taped excerpts from such sessions had become a commonplace of the nightly national television news programs. The 1973 Senate Watergate hearings, on the Nixon campaign break-in at the Democratic National Committee headquarters and the ensuing cover-up, attracted an immense national television audience and eventually helped to force President Richard Nixon out of office. The following year, the Joint Committee on Congressional Operations recommended live, gavel-to-gavel coverage of House and Senate floor sessions. The chairman of this committee, Democratic Senator Lee Metcalf of Montana, had opened its proceedings with this statement of the problem as he saw it:

A Congress unable to project its voice much beyond the banks of the Potomac—to be heard and understood only dimly outside Washington, D.C.— can be neither representative nor responsive. A Congress able only to whisper, no matter how intelligently, cannot check and balance the power of the Executive or safeguard the liberty of the individual citizen.[88]

The Senate did not accept the joint committee's floor-television recommendation. Twice, though, in the 1970s, the Senate did permit *spot* broadcasts of Senate proceedings. The 1974 swearing in, in the Senate chamber, of Vice President Nelson Rockefeller (appointed by President Gerald Ford) was carried live on television, and the 1978 Senate debate on the Panama Canal Treaty was broadcast live on radio (an earlier provision for televising this debate having fallen through at the last). But regular broadcasts of Senate sessions were still barred.

It was the House that welcomed television to its floor first. After authorizing only closed-circuit television and radio coverage of its sessions in 1978, the House installed its own complete television system the following year and began live, gavel-to-gavel broadcasts of House proceedings on a permanent basis.

The increased attention that House members got from this action eventually played a part in the Senate's decision to do the same on its side of the Capitol. But opposing senators continued to fret about how the Senate and senators would look on television—what, for example, the public would think about the empty seats. Would they realize that much of the Senate's business is done in committees which usually meet during Senate sessions? Senators also worried that the slow pace of the Senate would come across poorly on television, that broadcasters might play up the "kooky, the gimmicky shots" in order to entertain, and that the careers of "less glib, less photogenic" senators would be harmed.[89]

There were more serious institutional arguments, too, on both sides of the debate, in what proved to be the last sessions of struggle over Senate television. The leading foe of televising Senate sessions was Senator Russell Long, who earlier had also been a strong opponent of opening Senate committee mark-up sessions to the public. The leader of the other side was Senator Howard Baker (R., Tenn.) and, after Baker's retirement from the Senate, Senator Robert C. Byrd (D., W. Va.). The opposing contentions of these adversaries and those allied with them dealt with fundamental questions about what kind of institution the Senate was supposed to be.[90]

The legitimacy of the Senate, and public support for the body, would be enhanced by letting television into its sessions, Baker told a Senate committee in 1981.[91] He argued that the Senate was in danger of losing its equal standing with the House because it did not permit comparable television coverage, and that, similarly, "we cannot compete with the presidency unless we open up the deliberations of this body by television."[92]

Baker maintained that the Senate was intended to be a great deliberative body, a national forum for real debate among its members, as well as for educating the public. He wanted members to be drawn to the floor, for the public and senatorial focus to shift from the work of Senate committees to the deliberations of the full Senate, and he argued that admitting television to the Senate floor would accomplish this.[93]

Senator Long, on the other hand, argued that senators did their most important work off the Senate floor and in committees,[94] in a Senate with an inward focus, and that the outward-looking viewpoint that would be further forced on the Senate by televising floor sessions would interfere with its institutional efficiency in decision-making.[95] Long's Louisiana colleague, Democrat J. Bennett Johnston, maintained that the fact that the House had decided to allow their sessions to be televised did not mean that the Senate should follow suit, because the functions of the two bodies were intended to be different. The "intrusive eye of television," he argued, would make the Senate less deliberative and too quick to respond to swings in public opinion.[96]

Long had always agreed.[97] Baker was right in believing, Long thought, that televising Senate sessions would draw senators to the floor, but he feared that this, rather than improving the importance and quality of debate, would encourage senators to grandstand, to play to the voters, rather than act as "statesmen."[98] As he argued in 1986:

> Television will encourage us to use the Senate floor in ways that are unnecessary and ways that are not helpful to the decision-making process. It will encourage us to appeal to the electorate, to our constituents and to those we hope will vote for us in some future election. We can do that elsewhere. There is no need for us to do that on the Senate floor. It will be harmful to the institution of the Senate if we do.[99]

Both Baker and Long shared the belief that, since the 1950s, the Senate had become too individualistic, too inefficient in decision-making. Baker thought television would improve the situation; Long felt it would make things worse. Long and his allies also felt that televising Senate sessions would result in an increase in public and senatorial pressure for changing Senate rules—to restrict the filibuster and to require amendments to be germane to the measure to which they were offered.[100] These television opponents feared that the free-debate and free-amendment practices of the Senate would look bad to the viewing public. As Long's colleague, Senator Johnston, said:

> Unlimited debate, whether it is the classic filibuster or whether it is simply extended debate, is not a pretty thing to watch on television, and the public will never understand why it is important to this institution and to this Nation for the Senate to play the role of "the saucer where the political passions of the Nation are cooled.". . . It does not work efficiently; it is a messy, untidy, spectacle to watch, but I think it is vital to the Nation.[101]

As a matter of fact, before it was all over, the leaders for televising Senate sessions—Republican Majority Leader Robert Dole of Kansas, who led Senate Republicans after Baker's retirement, and Democratic Minority Leader Robert Byrd—tried with limited success to tie certain Senate rules changes to the television resolution. What they wanted was: to limit debate on a "motion to proceed," that is, to take up a measure for consideration, a motion that increasingly had become the subject of preliminary filibusters; to limit the introduction of nongermane amendments; and to limit postcloture filibusters, that is, filibusters through myriad amendments and parliamentary maneuvers after cloture.[102] The leaders did get a significant rules change to limit postcloture debate and delay, but the other suggested rules changes were rejected.

The Senate finally agreed to televise its sessions, first in a trial period.[103] Like the television the House had agreed to, it was to be gavel-to-gavel coverage, with the cameras owned and controlled by the Senate itself. Television stations and networks could use all or any part of it, live or taped. Senators and others could use taped excerpts for news, but not for campaign advertising.

What was the effect of televising Senate sessions? A report done for the Senate by the Congressional Research Service on the first month-and-a-half test of live television broadcasts found that there had been little impact on senatorial behavior or Senate floor proceedings. The number of "special order speeches"—senatorial speeches scheduled by agreement for a special time and which may be on any subject—had increased by 250 percent, but a new limitation had cut the average length of such speeches in half (from twelve minutes to six).[104] For Russell Long, the reported increase in the number of special order speeches was a confirmation of his earlier argument against television and a reason to continue his opposition. He told the Senate:

> What causes this Senator concern is the likelihood that television will lead to an increase in the number of speeches whose primary purpose is to appeal to the voters back home or to ensure one's reelection. In my mind the dramatic increase in the number of special order speeches during the test period shows the potential for this to occur in all areas of debate.[105]

Long's views were not those of the Senate majority, though. Alabama Democratic Senator Howell Heflin's opinion was more typical. Claims by opponents of televised sessions that senators would be encouraged to "grandstand and show off" had not been borne out, Heflin said, and "the business of the Senate has not and will not be substantially altered" by it. Acknowledging that he had earlier laughingly said that Capitol-area drugstores would probably enjoy "increased sales of hair spray and styling mousse and Ultra-Brite toothpaste" and that television had increased the use of charts and graphs and other visual aids on the Senate floor to such an extent that he had seen "more posters than at any time since I was in the fifth grade," Heflin concluded:

> But I really feel that . . . it has proven itself to be worthy. . . . I have noticed an interest in the legislative process and a knowledge of Government among my constituents in certain letters and inquiries which did not exist before we opened this chamber to the TV public.[106]

The Senate voted to approve television on a permanent basis. And, after the first year, one outside report found that the television cameras had already become an accepted feature of the chamber and that neither had the Senate been fundamentally transformed nor had it been made more efficient by television; the Senate, the report said, was "basically unchanged."[107]

Still, there can be no doubt that today's Senate is a more open and public institution, more exposed to view, more public-regarding and outward-looking. The televising of Senate sessions has contributed to this development. This development in turn has helped to individualize power in the Senate.

Increased Obstructionism

Increased individual power in the Senate has helped make it a less efficient body, because more decisions have been pushed to the Senate floor in recent years, and there senatorial tactics have become more obstructionist.

"If this is the world's greatest deliberative body, I'd hate to see the world's worst." Those words about the Senate by one of its senior Republican members expressed the growing frustration senators felt about their chamber as it entered the 1980s.[108] Oklahoma Democrat David Boren similarly decried Senate inefficiency, which he said resulted from a fragmentation, an individualization, of power in the Senate, coupled with the idiosyncrasies of Senate rules. "When you have this increasing fragmentation," he said, "and you add to that rules which allow the individual to exploit that fragmentation, you've got problems."[109]

Throughout the 1980s, the frustration level in the Senate remained high. David Pryor declared that the Senate suffered from procedural "gridlock," adding that "Over the past 20 years there appears to have crept in a decided inability to successfully complete the business of the Senate."[110] The Senate Committee on Rules and Administration found in 1988 that, among senators, "'Inefficient' floor procedures, characterized as spurts of activity punctuated by frequent periods of deadlock and inertia, is a major focus of complaint."[111]

Senate deadlock and inertia grew as, increasingly, issue fights went unresolved in committee; more and more individualistic, national-minded senators were inclined to "take it to the floor."[112] The number and percentage of bills that were the target of Senate-floor amendments, as well as the total volume of amendments offered on the floor, went up enormously. This Senate-floor amending activity peaked in the 1970s and then subsided somewhat, but never again to the lower levels of the 1950s.

Back in the Eighty-fourth Congress of 1955–1956, only about one-tenth of the bills that were reported to the Senate floor faced any floor amendments at all. By the Ninety-sixth Congress of 1979–1980, about one-third did.[113] During that same period the number of senators who offered a floor amendment to a bill increased from forty-seven to over ninety.[114] Sixty senators offered three or more floor amendments in the 1985–1986 session, compared with only fifteen who did so thirty years earlier.[115]

The growth in Senate-floor amending activity was just as evident in regard to contested amendments as it was in regard to uncontested amendments. Contested amendments are those that are voted on by roll call, with a 60–40, or closer, vote split. Less than a third of the senators (thirty-one) offered contested amendments, fifty-three such amendments in all, in the Eighty-fourth Congress of 1955–1956. By contrast, nearly three-fourths of the members of the Senate offered contested amendments in the Senate of the 1970s and 1980s: seventy-two senators offered 199 contested amendments during the Ninety-second Congress of 1971–1972; seventy-three senators offered 182 contested amendments in the Ninety-ninth Congress of 1985–1986.[116]

Senators became increasingly willing to offer floor amendments to bills reported by a committee on which they did not serve, thus deviating increasingly from the committee-deference part of the old reciprocity norm that said, in effect, "we'll support your committee on the floor if you'll support ours." Only one-third of all senators offered such amendments during the Eighty-fourth Congress, but eighty-six senators did during the Ninety-ninth Congress.[117] Floor amendments offered to bills by senators who were not members of the committees that reported them amounted to little over 48 percent of all amendments offered in the Eighty-fourth Congress but nearly 63 percent of all floor amendments offered in the Ninety-ninth Congress.[118]

Reforms for Greater Efficiency

Heightened floor amendment activity impedes and obstructs Senate action. It could be significantly curtailed by three Senate reforms that have been put forward with strong backing, one having to do with the germaneness of amendments, another with section-by-section offering of amendments, and the third with so-called "sense of the Senate" amendments.

Bills that come to the Senate floor are freely amendable (in the House, by contrast, the offering of amendments is frequently restricted and sometimes prohibited altogether). Senate amendments proposed to appropriations bills and to budget resolutions and reconciliation bills must be germane (pertinent or relevant) to the measure to which they are offered; so must amendments after cloture has been voted to cut off debate on a measure.[119] In most instances, though, senators are free to offer any kind of amendment, germane or not.*

Without a germaneness rule, senators, as Boren has complained, can "propose a national defense amendment to the agriculture bill, and vice versa."[120] The Senate Rules and Administration Committee has found that the consequences of the Senate's not having a germaneness rule are "profound," that this undermines and circumvents the authority of committees, plays havoc with scheduling, and reduces the impor-

---

* Unless they give up this right through a unanimous consent agreement that limits debate and amendments.

tance of floor debate by preventing its being focused. The Committee has declared:

> Committees can effectively veto bills by declining to report them, but supporters can offer the same proposals as non-germane floor amendments to other measures. The Majority Leader, acting for most Senators, may be able to control when measures reach the floor, but not the issues that Senators raise there as amendments. Any senator can compel his or her colleagues to vote on (or in relation to) virtually any issue at almost any time.[121]

And, since amendments need not relate to the measure under consideration, the work of the Senate can be obstructed by senators' bringing up the same issue over and over on the Senate floor, with no finality to votes.

The Democratic president pro tempore of the Senate, Robert C. Byrd, has proposed a new germaneness rule:[122] each day a measure is considered, the Senate could, by three-fifths of those present and voting (not the stiffer requirement of three-fifths of the whole Senate that is needed to vote cloture), require all noncommittee amendments to be germane.* This proposal should be adopted to expedite the work of the Senate.

The Senate should also adopt a proposal by senators David Pryor and John Danforth to limit "sense of the Senate" amendments, germane or not—amendments that simply express the opinion, or anger, or frustration of that body on some subject and do not have the effect of law. In 1987, for example, eighty-six amendments, many of them "sense of the Senate" amendments, were almost frenziedly added to that year's State Department authorization bill, causing one senator to protest: "We seem to create amendments, Mr. President, by reading yesterday's headlines so that we can write today's amendments so that we can garner tomorrow's headlines."[123]

As the Senate Committee on Rules and Administration has said, "sense of the Senate" amendments or resolutions give the Senate a "satisfying outlet," with immediate effect, without changing the law. But the Committee has pointed out their disadvantages:

> The principal objections to these amendments are (1) that they are pointless and a waste of time because they do not propose to make law; (2) that despite their lack of legal effect, they can provoke controversy and floor delays because of the subjects they address; and (3) that they can be drafted in ways that convey misleading impressions of the Senate or congressional opinion or that present Senators with awkward votes.[124]

Senator Pryor said in 1987 that in the preceding five years senators had proposed fifty-one "sense of the Senate" and "sense of Congress"

---

* After such a vote, the presiding officer's ruling that an amendment was nongermane and out of order could be appealed to the Senate floor, but a vote of three-fifths of those present and voting would be necessary to overrule the chair.

amendments, rarely with more than eight co-sponsors.[125] Pryor and Danforth would institute a kind of seriousness test for these amendments: none should be permitted to be offered on the Senate floor with fewer than twenty senatorial co-sponsors. Their proposed rules change would eliminate a lot of the extraneous considerations that presently interfere with the Senate's efficient operation.

So would another Pryor–Danforth proposal to require that amendments to a bill be offered in order, section by section, as the measure is being considered on the floor: an amendment to Section 7 of a bill, say, could not be offered while the Senate was considering Section 3; similarly, an amendment could not be offered to Section 3 after the Senate had approved this section and passed on to later sections. In arguing for section-by-section consideration of bills, the procedure used in the House and one that expedites orderly action, Senator Pryor once said: "In the House, each Member of the body accommodates his or her own individual schedule to the rule of the House. In the Senate, the rules and practices of the Senate accommodate each individual senator's schedule. One hundred years ago maybe that process worked. Today, Mr. President, it does not."[126]

The Senate Filibuster

Even more than the floor amendment, the Senate filibuster is the tool and cause of most obstructionism in that body. "Today, there are more opportunities to take many causes to the wall," Republican Senator Richard Lugar of Indiana has said.[127] And, he might have added, more senators are willing to do just that, through filibusters (and threats of filibusters) and other tactics of obstruction, to thwart the will of the Senate's majority.

Southern senators for many years used the filibuster as their main line of defense against, to begin with, the passage of federal antilynching laws, and later all federal civil rights laws. About the time the tide turned with regard to civil rights laws, however, Senate liberals began to employ the filibuster. Thereafter, so did the modern conservatives. Now there is no ideological cast to the filibusterers.[128]

From 1949 to 1959 the Senate's "cloture" rule was interpreted to mean that cutting off debate required a two-thirds vote of *all* senators. Liberal senators moved to amend the rule in 1959, and in the face of bitter opposition from southerners and other Senate conservatives reached a compromise to provide that debate could be cut off by a two-thirds vote *of those present and voting*.[129] In 1975 the rule was amended again to make it a little easier to get cloture—to require a three-fifths vote of the entire Senate, sixty senators.[130] (This is, in effect, a relaxed requirement because, on an important cloture vote, it is highly likely that all senators will be present.)

The old-style filibuster was one of talking a measure to death—in

1957 Senator Strom Thurmond of South Carolina holding the Senate floor for twenty-four hours and eighteen minutes in an effort to kill a civil rights bill, for example.[131] Now, filibusters entail more than just protracted talk, or as senators say, "extended debate." They often involve an inventively obstructionist manipulation of the rules to tie the Senate in knots—dilatory motions, unnecessary quorum calls, and repeated demands for roll-call votes. In addition to blocking action altogether, filibusters are used to force a change in, or a compromise concerning, a filibustered measure. They have also sometimes been used to hold one measure hostage until the Senate leadership agrees to work out a compromise on another matter or commits to bringing up some other measure.[132]

In the closing days of a session, when senators are in a hurry to get back home and when the number of senators necessary for a quorum may not actually still be in Washington, filibusters and other delaying tactics by a very few senators, sometimes even by one senator, can be quite effective in blocking Senate action. So can the mere threat of such tactics.

Extended-debate filibusters are much more numerous today than they once were, and some say that, rather than being wielded only in big ideological battles, they have become "trivialized."[133] In all the years from 1841 through 1959, for example, there were only forty-six filibusters. In the decade of the 1950s, there was an average of only one Senate filibuster during each two-year Congress. In the decade of the 1960s, the average number of filibusters tripled—to 1.8 a year. Then came even greater increases, to an average of nearly 5 filibusters a year during the 1970s to 6.6 per year during the first seven years of the 1980s.[134] During the last two Congresses of the 1960s (1967–1970), twelve cloture petitions were filed to stop or prevent filibusters, an average of three such petitions a year; in the seventeen years after that (through 1987), 202 cloture petitions were filed, an average of nearly twelve a year.[135]

Today the mere threat of a filibuster may cause a measure to be permanently shelved. The leadership now often calls a quick cloture vote in the face of such threat and, if cloture fails, lays the measure involved aside as unpassable. The increase in both actual filibusters and threats of filibusters is shown by the fact that, while the Senate only took thirty-eight votes on cloture during the first fifty years after its cloture rule was adopted (in 1917), there were forty-three such votes in the 1987–1988 session alone, far more cloture votes than there were filibusters.[136]

There are now three additional variations of the regular extended-debate filibuster: party filibusters, postcloture filibusters, and preliminary filibusters.

After the Republicans lost their six-year control of the Senate in the 1986 elections, as noted in Chapter 4, they began to wield a party-backed filibuster against Democratic proposals that they opposed—for example, public financing of Senate campaigns and a cap on campaign costs, an increase in the minimum wage, and others.[137] In 1988, Senate Democrats twice were able to secure a majority—fifty-three senators one time and fifty-six another—to cut off a filibuster and get a vote on a bill to raise the minimum wage, but they fell short of the sixty votes required.[138] Democratic leaders had to back off and eventually agree to a compromise calling for a lower increase in the minimum wage.

The party filibuster is not new in the history of the Senate,[139] but its use in modern times is new. And in a Senate with a narrow margin in party strength (fifty-seven Democrats to forty-three Republicans in 1993, for example), it can be—and is—used to great effect to prevent action by the majority.

The postcloture filibuster was more or less invented by the late Senator James Allen (1969–1978), a conservative Democrat from Alabama. It affords filibustering opponents of a bill a second chance to kill it.[140] When Allen first came up with the idea in 1976, the Senate's cloture rule limited each senator to one hour of debate after cloture, but the rule also provided that a senator could call up any amendments filed prior to the cloture vote and that time spent on parliamentary tactics did not count against the time limit. Allen, a dedicated student of Senate rules, knew this. On one memorable occasion, after his regular filibuster against an antitrust bill was ended by a cloture vote, Allen began to call up, one by one, the many amendments he had shrewdly filed prior to cloture. The result was, as former Senator James Abourezk (D., S.D., 1973–1978), an Allen opponent in that fight, has written:

> Not only did we vote on each amendment, which alone took fifteen minutes, but then we would vote on a motion to reconsider the vote, which took another fifteen minutes. If too many senators left the floor after the vote, Jim Allen would demand a quorum call. It would take another half hour or more to get the minimum of fifty-one senators back on the floor to fulfill the quorum requirement. If the required number of senators failed to reappear, Jim would make a motion for the sergeant-at-arms to bring them by force to the floor. This motion would require another vote. Because neither the motions nor the votes were charged against Jim Allen's time, if he had enough amendments and could physically stand it, he could go on all year.[141]

Allen's postcloture filibuster eventually forced a compromise on the antitrust bill he was blocking.[142]

It was Abourezk, himself, though, together with another liberal Democrat, Howard Metzenbaum of Ohio, who took the postcloture filibuster to its farthest extreme. In 1977 Abourezk and Metzenbaum started a filibuster against a gas deregulation bill supported by President Jimmy Carter. When it became clear that cloture would be voted, they filed 508 proposed amendments. After cloture they began slowly to call up each one, using as much time as possible, as was their right under the rules. Controversial rulings by Vice President Walter Mondale, who was then presiding over the Senate, eventually ended this particular postcloture filibuster.[143]

Reaction to the Abourezk–Metzenbaum postcloture tactics finally caused the Senate, in 1979, to amend its rules to provide for a total time limit after cloture of one hundred hours, including all time spent on amendments, motions, and roll calls. Later, in 1986, as a part of the deal to get the Senate to vote to televise its sessions, Senator Byrd was able to get the Senate to reduce the total postcloture time to thirty hours.[144]

Soon after they voted limits on the postcloture filibuster, however, senators came up with the preliminary filibuster—a stall on the previously routine motion to proceed to consideration of a measure. Opponents began to filibuster on the question of even taking up a bill; if they lost they could still filibuster on the measure itself. Senate leaders tried to short-circuit the preliminary filibuster by making their motions to proceed during the "morning hour," the first part of each Senate day when motions to proceed are not debatable.

But that did not work. Opponents came up with a creative alternative, as when, for example, Senate Republicans found new ways to block consideration of the 1987 defense authorization bill. The Republicans in the Senate were opposed to that bill because, as reported by the Democratically controlled Armed Services Committee, it would have limited President Ronald Reagan's ability to press ahead with his so-called Star Wars antimissile defense system. The *New York Times* gave the details of this particular Republican preliminary filibuster as follows:

> The filibuster of the military programs bill began when Senator John Warner of Virginia . . . refused to vote on the usually routine motion to approve the journal of the previous day's proceedings in the Senate. Senator Warner said he could not vote because he had not read the journal.
>
> Under the rules, the full Senate must vote to excuse a senator from voting. A roll-call vote was ordered. In a carefully orchestrated plan, another Republican, Dan Quayle of Indiana, then refused to vote on excusing Senator Warner, taking the position that no senator should vote to compel another to vote.
>
> Next, a roll-call was ordered on Senator Quayle's refusal. The parliamentary web was thus spun faster and tighter, all within the rules, and within hours the Senate was plunged into gridlock.[145]

There is no way, of course, to stop the use of dilatory tactics of various kinds in the Senate so long as the norm of reciprocity remains weak. No serious proposals have been made in recent times to make it easier for the Senate to invoke cloture on a regular filibuster, say, by allowing debate to be cut off by a simple majority vote. There are no efforts to cut the time limit on postcloture debate below the presently permitted thirty hours.

A reform aimed at the preliminary filibuster has been proposed, though, by senators Pryor and Danforth. It would limit debate on a motion to proceed to a total of one hour. In support of this proposal when he first introduced it, Pryor noted in 1987 that there had been twelve instances during the preceding Congress (the Ninety-ninth) when legislation had been delayed by filibusters on a motion to proceed and that, halfway through the 1987 session, there had already been seven unsuccessful attempts to cut off debate on motions to proceed to the consideration of two bills.[146] The efficiency of the Senate would be improved by enactment of the Pryor–Danforth limit on debate on motions to proceed.

A nationalized environment and resultant changes within the Senate have led to institutional changes—structural and procedural. The Senate, and senators, have become staff-rich, even "staff-saturated," and this has helped to make the Senate a more national and individualistic institution while, somewhat paradoxically, allowing more attention to be devoted to local constituents, making the Senate less efficient in lawmaking.

The proliferation of Senate subcommittees and the increase in the number of committee seats, especially on the most influential panels, have further diffused Senate power. These developments have also caused numerous scheduling conflicts for senators and slowed Senate decisionmaking.

The Senate is a much more open and accessible body than it was in the 1950s; its committee mark-up sessions are open to the public, and its sessions are televised. These developments, too, have made the Senate more public-regarding and more outward-looking, as well as more individualistic in the way power is wielded internally. Obstructionism in the Senate has reached "gridlock" proportions.

Increased senatorial frustration in recent years, public pressure for action, and, perhaps, Senate concern about the poor Senate image that may be projected by the present televising of floor action, make these and other reforms increasingly practical.

In sum, then, the Senate has become less efficient in law-making, and certain needed reforms could improve its operations and ability to act. It should be noted, too, Nelson Polsby has written, that the Senate has changed somewhat in function, from a kind of council of review and revision on the president's program to a pro-active, "publicity-seeking, policy-incubating, interest-group-cultivating body."[147] The Senate has become "a grand arena where a lot of policy gets proposed and debated"; not being content any longer at simply being the saucer that cools the coffee, or tea, of the House, senators are now "brewing more of the tea than they're cooling."[148] As the Senate has become more national and individualized, it has also become "more reflective of the preferences of the entire membership and more responsive to public opinion," as Barbara Sinclair has put it, "a superb forum for the articulation of issues."[149]

6

Nationalized Parties, National Conflict

When President George Bush nominated a conservative and pro-military fellow Texan, Republican John G. Tower,* to be secretary of defense in 1989, most Washington observers supposed that Senate confirmation of Tower's appointment would be swift and uneventful.[1] After all, only eight presidential cabinet nominations had ever been rejected by the Senate, and only two of those during the whole of the twentieth century, the last during the Eisenhower administration.[2] What was more, Tower was a former member of the Senate, having served in that body from 1961 to 1985. As a matter of fact, he had served as a longtime member of, and had once even chaired, the very committee—the Senate Armed Services Committee—that now had jurisdiction over his confirmation. It was thought wildly unlikely that senators would refuse to confirm a person who had once been a member of their own "club."

The Tower nomination was different. It ran into trouble almost immediately. The FBI investigative reports on Tower, furnished in confidence to the members of the Armed Services Committee, showed a history of serious "problem drinking" and indiscreet "womanizing." Bush officials at first tried to downplay the import of these reports. Then, after a delay during which leaks and rumors multiplied, the White House finally ordered new FBI investigations. These were of no help. When a

---

* Later killed in a private-plane crash.

noted conservative leader openly questioned Tower's moral fitness in testimony before the Committee, saying aloud what nearly everyone in Washington was saying privately, the rumors became public issues that could not be ignored.

These were not the only problems the Tower nomination faced. From the first, some senators had expressed concern about the fact that Tower, immediately upon resigning his position as a U.S. arms negotiator (an executive post he had accepted after leaving the Senate and the Armed Services Committee), had unethically profited from his former position, earning huge fees as a consultant to arms manufacturers.[3]

As the controversy over Tower's confirmation increased, the Bush people were surprised to learn that, while Tower had indeed served many years in the Senate, his friendship account there had long since been overdrawn. As Elizabeth Drew, national correspondent for the *New Yorker*, wrote, the diminutive Texan actually "had few friends among senators, a large proportion of whom regarded him as a mean and petulant man, one who bullied his colleagues" when he was Armed Services Committee chair.[4] (Years earlier, one of Tower's Senate colleagues had probably capsulized senatorial sentiment about Tower when he said, "John Tower is the only man I ever knew who could strut sitting down."[5])

Tower's confirmation was made more difficult, too, by the fact that the Senate—and the country—had changed considerably during the preceding twenty-five years:

> When Tower entered the Senate in 1961, it was a cozier place, where members ducked into hideaways for drinks and conversation, tolerated each other's excesses and lapses and protected one another from the outside world, often without regard to party. Journalists were less intrusive, too.
>
> When Tower left the Senate in 1985, the pace had quickened and the Senate was becoming less collegial—a more uptight and frenetic place. Senators were more likely to dash off the floor to a fund-raiser or to National Airport than to stop for a drink in someone's office. Public attitudes toward alcohol and the role of women had changed. Cynicism toward political ethics was running deep, and politicians were running scared. The last of the old barons were going, and younger, better-educated and more news media-conscious members were taking their place.[6]

As the damaging evidence, public and private, began to build against John Tower, Armed Services Committee Chairman Sam Nunn came increasingly to feel that Tower should not be confirmed for the highly sensitive and strategically important position of secretary of defense.[7] As Nunn's doubts about Tower's fitness mounted, though, he was characteristically careful to consult and work in tandem with the ranking Republican on the committee, Senator John Warner of Virginia. He was careful too, at first, together with Warner, to keep his doubts and objections out of the press and to express them only privately, in three separate meetings with White House officials.[8]

Nunn clearly hoped that President Bush would decide to withdraw the Tower nomination and name a more suitable person. But the president and his advisers would not back down. They angered Nunn when they arranged a separate back-stiffening meeting and briefing exclusively for the Republican members of the Armed Services Committee.[9] When the vote in the Committee finally came, committee members divided strictly along party lines, all nine Republicans voting for Tower's confirmation, all eleven Democrats voting against it.[10] The Committee recommended that the Tower nomination not be confirmed by the Senate.

That should have been the end of it, with the president sending up another name. There were two schools of thought within Bush administration circles, however, about what to do next. According to an unidentified White House official, one school held that while "this is unfortunate, let's nominate someone else and work with Congress," but another school felt that "this is the first shot in a tug of war between the Democratic Congress and the Republican executive branch."[11]

The president and his advisers decided stubbornly to press on, to carry the fight to the Senate floor. That was a mistake. Enlisting Senate Republican leader Robert Dole of Kansas as their point man, they also decided to go with the second school in the Bush camp—to see the fight as a partisan one, indeed to make it one, while accusing the other side of doing so, and to launch a personal attack on Senator Nunn, ascribing partisan and "power-grab" motives to him. These were additional mistakes.

The fact was that Nunn, a moderate southern Democrat, was not thought of as strongly partisan. As one observer pointed out, he was certainly "less partisan than most of his Democratic colleagues."[12] Furthermore, characterized by Senate watchers as one of the remaining Washington politicians "who act with honor and play it straight,"[13] and as "sober, unflashy, a cool craftsman of military policy,"[14] Nunn, probably second only to Dole, was one of the most highly respected and influential members of the Senate when the Bush people decided to make the Tower nomination a partisan, anti-Nunn contest.[15] Even the ranking Republican on the Armed Services Committee, Senator Warner, felt constrained to defend Nunn's motives. For example, when Dole attacked Nunn during a joint Dole–Warner press briefing on the Tower nomination, Warner quickly interjected, somewhat to Dole's consternation: "Sam Nunn happens to like Tower personally, and it's not a power grab of trying to run the Pentagon from Capitol Hill. That much I can tell you."[16] But, by then, the skirmish had already become a partisan brawl.

The personal attack on Nunn backfired. A number of conservative, pro-military Democrats, like Charles Robb of Virginia and David Boren of Oklahoma, who might otherwise have voted with the White House, felt compelled by their respect for Nunn's ability and integrity to stand with him against confirmation. And Nunn, once he had taken a stand, was determined not to be overridden; he went to work, talking to senators one by one, laying out the evidence against Tower as he saw it.[17] The partisanship of the pro-Tower side backfired, too. Republican partisanship begat Democratic partisanship. The Senate Democratic leader, George Mitchell of Maine, joined in the quarrel, publicly defending Nunn and the committee majority on the floor of the Senate as well as helping to count noses and solidify the Democratic ranks in private.

President Bush asserted that Tower should be confirmed because, as he put it, "of the right of the President to have—historical right—to have who he wants in his administration."[18] Dole made the same argument. "President Bush selected John Tower. He has the right to have people [he wants] in his Cabinet."[19] Democrats were quick to point out that senators have the full constitutional power to reject presidential nominees for whatever grounds they choose, citing for this principle the words of Alexander Hamilton himself, who, in *Federalist 76*, wrote that the exercise of the Senate's confirmation powers "would tend greatly to preventing the appointment of unfit characters."

More tellingly, confirmation opponents made much of John Tower's own earlier senatorial opposition to various presidents' cabinet and subcabinet appointments, including Jimmy Carter's nominations of Paul C. Warnke to be head of the Arms Control and Disarmament Agency and of Ray Marshall to be secretary of labor, Gerald Ford's nomination of William J. Usery to be secretary of labor, and John Kennedy's nomination of Paul H. Nitze to be secretary of the navy. Opponents quoted Tower's own words on Senate confirmation powers and responsibilities, spoken in 1977 when he opposed the Warnke nomination: "The suggestion has been made that the people should trust the president to make the right appointment. I think that ordinarily we do around here. But, after all, the Constitution has vested in us the responsibility for advice and consent, and it is one that we should exercise."[20]

In the end, when the Senate vote came, each side was able to hold its own partisans pretty much in line. The Democrats won, of course— they were the majority party. It fell to the Republican Vice President Dan Quayle (who was in the chair to vote for Tower's confirmation in case of a tie) to announce in traditional words a nonroutine result: "The nomination of John Tower to be secretary of defense is not confirmed. The President is to be notified of the Senate vote."

The 53–47 defeat of Tower's confirmation resulted from a near party-line vote.[21] The Republicans held all but one of their members in favor of the nominee, and only three Democrats left their party's ranks to vote for him.* There was reason to believe that of these four, the lone

---

* The Republican was Nancy Landon Kassebaum of Kansas, who later said she had voted against Tower mainly because she was concerned about his consulting activities after leaving the Senate. The Democrats were Lloyd Bentsen of Texas, Tower's home state; Howell Heflin of Alabama, who was in a tough reelection campaign; and Christopher

Republican defector and one of the Democrats would have stayed in party ranks and voted the other way if their votes had appeared crucial to the outcome.[22]

It had been an unusually bruising battle. The fight caused some heated and personal words to be said on the Senate floor. The minute the fight ended, though, Democrats began trying to smooth things over. "This was a unique case," Senator Nunn said, adding that he did not think the Senate would in his lifetime ever consider another nomination "with so many allegations and such a degree of controversy."[23] (Nunn was wrong. The even more scalding debate over the 1991 nomination of Clarence Thomas to serve on the Supreme Court will be discussed momentarily.) Majority Leader Mitchell declared, soothingly, "This vote is not and should not be interpreted as a vote to harm the President."[24] But the Senate Republican leader was having none of it. Just before the vote Senator Dole declared, "It was a pitched partisan battle and has been for weeks," and Tower "knows this is politics. He knows he's being shot down because he's a Republican and there are more Democrats than Republicans."[25] Dole remained on the attack even after the defeat. In a CBS television interview, he said that the opposition to Tower had been "Nunn-partisan—N-U-N-N." Tower was "maligned and criticized with leaks and innuendos," he charged, adding bitterly that "We're not going to forget that easily."[26]

The increased partisanship of the Senate was shown again in the 1991 fight over the Supreme Court nomination of Clarence Thomas.[27] Like retiring Justice Thurgood Marshall, the man he was named to replace, Thomas was an African-American, but his judicial philosophy was decidedly more conservative than Marshall's, and liberal Democratic senators soon lined up against him.

Through the years, the Senate has rejected 28 of the 139 nominations to the Supreme Court,[28] and many of these confirmation fights have been highly partisan. But in recent years, with the exception of the 1987 disapproval of President Ronald Reagan's nomination of conservative Robert Bork to serve on the Supreme Court, opponents have not usually been able to muster enough Senate votes to reject the nominees on the basis of judicial philosophy alone.[29] That was the trouble Senate Democrats at first confronted in trying to defeat the Thomas nomination.

Democratic senators questioned Thomas severely when he came before the Judiciary Committee, especially concentrating on his views concerning affirmative action and abortion. But they were unable to tie the nominee down very much, and at first nothing was brought out publicly to damage his personal reputation. The Republican members of the

---

Dodd of Connecticut, who felt somewhat in Tower's debt because, twenty-two years earlier, Tower had opposed a Senate vote to censure Dodd's senator father, Thomas Dodd, for financial misdealings.

Judiciary Committee solidly defended Thomas and, together with the "handlers" provided him by the White House, aided Thomas in presenting himself in the best light.

When the fourteen-member Committee voted, senators split almost exactly along party lines. Democrat Dennis DeConcini of Arizona was the sole exception to the party-line vote, and this produced a 7–7 tie. The Thomas nomination was sent without recommendation to the Senate floor where, despite the split committee vote, it seemed headed for sure approval.

Then details of a sexual harassment charge against Thomas were leaked to the press, and a nasty and highly partisan controversy exploded. The Senate delayed the confirmation vote, and the Judiciary Committee reopened its hearings. Republicans immediately decided on a strategy of attacking Thomas's accuser, University of Oklahoma law professor Anita Hill, and they pursued it with a vengeance.[30] Thomas denied the sexual harassment allegations and went on the offensive, too, claiming that he was the victim of an underhanded and unfair conspiracy. Raising the race issue, he charged that the Senate hearings amounted to a "high-tech lynching for an uppity black."[31] Committee Democrats, by contrast, sought to appear neutral and evenhanded in the hearings—and their side lost the television battle for public opinion. Polls showed that a substantial majority of Americans believed Thomas, not Hill.[32]

The final vote in the Senate—fifty-two to forty-eight for confirmation—was the closest vote on a Supreme Court nominee in over a century. The division was quite partisan. Eighty percent of Senate Democrats—forty-six senators—voted against confirmation; eleven voted in favor. Ninety-five percent of the Republicans—forty-one senators— voted in favor of confirmation, with only two defectors.

In noting the high degree of partisanship that characterized the Thomas fight, Senate Majority Leader Mitchell said that the process of confirming justices "has taken on the trappings of a political campaign. Indeed, in the eyes of many Americans the process has become confused with electoral politics."[33]

Increased Senate Partisanship

Interestingly, senators have become more partisan precisely during the time that the people of the country who elect them have become less partisan—a kind of *less partisan voter–more partisan senator paradox*. Interestingly, too, each of the party conferences, or caucuses, in the Senate (the Senate Democratic Conference, made up of all Democratic senators, and the Senate Republican Conference, made up of all Republican senators) has grown stronger and more cohesive during the very period that senators and the Senate have become more individualistic—a kind of *individualistic senator–stronger party conference paradox*.

How can these paradoxes be explained? Part of the problem, especially with regard to the first paradox, is in the imprecision with which many people think, and write, about political parties in this country.

American political parties are different from those in Europe.[34] Party organizations here do not have members, as such; party affiliation is by self-identification only. Party organizations are decentralized and fragmented, like the levels and branches of our governmental system itself. And they do not control party nominations for office; instead, party candidates are designated by the party voters in primary elections. Little wonder, then, that the New Mexico Republican State Central Committee, say, or the Republican National Committee, has virtually no power to pressure New Mexico's Republican Senator Pete Domenici on a vote. Domenici, after all, was not nominated by either of these party organizations.

Are American political parties in decline? *Washington Post* columnist David Broder, in his 1971 book *The Party's Over*, argued that they were, maybe permanently.[35] Political scientist Larry Sabato counterargued in his 1988 *The Party's Just Begun* that America is experiencing a party resurgence, which can be encouraged and augmented.[36] Who is right? To answer we have to ask another question: What is meant by "party"? There are three possibilities, because American political parties appear in three distinct guises or aspects: (1) the *party in the electorate*, the potential voters who identify with one or the other of the parties; (2) the *party organization*, such as the Republican National Committee or the New York Democratic State Central Committee; and (3) the *party in government*, officials elected to office under one or the other of the party labels.

In trying to understand what has happened in regard to partisanship and party in the U.S. Senate, then, and whether or not there has been a "party decline" in America, we must start with a consideration of the Democratic party and the Republican party in the electorate.

A Changing Party in the Electorate

Changes have occurred in the party in the electorate of both major parties in recent decades. First, let us look at the horse-race standings: the relative percentages of Americans who call themselves Democrats or Republicans. Has there been a party "realignment" in the 1980s, like the one that made the Democratic party the front-runner party after the election of President Franklin D. Roosevelt in 1932, this time in favor of the Republicans?

No, the Republican party has not become the front-runner party in the electorate. It is clear, though, that since the 1960s there *has* been a realignment in the South as well as in some of the border states. The percentage of Republican identifiers, particularly among whites, has increased enough at least to put the Republicans on a parity footing with the Democrats there.[37]

Nationwide, Democratic identifiers still lead, but the gap between Republicans and Democrats has narrowed in recent years. During the 1950s, 1960s, 1970s, and the first four years of the 1980s (1980–1983), when asked their party affiliation by the Gallup poll, the percentage of Americans who called themselves Democrats consistently averaged around 45 percent or better.[38] During the same years, the percentage of Americans who called themselves Republicans declined in percentage from the low 30s to the mid-20s while those who identified themselves as Independents increased from the low 20s to about 30.[39]

Some people seem to be sort of party "floaters," though. Their party identification is changeable, affected by the momentary popularity or unpopularity of the presidential officeholders or candidates of the two parties. For example, back when Democratic President Harry Truman's approval rating hit bottom in 1946, so did the percentage of the nation's potential voters who identified themselves as Democrats; a Gallup poll found that the percentage of Americans who called themselves Democrats had dropped to 39 percent that year. But with a vigorous campaign against Republican Thomas E. Dewey in 1948, Truman turned things around and won reelection; correspondingly, that year, too, the percentage of Democratic identifiers was found to have increased to a more normal 45 percent. The Democratic percentage rose to a zenith of 53 percent in 1964, the year that Democrat Lyndon Johnson defeated Republican Barry Goldwater for the presidency. It dropped to 43 percent in 1972, the year that Republican President Richard Nixon swamped Democratic challenger George McGovern to win reelection.[40]

This tendency on the part of some potential voters to choose their party label on the basis of the popularity or unpopularity of national leaders is, undoubtedly, a principal reason, at least outside the South, for the narrowing of the gap between those who identified themselves as Republicans or Democrats after President Ronald Reagan's reelection in 1984. From 1984 through 1989, Republican affiliation in the whole country rose to an average of 31.4 percent, according to the Gallup poll, and the 1989 average was 32.5 percent.[41] During the same years, the average for the Democrats declined to 40 percent (with an average of 40.2 percent in 1989), and the Independent average declined a little, too, to 28.5 (with an average of 27.25 percent in 1989).[42] The Democratic advantage of earlier years was seen to shrink even further when nonvoters were left out.[43]

Democrats had not had one of their own in the White House since Jimmy Carter (1976–1980), and they had no doubt suffered in the party-affiliation surveys during the time when a hapless Carter, presiding over, among other things, a worsening economy, was remembered as the personification of the Democratic party, while a blessed and charming Ronald Reagan was seen as the personification of the Republican party.[44] In addition to the matter of the narrowing Republican–Democratic gap, this "leader comparison" factor may also help account for the fact that 27 percent of the Democrats in a 1986 survey said that they were less committed to their party than they had been five years earlier, whereas 43 percent of the Republicans said that they were more committed.[45]

The net gain in Republican identifiers during the Reagan years and afterward involved, though, some of the "least politicized, least sophisticated" people in the electorate, whether young or old.[46] These may well be the most volatile and changeable in their party affiliations.[47] In fact, among the youngest of the young voters (those who remember only the Reagan of his last years in office), the shift toward a Republican identification was not nearly so evident.[48] On the other hand, in the electorate at large, political scientist Warren Miller found that, while the increase in Republican identifiers during Reagan's first years came from the ranks of those whose party affiliation was usually most subject to change, those who increased the Republican percentage later on were better educated and more conservative, indicating that these changes, at least, might prove more permanent.[49]

In any event, we can surely say that there has not yet been a nationwide realignment in party identifier percentages—at least to the extent of an overall, net change in the answers to the horse-race questions: Who's ahead? Who's behind? The Republican party has not become the front-runner party in the electorate, although a realignment in the South and a recent narrowing of the gap between Democrats and Republicans generally has made the Republican party more competitive virtually everywhere.

It must be conceded, though, that there has been a decline in the electorate's party loyalties, a trend some call "dealignment."[50] The percentage of Americans who feel that neither party matters in terms of dealing with major problems, rose to 44.6 percent in 1988.[51] Similarly, while in 1952 only 10 percent of Americans were neutral—that is, neither negative nor positive toward the Republican and Democratic parties—by 1988, 30 percent were.[52] Since the 1950s there has been an increase in the percentage of Americans who identify with neither of the two major parties, but call themselves "Independents."[53]

Another way to gauge the decline in party loyalty is to look at "ticket-spitting"—voting for candidates of different parties for various offices in the same election. Split-ticket voting has increased. More than 90 percent of American voters questioned in a 1986 survey agreed with the statement, "I always vote for the person who I think is best, regardless of what party they belong to," while only 14 percent agreed with the statement, "I always support the candidates of just one party."[54] Between 1952 and 1988 there was an increase in reported split-ticket voting: the proportion of voters surveyed who said that they had split their tickets between presidential and House candidates increased from

12 percent to 25 percent; during the same period, there was an increase from 9 to 27 percent in those who said that they had split their tickets between House and Senate races.[55]

Yet party identification is still important in the way people vote. As a matter of fact, the extent of decline in party loyalty has probably been overestimated,[56] and most of the decline has likely come from the ranks of nonvoters, so that the trend should be called "*non*alignment" rather than "*de*alignment."[57] About 80 percent of Americans still identify themselves as either Democrats or Republicans.[58] The trends toward increased split-ticket voting and split-ticket outcomes, as well as the trend toward increased voter neutrality toward the parties, were all slowed in the election years of 1984 and 1988.[59] One authority has found that Americans in the 1980s voted as willingly along party lines in presidential elections as they did in the 1950s.[60]

In the 1988 elections, 81 percent of the voters cast their ballots for their party's presidential candidate; 72 percent voted for their party's Senate candidate that year, and 75 percent in 1990.[61] In any event, even in elections in which a third of the voters split their tickets, the other two-thirds cast straight, party-line votes for the entire ticket of their party.[62] Too, partisanship plays a greater part in Senate and presidential elections than in House elections, providing, as two political scientists have concluded, "a strong cue for many voters deciding how to cast their Senate as well as their presidential votes."[63]

What is probably more significant than party dealignment and more important than realignment (in the net, horse-race percentages) is the fact that, within each party in the electorate, there has been a considerable measure of *ideological* realignment. A big part of this has taken place in the South.[64] Many white Southerners have left their traditional Democratic affiliation for a new alignment with the Republican party.[65] At the same time (and a major reason for the white Southern conversion), there has been a huge increase in the numbers of African-American voters in the South (a consequence of the Voting Rights Act of 1965), and these new black voters are overwhelmingly Democrats.[66] Rural and conservative Democrats in the South, "the old Southern establishment—opposed and ultimately defeated by their own national party on the central issue of civil rights and, thus, 'liberated' to vote their pocket books—began to drift toward the GOP once the Civil Rights and Voting Rights Acts had passed."[67] Ideological differences were involved in these switches (and in the North, too), so that, as political scientist John Petrocik has concluded:

> The Republicans, even without an enlarged white Southern constituency, would have become more conservative on both race and welfare issues by 1984. The increased representation of white Southerners in the party made a mass base of the GOP slightly more conservative on welfare issues. The greater liberalness of the Democrats, on the other hand, has depended significantly upon the declining contribution of the white South to Democratic identification.[68]

Nationwide there is a growing racial gap between Democrats and Republicans. A slightly greater percentage of whites favor the Republican party than identify with the Democratic party (46 percent to 41 percent), whereas blacks are overwhelmingly Democratic, and more of them (10 percent) call themselves Independents, even, than identify with the Republican party (4 percent).[69]

There is an increasing income-level gap between party identifiers, too—more Democrats being at the lower-income levels, more Republican identifiers at the higher-income levels.[70] Democratic identifiers lead Republicans by a huge margin of 37 percent among Americans whose families earn under $12,000 a year, but drop below Republicans by a margin of 11 percent among those whose families make more than $42,000.[71]

These racial and income-level gaps between the two parties have paralleled a growing ideology gap. The electorate as a whole has not become more conservative in recent years, but Republican identifiers certainly have, and this change has occurred at the same time as those who identify with the Democratic party have become more liberal. The result is more pronounced ideological and programmatic differences between Democratic and Republican identifiers.[72] Overwhelmingly, now, those who call themselves liberals also call themselves Democrats, and those who call themselves conservatives also call themselves Republicans.[73] Looking at all Republican and Democratic identifiers, we find that 52 percent of Republicans consider themselves conservative, whereas only 22 percent of Democrats do.[74] Furthermore, the Democratic party attracts a greater percentage of moderates than does the Republican party.[75]

Issue differences stand out, too, so that, as one authority has found, "party loyalty now is more closely tied to policy concerns than in the past."[76] In another study, nearly 52 percent of Republicans said that the government was spending too much on welfare, compared with only 37 percent of Democrats who said that; nearly 50 percent of Democrats thought that too much was being spent on the military, while only 28 percent of Republicans agreed.[77] A 1988 *New York Times*/CBS News poll found that 59 percent of Republicans (compared with only 33 percent of Democrats) preferred small government with fewer services, whereas 56 percent of Democrats (compared with only 30 percent of Republicans) preferred big government with more services.[78]

Surveys show that more Americans now perceive significant issue differences between the two parties than in earlier times. During the years 1952 to 1976, between 46 and 52 percent of Americans thought that there were important differences in what the two parties stood for, a measure that rose to between 58 and 63 percent during the years 1980 to 1988.[79]

Each party in the electorate, then, has become more homogeneous, more ideologically coherent, and simultaneously more unlike the other party[80]—the average Republican being a conservative, the average Democrat a moderate to liberal.

These ideological and issue differences between the two parties in the electorate are greatest among their activists and elites.[81] When white activists in the South switched to the Republican party, for example, their conversions were closely related to their conservative positions on issues.[82] Nationwide delegates to the 1988 national conventions of the two parties were found to be strikingly different in their policy preferences.[83] For example, 67 percent of the Democratic delegates polled agreed that large corporations were too powerful for the good of the country, whereas only 14 percent of the Republican delegates did. Eighty-two percent of the Democratic delegates surveyed thought that the government should institute a national health care program, compared with only 15 percent of the Republicans. Eighty-three percent of the Republican delegates polled said that they thought that the government had no role in seeing to it that everyone had a job; 70 percent of the Democrats thought just the opposite.

Overall, nationwide, "Activist Democrats are more liberal than rank-and-file Democrats, and activist Republicans are more conservative than GOP voters,"[84] and these differences widened significantly during the especially polarizing Reagan years (1980–1988).[85] Party activists are the people who are likely to be most influential in party nominating contests and their outcomes, which tends to produce more polarized nominees for the U.S. Senate: moderate to liberal Democratic nominees and conservative Republican nominees.[86]

To summarize, changes have been taking place in the parties in the electorate. The percentage difference between Americans who call themselves Democrats and those who call themselves Republicans has narrowed in recent years. Party competitiveness has become nationalized, with a more two-party South and a more two-party country. There has been some party dealignment in America, a decrease in party loyalty, and, in that sense, some party decline. But dealignment has been accompanied by an ideological realignment among the party identifiers, so that Democrats are now more likely to be moderate to liberal on issues and Republicans to be conservative. This difference is most pronounced among the activists and elites of each party, those who have greatest influence on party nominations. As a result of this ideological realignment, there is now a greater likelihood that each party's nominees—including nominees for the Senate—will be similarly different from each other.

Stronger Party Organizations

Most people have the feeling that America's national "political parties" were stronger in the 1950s than they are now. But even then, some thinking about national parties was based on myth. Donald Matthews wrote that the "popular mythology" of the time was that candidates ran on party-formulated "principles," that the voters chose the party that best represented their opinions, and that the winning party then enacted its program in accordance with its popular "mandate," or if it failed to do so, was turned out of office at the next election.[87]

The 1950s reality, though, was something quite different. Matthews reported then that "the parties are loose coalitions of state and local organizations with divergent policy aims, leadership groups, and electoral followings. Senators who belong to the same national party are elected at different times, by different electorates, on different platforms."[88] Similarly, President Dwight Eisenhower (1952–1960) said at the time, "Now, let's remember, there are no national parties in the United States. There are forty-eight parties. . . . The most I can say is that in many things they do not agree with me."[89]

What people mean, then, when they talk about parties having been stronger in the past is that state and local party organizations—in many cases, boss-ridden party machines, really—were stronger, with their power springing largely from their considerable ability at one time to control nominations, jobs, contracts, welfare, and other tangible benefits. The direct primary election, which legally took party-nomination control away from party organizations and gave it to the party in the electorate, had long since begun to weaken the old party machines. A more educated citizenry and a rising standard of living, television (which allowed candidates to go over the heads of the party leaders directly to the people), civil service and merit systems, and contract and welfare reforms helped to complete the job.[90]

The old machines died away, and by the early 1960s state party organizations, too, had been weakened in most states for the same reasons.[91] Then new challenges to the state parties developed, particularly the weakening of party loyalty in the electorate; the advent of candidate-centered, personal campaigns; and the rise of political action committees (PACs).[92] But state party organizations were able to survive and adapt. More than that, according to political scientist John Bibby, they have in recent years enjoyed a comeback, as evidenced by the fact that many now have permanent party headquarters, professional leadership and staff, adequate budgets, and programs for assistance to local party units and party candidates and officeholders.[93] The result, Bibby concludes, is that although American parties are never going to be as strong as European parties,

> The prophets of party demise . . . have been proven wrong. Since the 1960s American state party organizations have become more professionalized and organizationally stronger in the sense that they can provide campaign services to their candidates. They do not control nominations because of the direct primary, but many are playing an important role in campaigns. The state parties have also become more closely integrated with the national party organizations and are being utilized by those organizations in presidential, senatorial, and congressional elections.[94]

This resurgence in the strength of state party organizations was paralleled at the national level: the national committees and the congressional campaign committees also gained vitality. Political scientist Joseph Schlesinger believes that this party-organization resurgence has actually been a result of loosening party-loyalty ties in the electorate and increased party competitiveness.[95] Schlesinger has written that, today, especially with changes in the South, there are virtually no states that are not two-party, competitive states. This fact, taken together with loosening partisan identification in the electorate, which has made the electorate more unstable and less predictable, has created a greater sense of electoral insecurity in candidates and officeholders. "Candidates need all the help they can get; they are finding that the best place to get it is from their fellow partisans," Schlesinger has written, arguing further that this has produced a party-organization revitalization, in a continuing cycle: "More organizational effort leads to more intensely competitive contests, which in turn make organizational effort ever more worthwhile."[96]

The Republican National Committee (RNC) and the Democratic National Committee (DNC) are national party organizations that are sometimes called the presidential wings of the parties. They may be more interested in presidential than congressional elections, although they now sponsor national party advertising, give assistance to local and state parties, and provide candidate services that have considerable impact on Senate (and House) races. The four House and Senate party campaign committees—including the National Republican Senatorial Committee and the Democratic Senatorial Campaign Committee—are national party organizations in the congressional wings. All of these national party organizations are stronger today than ever before and are, as one authority has found, "financially secure, institutionally stable, and highly influential in election campaigns and in their relations with state and local party organizations."[97] The national party organizations, he reports, are "no longer in decline, but they are in fact flourishing in today's cash-oriented, technologically sophisticated world of campaign politics."[98]

The RNC began the revitalization process much earlier than the DNC. The RNC pioneered, for example, a direct-mail financial appeal to small givers that by 1963 was raising more than a million dollars a year.[99] In the mid-1970s, spurred by concern over Watergate-era defeats at the polls, the RNC further institutionalized and expanded its fundraising efforts and multiplied its staff numbers, as well as its candidate and campaign services. In turn, the DNC finally, after Democratic President Jimmy Carter had lost his reelection bid in 1980, began to augment its own activities and endeavors in an impressive, though always lagging, effort to catch up.[100]

In the presidential years from 1976 through 1988 the RNC's total receipts increased three and a half times—from $29 million in 1976 to $78 million in 1980 to $106 million in 1984 and then down a little to $91 million in 1988. In the same years the DNC, though it had fewer potential backers among the potential big givers, increased its total receipts from $13 million to $15.4 million to $46.6 million and then to $53.3 million, a fourfold increase.[101]

The RNC and the DNC each had staffs of only thirty people in 1972. By 1976 the RNC staff had exploded to 200, and by 1990 (with some decline after 1984) there were 400 people working at the RNC.[102] DNC staff numbers jumped first in the presidential year of 1984—to 130 people—increased in 1988 to 160 and leveled back down to 130 in 1990.[103] Each of these national party organizations is now housed in its own imposing national headquarters building in Washington, D.C.

Party nationalization has grown with the revitalization of the RNC and the DNC. These national party organizations have greatly increased their influence over their own state and local party organizations and have used their funds and legal authority to nationalize party campaign efforts.[104] The national committees sponsor party-building advertising and finance state and local registration and get-out-the vote campaigns.

Now let us turn to the growth in the importance and influence of the two Senate campaign committees—the National Republican Senatorial Committee (NRSC) and the Democratic Senatorial Campaign Committee (DSCC)—growth that can hardly be exaggerated. Formed in 1916, as mentioned in Chapter 2, after the Constitution was amended to provide for the direct election of senators (counterpart House campaign committees had been formed at the end of the Civil War), the NRSC and the DSCC have the highly specialized job of helping to elect, and reelect, their own partisans to the Senate.

Each of these Senate campaign committees now has its own impressive national headquarters (the DSCC in the DNC building) and its own large and professional staff. Each had only four staff members in 1972; by 1990 the DSCC staff had increased twelvefold, to forty-eight, and the NRSC staff had multiplied twenty-five times, to one hundred.[105]

The Senate campaign committees also raise tremendously increased sums of money now. Funds received by the two committees, from individuals and PACs, have multiplied from no more than around $2 million, or less, for each of them as late as 1976 to $16.3 million for the DSCC and $66 million for the NRSC in 1988.[106] Under the law, each committee is allowed to contribute $17,500 to each of its Senate candidates during each election cycle and also to make "coordinated expenditures" (for polls, advertising, and direct-mail fund-raising appeals, for example) on behalf of each of its candidates in amounts ranging from $100,000 to $2.1 million. In 1990 the NRSC contributed $859,140 to its Senate candidates and spent $7.7 million on coordinated expenditures on their behalf, while the DSCC contributed $510,133 to its candidates and laid out $5.19 million in coordinated expenditures.[107]

The campaign committees also do a lot of other significantly important things, these days. They recruit candidates in key races; run campaign schools; provide exceptionally sophisticated campaign management, communications and advertising, polling, fund-raising, and other services; help corral PACs oriented toward their parties and candidates and coordinate their giving; and cultivate a hothouse of successful and distinctly partisan national campaign consultants and match them up with candidates who need their services.[108] In short, what the Senate campaign committees do has been vastly expanded in recent years—a development recognized and immensely appreciated by senators and Senate candidates.[109]

Party organizations are *not* still suffering from a decline, then. They are stronger and more important than ever. Their increased efforts and activities, as we saw in Chapter 2, have helped to nationalize Senate campaigns. They have also helped to make Senate campaigns more nationally partisan—thus helping to produce more nationally partisan senators and, within each of the two Senate parties, more internal cohesiveness on issues.[110] (In fact, some think the revival of national party organizations and their increased advertising and other activities might even have the potential of reversing the trend toward dealignment, or party decline, in the parties in the electorate.[111])

The resurgence of party organizations, especially the national party organizations, taken together with changes that have taken place in the parties in the electorate (particularly, increased two-party competitiveness in the South and elsewhere and the ideological realignment among party identifiers and especially among party activists), has generated changes in the parties in government in the Senate, helping to produce more partisanship in that body.

Greater Senate Partisanship

Eminent political scientist James MacGregor Burns wrote in the 1960s that America was suffering from a "four-party system"—that American government, including the Senate, was often deadlocked because the two parties in government, the Democrats and Republicans, were each split into internally warring liberal and conservative factions.[112] Today the Senate has become much more a two-party body (and that, as we shall see, has sometimes produced a different kind of deadlock).

As each party's set of identifiers and activists in the electorate has become more homogeneous and more unlike that of the other party—Republican identifiers and activists more conservative, Democratic identifiers and activists more moderate to liberal—these differences have been carried into, and have colored, the nominating processes for the U.S. Senate. Thus the Republicans have become more likely to nominate conservative Senate candidates, the Democrats more likely to nominate moderate to liberal Senate candidates.[113]

This has been true, of course, in the South. There, Democratic nominees for the Senate have increasingly been "nationalized"—that is, have become more like the Democratic nominees in other parts of the country;[114] and Southern Democratic Senate candidates are less likely, now, to be able to secure their party's nomination without the strong support of liberal-leaning labor unions, teachers, and, notably, African-Americans.[115] Southern Democratic senators elected from 1965 on were significantly more liberal than those elected earlier, and not just on racial issues.[116] Five new Democratic senators elected in the South in 1986 were significantly more liberal on all issues than their predecessors; indeed, their rejection of President Reagan's nomination of Robert Bork for the U.S. Supreme Court was only one indication, though a spectacular one, of this fact.[117] In the 1991 confirmation of Clarence Thomas for the Supreme Court, seven of the eleven Democrats who voted for Thomas were from the South, but their actions were more explained by the fact that black voters in their states backed Thomas than by old-style Southern conservativism.[118]

Nationwide, the "sharpest ideological division" among Democratic and Republican senators was seen among those who were first elected in the 1980s, so that, as one observer noted, "the politics of the 80s is producing very conservative Republicans and very liberal Democrats."[119]

The ranks of what might be called Senate "countertypes"—liberal Republican senators and conservative Democratic senators—have been noticeably thinned out. "First to fade," one authority has written, "were the progressive Republicans"—liberal Republican countertypes like Senator Clifford Case of New Jersey or Edward Brooke of Massachusetts. "Their counterparts, the conservative Democrats," he added, "have been vanishing as well, although more slowly"—senators like John McClellan of Arkansas and Richard Russell of Georgia.[120]

Countertype decline in the Senate occurred in four ways: when a liberal Republican was replaced by a conservative Republican (Alfonse D'Amato replacing Jacob Javits in New York, for example); when a conservative Democrat was replaced by a conservative Republican (Thad Cochran replacing James Eastland in Mississippi, for example); when a liberal Republican was replaced by a liberal Democrat (Barbara Mikulski replacing Charles McC. Mathias, Jr., in Maryland, for example); and when a conservative Democrat was replaced by a liberal or moderate-liberal Democrat (John Breaux replacing Russell Long in Louisiana, for example). Three of the five most conservative Democrats (with the highest 1986 "opposition to party" scores)—Russell Long of Louisiana, Edward Zorinsky of Nebraska (replaced by liberal Democrat Robert Kerrey), and John Stennis of Mississippi (replaced by conservative Republican Trent Lott)—as well as two of the three most liberal Republicans (again, with the highest 1986 "opposition to party" scores)—Mathias of Maryland and Mark Andrews of North Dakota (replaced by moderate-liberal Democrat Kent Conrad)—did not, because of death, defeat, or retirement, return to the Senate after the 1988 elections.[121]

Reflecting what has taken place among identifiers and activists in the party in the electorate, then, Senate Democrats and Republicans have each become more internally homogeneous and each more unlike the other.[122] This, more than increased party discipline or anything else, has produced more Senate partisanship and party-line voting.

In the 1960s and 1970s, Senate Democrats had found it virtually impossible to get together on issues because of the wide ideological splits within their own ranks, and for that reason, little attempt was made to do so. "The spread between Kennedy [Edward M. Kennedy of Massachusetts] and Long [Russell Long of Louisiana] was too great to bridge," is the way New Jersey Democratic Senator Bill Bradley put it.[123]

Senate Republicans had always been more homogeneous, more cohesive, than Senate Democrats, but today some Republican senators feel that the Democrats have surpassed the GOP in the ability to vote together as a party. For example, Kentucky Republican Senator Mitch McConnell has noted:

> It seems like the Democrats are more together because there are really not any conservative Democrats any more. There has been a realignment in the South. On defense, the Democrats of the South still tend to be more con- servative, but on most other issues, they are like [the rest of the Senate Dem- ocrats].[124]

Similarly, Senator Kassebaum has said, almost wistfully: "Do the Democrats, I wonder, do it [get consensus] more? I guess I'm glad I'm not pushed too hard, but I wish we had a little bit more of it."[125]

Internal homogeneity has meant more cohesiveness on issues. Toward the end of the 1980s, Senator Byrd, then majority leader, was able to say that Senate Democrats were "more unified than at any time in my 36 years in Congress."[126] And, studying both Senate parties, polit- ical scientist Samuel Patterson has concluded that "On important mat- ters of national public policy, the Senate usually divides mainly along political party lines."[127]

Consider "party-unity voting" in the Senate—roll calls on which a majority of Democrats oppose a majority of Republicans. There has been a steady, overall upturn in this measure of Senate partisanship since 1970.[128] There was more such partisanship in the Senate in 1990 than in any year since 1961.[129] And party-unity voting was even slightly higher in the Senate in 1991, when an average of 80 percent of the Democrats and 81 percent of the Republicans voted with their parties on roll-call votes where a majority of one party lined up against a majority of the other.[130]

Similar, and even higher, Senate partisanship shows up when only important or major roll calls are taken into account, leaving out, as Sam- uel Patterson puts it, votes on things like "Made in America Month" or "National Digestive Diseases Awareness Week." On the more impor- tant roll calls, a majority of one Senate party lined up against a majority of the other party 75 to 80 percent of the time throughout the 1980s, the highest levels of Senate partisanship on such key votes in the postwar era.[131] That trend continued into the 1990s.[132]

Consider "party-unity scores" of individual senators—the percentage of the roll calls on which they sided with their party when a majority of Democrats and a majority of Republicans voted against each other. These partisan scores have risen, too. The average party-unity score for all Republican senators, which was 68 percent in the 1960s, rose to 75 percent in 1990 voting and 82.9 percent in 1991; similarly, the average party-unity score for all Democratic senators, which was 65.7 percent in the 1960s, climbed to 80 percent in 1990 voting and 82.8 percent in 1991.[133] And one study has shown that the levels of party unity inside Senate committees generally exceed those on the floor.[134]

In addition to internal party homogeneity, other factors have played a part in creating heightened Senate partisanship. One such factor has been increased two-party competition: nationwide in Senate elections, as well as internally for control of the Senate.

Two-Party Competition

We have already discussed how regional and other changes in the electorate have meant that, as one authority noted, "two-party competition has spread to every region of the country" and "no state can be considered safe for either party."[135] Similarly, according to another authority, "Both parties have an equal chance of winning any seat in the Senate."[136] The extent of this two-party competitiveness for Senate seats is seen in the fact that by 1980 fully half the states were electing "mixed" Senate delegations, that is, sending both a Democratic senator and a Republican senator to represent them,[137] and the number of such mixed delegations has remained high (twenty-one such states in 1993).

One political scientist, Joseph Schlesinger, maintains that greater cohesiveness within each party is an inevitable result of increased party competition. "Not only has the spread of competition to all regions meant greater voting cohesion within the Congress; it has also reduced the range of ideological differences within the parties."[138]

Heightened nationwide party competition has produced closer party margins in the Senate. The Democrats enjoyed comfortable Senate majorities all through the 1960s and 1970s (after the close partisan balance of the 1950s). But beginning in the early 1980s, the margin between the two parties has narrowed to a degree that makes it quite possible for control of the Senate—and thus the majority leader's position and that of all committee and subcommittee chairs, as well as authority over the Senate's schedule—to pass from one party to the other with almost any biennial senatorial election.[139] Control *did* shift in 1980, to the Republicans, when, with Ronald Reagan's election to the presidency, the Republicans also gained a majority in the Senate for the first time since the 1953 to 1955 period. In 1986 majority control swung back to the Democrats.

Intensified competition for control of the Senate, which is likely to continue in the future,[140] has played a part in the escalation of Senate partisanship. "We [Democrats] figured we had the majorityship by the divine right of kings," Democratic Senator J. Bennett Johnston of Louisiana has said. "Nobody really, seriously, considered a threat by the Republicans to be credible. Now it is, and we're more careful."[141] Democratic Senator Bill Bradley of New Jersey made a similar point: "Until we lost [control] in 1980, the necessity of consensus was not felt; we had plenty of members."[142] On the Republican side, Pete Domenici of New Mexico has also linked increased Senate partisanship and closer party margins. "Recently, the two parties have been relatively equal in strength, and we're moving more toward making party more important here."[143]

When Robert C. Byrd was first elected to lead Senate Democrats in 1977, his party enjoyed a 62–38 margin over the Republicans. Contrasting those earlier times with what was to follow, Byrd has said:

> We had the numbers then. We could afford to be a little [individually] independent. But being in the minority for six years . . . we learned that we don't have the luxury of each going his own way, and we do have to look at things together as a party. Because, while a given senator may be reelected on his own, quite independent of his party, at the same time he is a committee chairman or a subcommittee chairman because the party has control of the Senate.
>
> So, it is important that our party stay in control. Otherwise, we go back to being the ranking members [of the committees and subcommittees] and having less influence and impact on national programs and the direction of our country. That's the penalty if we don't stick together.[144]

A number of senators feel that, as a result of the now biennially close contests for Senate control, both parties in the Senate sometimes press for roll-call votes for partisan, electoral reasons. According to Republican Senator Steven D. Symms of Idaho, the Republicans started this. "In the 1970s, conservative Republicans—Helms, Laxalt, and McClure— offered amendments and forced votes. This gave me a record to run against [former U.S. Senator Frank] Church and beat him; it made it possible to polarize the electorate in 1980." Senate Democrats then took up the same tactic: "Now, Bobby Byrd [Democratic leader Robert Byrd of West Virginia] is a genius over there on the floor. He has forced Republicans to vote on a lot of issues. . . . The Democrats have sometimes offered amendments to embarrass the Republicans, and the Republicans voted consistently as a block."[145]

Similarly, as the 1988 Senate session was winding down to a close, Republican Don Nickles declared that, "Right now, at the tail end of the session, a lot of things are used [by the Democrats] for political purposes—such as minimum wage and parental leave and, earlier, plant closings."[146]

The party-polarizing presidency of Reagan also helped make the Senate a more partisan place. There is nearly always a greater degree of partisanship in the Senate (and in the House, too), more party-line voting, when the issue in dispute is one on which the president is active. Oregon Republican Senator Bob Packwood made that point when he said, "A presidential veto causes a party effort, or an issue where the President is involved (although there was less of that for the Democrats when [President Jimmy] Carter was in)."[147] In such cases, the president's partisans in the Senate are more constrained to stand up for him, and senators of the other party to line up against him. (We saw this, of course, in 1989 when President Bush tried unsuccessfully to push through the confirmation of John Tower to be secretary of defense as a partisan matter.)

Ronald Reagan was an especially partisan president, and this had a highly polarizing effect on the parties in the Senate (and House), producing a kind of "revitalization of partisan conflict."[148] The Republican takeover of Senate control when Reagan was first elected also had immediate and long-lasting party-polarizing consequences, especially on Senate Democrats. As former Senator Long has said:

> For as long as I can remember, the Republicans would meet behind closed doors, over lunch . . . [and] come in with solid, party line votes. And that eventually forced Democrats to offset that. So, under Bob Byrd, after President Reagan came in, we began having a once-a-week meeting to determine what was going on, our position on various issues.[149]

Another former Democratic senator, Lawton Chiles of Florida, has made the same point, emphasizing its continuing effect:

> In the old days, no position [on issues] was taken in the [Democratic] caucus. We carried that on for a while [after the Republicans gained the majority in 1980]. The Republicans were disciplined and working us over pretty good. It took us a long time to realize we were a minority. We kept operating like that until, finally, we began to discipline ourselves and find those areas where we had agreement. On a budget vote once, we got a one hundred percent Democratic vote. The discipline carried over, and we worked on it harder after we became the majority again.[150]

Thus, a partisan Republican majority from 1980 to 1986 helped cause greater partisanship in the Democratic minority during those years as well as in the Democratic majority in the years that followed. This in turn caused Republicans to be more partisan. According to Republican Senator Richard Lugar of Indiana: "The Republicans sort of went along for a while and were surprised the Democrats were voting in such numbers together. The shock of this occurring for four months finally led Republicans to say, Well, now we have to pull up our socks on a couple of issues."[151]

Each Tuesday noon while the Senate is in session, now, all work stops so that both parties can hold separate meetings of their party conferences. Many additional such party caucuses are held at other times as well. Especially for the Democrats, this is a huge change from earlier times, as former Democratic Senator William Proxmire of Wisconsin said in a 1988 interview:

> When I started in the Senate [1957], there was only one caucus at the beginning of the Congress for the "LBJ [Senate Majority Leader Lyndon Johnson of Texas] State of the Union" and another one two years later. Now, we have one every Tuesday. At the end of a session, there are three or four a week. Byrd [former Majority Leader Robert Byrd of West Virginia] and the other leaders take their lead from the consensus. The Republicans *always* operated more on the basis of consensus.[152]

In addition to more partisan voting and greatly increased party caucusing, today's increased Senate partisanship is evident in other ways, too. There have been party retreats as well as the appointment of party task forces to study and recommend partisan positions on particular issues.[153] The Senate is represented, and negotiates, by party in "summit" meetings involving the House, the Senate, and the White House, to work out agreements on the federal budget and deficit reduction.[154]

Significantly, after the Republicans lost control of the Senate in the 1986 elections they began to use party-line filibusters and related stall tactics to stymie the Democratic majority. The Democrats, in turn, resorted to attempts at near party-line cloture votes to facilitate action on their agenda. As we noted in Chapter 5, the partisan filibuster is not new in the history of the Senate, but the way it is used in modern times is new. With a narrow margin in party strength (fifty-seven Democrats to forty-three Republicans in 1993, for example), the party filibuster can be especially effective to prevent action by the majority.

In 1987, the first year after the Democrats regained control of the Senate, there were unprecedented, and successful, Republican party-line filibusters to block that year's defense bill, as well as the supplemental appropriations bill, until figures more to the liking of the Republicans could be negotiated.[155] That same year, a Republican filibuster prevented consideration—and otherwise sure passage—of a Democratic campaign-finance reform bill, when Democrats repeatedly came up short of the sixty votes needed for cloture to cut off debate during that session (as well as in the next). In 1987, too, a Republican filibuster paralyzed Democratic efforts to pass a new minimum-wage bill.[156]

Senate Republicans continued in 1988 to use party-line filibusters and obstructionist tactics to thwart the Democratic majority, thereby blocking such measures as child-care legislation and, again, the minimum-wage bill. The Democratic leadership took to filing cloture petitions almost as soon as there was a hint of a Republican filibuster—a move that, if thereafter voted by three-fifths of the Senate, would cut off debate and restrict the consideration of amendments to germane measures already filed. Democratic Majority Leader Byrd introduced forty-three cloture petitions (against both party-line and individual filibusters) in the 1987–1988 session.[157]

At first, after George Mitchell succeeded Byrd as majority leader in 1989, he operated more cooperatively with the Republican leader, Robert Dole of Kansas, than Byrd had. Mitchell and Dole worked together to pass clean air legislation and were able to agree on procedures for taking up controversial measures dealing with flag burning, family leave, and crime. But the "era of good feeling" exploded in pent-up partisan rancor in 1990, after Mitchell was able to obtain cloture and cut off a Republican filibuster (with the help of eight Senate Republican defectors) before pushing a new Democratic-backed civil rights bill through the Senate.[158] The Republicans, under Dole, had long been engaging in a variety of party-line stalls on Democratic measures, and the civil rights cloture petition marked the eighteenth time in 1990 that Mitchell had used that device to end a real or threatened (party-line or individual) filibuster.

This time, Dole blew up. "If we are going to be treated like a bunch of bums on this side of the aisle, say so," he said. "There will not be any more time agreements or any agreement at all on anything until we have some understanding that we are going to run this place in a civil manner."[159] And the next day, to underscore his point, Dole objected to and thus blocked a routine request by Senator Edward Kennedy to modify his own amendment. Dole declared, "Since we've been deprived now of our rights . . . we are supposed to turn the other cheek. We are not going to do that."[160]

Partisan Senate Leaders

The positions of majority leader and minority leader in the Senate were created only some seventy years ago.[161] Even as late as the 1950s, when Democrat Lyndon Johnson of Texas served as majority leader (1955–1961) during the Republican presidential administration of Dwight Eisenhower, Johnson viewed his role as that of *Senate* leader, the faithful agent of a cross-party conservative coalition,[162] rather than as a *party* spokesman and open opposition leader.[163] "Johnson didn't see any percentage in open dissent against Eisenhower, a popular president, a national hero," Horace Busby, once a Johnson aide, has said. "Why harm the [Democratic] party by taking him on all the time?"[164] But since those days, as each Senate party's members have become more ideologically in accord with one another and more partisan, it has been easier for each party's leaders to be partisan and to be committed and active on issues. This, coming full circle, itself has helped to produce more Senate partisanship.

It happens that both party chiefs in today's Senate—Republican

Robert Dole and Democrat George Mitchell—have strong national partisan backgrounds. Dole is a former chair of the Republican National Committee (and was the vice presidential nominee of his party in 1984), and, according to one of his fellow Republican senators, "enjoys more partisanship."[165] George Mitchell, before coming to the Senate, chaired the Maine Democratic State Central Committee and served as a member of the Democratic National Committee and its Executive Committee (and was an unsuccessful candidate against Robert Strauss to chair the DNC). His popularity among Senate Democrats blossomed as a result of his effectiveness as the "partisan and combative" head of the Democratic Senatorial Campaign Committee in the 1986 elections—elections that won back control of the Senate for the Democrats.[166]

Elected majority leader in 1989, Mitchell used the Senate Democratic Policy Committee, as well as party task forces, to develop Democratic programs. He announced a Senate Democratic Agenda for the 101st Congress that had been approved by the Senate and that ranged from deficit reduction to national security to expanded education, health, and environmental efforts.[167] As majority leader, Mitchell has acted as a legislative agenda-setter and a national party spokesperson.[168] Indeed, it was the perception by many of his fellow Senate Democrats that he would be a good national party spokesman that helped gain him the election as majority leader in the first place.[169] With Mitchell as the new party chief, the previously moribund Senate Democratic Policy Committee drafted a legislative agenda that was adopted by the Democratic Conference, began to supply Democrats with substantive "issue alerts" and other information and materials, and began to sponsor regular, well-attended Thursday luncheons to discuss party issues.[170] By the middle of his second year in the leadership position, Mitchell had become the target of choice for partisan Republican attacks, especially by House Republicans and certain officials within the Bush White House.[171]

We have looked at contemporary American political parties. Now we can turn our attention more directly to the resolution of the two Senate-party paradoxes that were mentioned earlier.

Party Paradoxes Resolved

The first of the puzzles about Senate parties was the *less partisan voter–more partisan senator* paradox. It is clear that the American people have become somewhat less partisan in recent years. But at the same time, each party's identifiers, and especially each party's activists, or "nominators," have become more ideologically different from the other's. There has been a resurgence of national party organizations, especially of the two Senate partisan campaign committees, and this has helped to produce more nationally partisan campaigns—and senators.

Greater Senate partisanship became *possible* as each Senate party grew more internally homogeneous ideologically. Greater Senate partisanship became more *probable* as each Senate party grew more ideologically different from the other. The development of nationwide two-party competitiveness in Senate elections, as well as two-party competitiveness for control of the Senate, augmented partisan senatorial trends. And so did the party-polarizing years of the Reagan administration and the partisan shock Senate Democrats experienced when, for six years, they lost their Senate majority. Lastly, each party's leaders in the Senate began to feel freer to be partisan and issue oriented.

The result: while the people have become somewhat less partisan, the Senate has become more partisan, more nationally partisan.

What about the other paradox—the *individualistic senator–stronger party conference* paradox? This is not as much of a puzzle as it might at first appear. It is true, of course (as we have seen in Chapters 4 and 5), that senators (and the Senate) have become more national in outlook and more individualistic, less willing to follow their leaders, less inclined to go along in order to get along. How is it, then, that these more individualistic senators are willing to give more importance and strength to their party conferences?

The answer has to do with whether the party conferences are seen as the tools of the Senate party leaders or as the agencies of the individual members. Some party conferences may be both to some extent, of course. Primarily, though, as senators became more individualistic, they began to use party conferences as their own collective instruments. Earlier the conferences were the tools of the party leaders (when those leaders permitted conferences to be held at all). In the case of the Democratic Conference, this becomes very clear from remarks made in personal interviews:[172]

> [The Democratic leader] tests the waters and figures where the significant majority is as a party and uses that to decide on the agenda, what to bring up on the floor and force to a vote. (Jeff Bingaman of New Mexico)

> They [Democratic caucuses] do help develop a consensus, or the leader has a chance to hear what members think, completely off the record, and some things he plans on doing he decides not to. (J. James Exon of Nebraska)

> There are votes [in the Democratic caucus], kind of straw votes, not binding, but to find out where the Democrats are. Byrd [then majority leader] makes an effort. He is increasingly concerned and considerate about getting consensus, putting aside his own position. (James Sasser of Tennessee)

> Caucuses are much more important than they used to be. The leaders won't risk a caucus revolt. Sometimes the leaders will orchestrate a caucus move, but mostly it is group leadership, rather than individual leadership, now. . . . The leaders are less secure in their willingness to lead. (David Boren of Oklahoma)

This bottom-up, rather than top-down, leadership style is a change from the 1950s. Back then, as the heads of their respective Senate parties,

Republican Robert A. Taft of Ohio and Democrat Lyndon B. Johnson of Texas instituted strong, centralized party leadership.[173] Johnson operated virtually without any formal guidance from his party's caucus. He allowed the caucus to meet only five times during his first six years in the leadership position; under pressure from a growing number of liberal Democratic senators he did grudgingly call the Democratic Conference together six times during his last two years.

On his 1960 election as vice president, Johnson was succeeded in the position of majority leader by Mike Mansfield of Montana. Partly in reaction to senatorial resentment of Johnson's heavy-handedness, partly because of the new individualism that was already building in the Senate, and partly because of his own personality and style, Mansfield was a shepherd, not a drover. "I'm not the leader, really," he said. "They don't do what I tell them. I do what they tell me."[174] While this may have been a little too self-deprecating, there was a lot of truth in Mansfield's words. Things had changed. As Democratic leader, Mansfield agreed to reforms that allowed Democratic senators to call party conferences themselves, previously a prerogative of the leader.

On the Republican side of the aisle, Robert A. Taft was followed by two other strong GOP leaders, William Knowland of California and Everett Dirksen of Illinois. Then the Republican leadership changed too, to a kind of consensus leadership, responding more to party members rather than steering them. The leadership style of Hugh Scott of Pennsylvania, who served as Republican leader from 1969, upon Dirksen's death, until 1977, paralleled that of Mansfield. And party leadership has continued in that style ever since.

Robert Byrd, who was elected Democratic leader when Mansfield left the Senate in 1976, built his support on deference to, and doing favors for, his Democratic colleagues. One of his Democratic contemporaries in the Senate once said that if he had ever taken a pencil out of his pocket in Byrd's presence, the West Virginian would probably have offered to sharpen it.[175] As we have already noted, during the time when the Democrats were out of power in the Senate (1980–1986), Byrd began to turn to an invigorated Democratic conference for direction.

Republican leader Howard Baker of Tennessee, who served in that capacity from 1976 through 1984, was also a consensus leader. Dole has been too, although a somewhat more aggressive one.

The Republicans in the Senate have long divided up their leadership positions, so that the four principal Republican party positions— Republican leader and chairs of the Republican Conference, Policy Committee, and Committee on Committees—have been occupied by four different Republican senators.

Until recently the Democrats were different: the Democratic leader held all the Democratic counterparts of those posts (although the Democrats call their committee on committees the Steering Committee). With the election of Mitchell as majority leader in 1989, though, that changed. Mitchell named a then very junior senator, Tom Daschle of South Dakota, to co-chair with Mitchell a revitalized Senate Democratic Policy Committee, and he appointed Daniel Inouye of Hawaii (one of the Democrats he had defeated for the leadership position) to chair the Senate Democratic Steering Committee. In these and other ways, Mitchell further expanded the concept of consensus leadership, with the party conference being the collective instrument of the membership, more than a tool of the leader.[176]

In a fragmented, individualistic Senate, then, the two party conferences have begun to serve as overarching mechanisms. They are much stronger now than before. But this is not a sign of increased central control. Instead it results primarily from the fact that party conferences and policy committee meetings have become, as one political scientist has put it, "important opportunities for rank-and-file members to influence party and committee leaders' decisions about scheduling, political tactics, and policy substance."[177]

The result: while senators have become more individualistic, their party conferences have become stronger.

The Effect of Greater Partisanship

Has the fact that the Senate has become more partisan, its party conferences stronger, been a good development? Has it fostered better government?

Political scientists of earlier times decried the stalemate that develops in American national government when parties are not internally cohesive and cannot take a common stand for which they can be held accountable. We have already mentioned James MacGregor Burns in this connection. It was that kind of deadlock, too, that was one of the concerns of the American Political Science Association when in 1950 its Committee on Political Parties, headed by E. E. Schattschneider, issued the landmark report, "Toward a More Responsible Two-Party System."[178] The writers of the report wanted more accountable government, through a system by which voters could hold not only individuals but also parties responsible for their actions, once in office. The committee declared, in part:

> Historical and other factors have caused the American two-party system to operate as two loose associations of state and local organizations, with very little national machinery and very little national cohesion. As a result, either major party, when in power, is ill-equipped to organize its members in the legislative and executive branches into a government held together and guided by the party program. Party responsibility at the polls tends to vanish! . . .
>
> An effective party system requires, first, that the parties are able to bring forth programs to which they commit themselves and, second, that the parties possess sufficient internal cohesion to carry out these programs.[179]

As we have seen, things have changed since that statement was issued in 1950. Parties now *are* more internally cohesive—in the electorate, in party organizations, and in government. But because of another development, this cohesiveness can also result in deadlock.

As political scientist James Sundquist has pointed out,[180] those who wrote the "responsible parties" report seemed to assume that the same party would usually control both the presidency and the Congress. Thus if the people elected to the two branches had a program and stuck to it, they could offer a clear alternative to the other party and, most important, could enact and implement their program.

The assumption that there would usually be same-party government in America was firmly based in history. Until Dwight Eisenhower's second election in 1956, no newly elected president in this century faced a Congress in which the opposition party controlled even one house of Congress. The president–Congress norm during the first half of the 1900s and earlier, then, was a president and a congressional majority of the same party. In the second half of the twentieth century, by contrast, divided government became the norm.

In five of nine presidential elections in the second half of this century—1956, 1968, 1972, 1984, and 1988—the people gave one party the presidency and the other party a majority in at least one house of Congress (and in three of those elections majority control of *both* the House and Senate went to the opposition party). If off-year congressional elections are factored in, the modern reality of divided government is even more apparent. At the end of President George Bush's term in January 1993, the Republicans had controlled the presidency for twenty of the last twenty-four years, while the Democrats had controlled the House during all those twenty-four years and the Senate for eighteen of the twenty-four. Strong party cohesion and responsibility in a situation like that lessens government unity and action, rather than enhances it.[181]

In modern times, when their party has controlled Congress presidents have generally enjoyed a high success rate, an average of 75 percent or better, in getting their program enacted into law (a phenomenal 93 percent for Lyndon Johnson in 1964); but their success rates have generally dropped considerably with divided government.[182] In President Reagan's first year in office, when the Republicans controlled the Senate, his success rate was 80 percent; after the Democrats regained Senate control in 1986 Reagan's success rate sank to 43.5 percent (although it should be noted that usually there is a decline in congressional success rates as administrations age).[183] President George Bush's third-year, 1991 congressional success score was 54 percent.[184]

Internal party cohesion and responsibility in the Senate have increased as a result of forces that began outside the Senate. This does provide more national party responsibility, giving the voters an improved opportunity to hold parties, as well as officeholders, accountable. But increased national partisanship came hand-in-hand with a trend toward divided government, and the two together increased chances for heightened national conflict between the president and Congress.

Party responsibility of the type some political scientists long for, that bridges the branches, will most likely come with the election of a Democratic president (and a Democratic congressional majority at the same time). In other times, party conflict within the national government will continue to be the norm.

One way to change that prospect would be through conversion to a parliamentary system, allowing the majority party in Congress to elect the president. But even that wildly unlikely structural revision of our system would not assuredly solve the problem, unless such a transformation was expanded to include the adoption of a unicameral national legislature. Otherwise, separate elections for the House and Senate could easily produce a two-house Congress in which at least one house was controlled by the opposition party.

A more practical way to reduce partisan conflict in America's national government, and encourage cooperation, would be the greater use, and institutionalization, of the "summit" device—like that now commonly used for producing a federal budget—a device that permits and facilitates negotiation across three kinds of lines: party, house of Congress, and branch of government.

There is, of course, no guarantee of success with such an instrument, as the 1990 muddle and confusion over adopting a federal budget dramatically and painfully illustrated. It should be noted, though, that one of the main problems in that case was the reappearance of the earlier type of party deadlock: lack of cohesion within President Bush's own Republican party, as a majority of House Republicans angrily rebelled against the summit's first carefully crafted compromise.[185]

The potential of heightened national partisanship in the Senate to create stalemate could be reduced by restricting the opportunities now allowed under Senate rules for obstructionism generally—the kinds of internal Senate reforms, including reforms aimed at limiting the filibuster, that we have already discussed in Chapter 5. The filibuster and related obstructionist tactics are especially deadly in the Senate when wielded by a party rather than by a group of individual senators, and when the parties are nearly equal in strength, the majority party not having the three-fifths majority necessary to vote cloture by itself. Generally cutting down the effect of the filibuster and related obstructionism would lessen the deadlocking potential of the party filibuster as well.

The Senate has become a more partisan place. Each party in that body has become more internally homogeneous and less like the other party. Each has also become more internally cohesive on the issues, and party-line voting, especially on the most important votes, has increased.

While the voters have become somewhat less partisan, the Senate has become more partisan. And while senators have become more individualistic (as detailed in earlier chapters), they have increasingly turned to their party conferences for overarching group action. The party conferences have become stronger, but are more the instrument of their members than of their leaders.

As Senate parties were becoming more internally cohesive, another trend was developing, that of divided government. In such a situation party cohesion, rather than reducing the possibilities of government stalemate, can increase them. The Senate should press for the institutionalization and greater use of the summit device to encourage cross-branch and cross-party cooperation. The Senate should also adopt reforms to limit the force of the filibuster and related obstructionism; this would reduce the deadlocking potential of the party filibuster, also.

Increased national partisanship and national conflict in the Senate, as well as the other Senate trends we have previously discussed—the growth in national interest-group activity and influence and the power of money, as well as the development of a much more nationally oriented and individualistic Senate—have made the Senate less efficient in the making of national policy.

IV

NATIONAL POLICYMAKING IN THE SENATE

7

Making National Budget Policy

In 1990 Congress held its longest election-year session since World War II. The reason was the federal budget. Congress spent ten months trying to put together a multiyear deficit-reduction package of taxing and spending measures on which both houses of Congress and the president could agree.

By the time a budget deal was finally struck, Washington's ability to govern was being more widely questioned than at any time since the late 1960s,[1] President George Bush's approval rating had gone into a virtual free-fall,[2] the budget debate had grown sharply partisan, and the Democratic party had won a resurgence in public standing at the expense of the Republican party, which had been made to wear the label of the "party of the rich."[3]

President Bush started the annual funding process by sending his proposed fiscal 1991 budget to Congress in January 1990, as required by law.[4] He called for expenditures of $1.2 trillion and forecast a $64.7-billion federal deficit. Some people called it a "Darmanesque" budget, after the President's politically slick director of the Office of Management and Budget, Richard Darman. This budget was designed only to meet legal requirements and start the negotiation process. It avoided the difficult decisions—especially on the matter of a tax increase—that were required to meet the deficit-reduction targets dictated by the Gramm–Rudman–Hollings law.*

---

* Officially, the Balanced Budget and Emergency Deficit Control Act of 1985. It provided for deficit limits and mandatory across-the-board cuts if Congress could not voluntarily comply with the limits.

191

Contesting for the presidency in 1988, George Bush had made the centerpiece of his campaign a "read my lips, no new taxes" pledge, and he did not now want to be seen as backing away from such an unequivocal commitment. Thus, Bush's fiscal 1991 budget was based on economic projections, and federal tax revenues, that were much too optimistic, as well as on estimates of savings-and-loan bailout costs that were clearly too low. It underestimated the year's deficit by $61 billion, according to the Congressional Budget Office.

Congressional Democrats were not about to take the initiative—and the heat—themselves for raising taxes to cut the deficit. They correctly maintained that no Congress could pass a tax increase without the vigorous support of the president.

In April the House Budget Committee thus recommended a fiscal 1991 $1.2-trillion budget based on the same, too-optimistic projections that had come from the White House, and this House budget, which provided higher funding for domestic programs and lower funding for defense than the president wanted, was adopted on May 1 by the full House of Representatives in a near party-line vote (218–208). The next day the Senate Budget Committee quickly approved its own $1.2-trillion budget proposal, with even steeper defense cuts. The time for bargaining was ripe.

President Bush had wanted a budget "summit" all along, the kind of unofficial House-Senate-presidential negotiating session that had produced a two-year budget agreement in 1987, toward the end of President Reagan's final term. Majority Leader Mitchell and the chair of the Senate Budget Committee, Jim Sasser of Tennessee, on the other hand, resisted the summit idea.[5] Mitchell, Sasser, and other Democrats figured that President Bush wanted to use the summit device to get out of his "read my lips, no new taxes" pledge while blaming the Democrats for the increase, and they knew also that Bush still wanted a cut in the capital-gains tax rate. What was there to negotiate?

President Bush called the foremost congressional leaders of both parties to the White House for an unusual Sunday meeting on May 6. Worried about high interest rates and the country's weakening economy, he formally, and publicly, proposed a House-Senate-presidential budget summit with "no preconditions"—code words meaning that tax increases could be discussed by the summit. In other words, taxes would be "on the table," in the Washington phrase of the time.

The Democrats soon grew wary again, though, when "a senior White House official" told reporters "on background" (that is, not for attribution) that Bush's no-preconditions statement simply meant that the Democrats could propose taxes but that the President would veto them. This senior official was, in fact, White House Chief of Staff John Sununu, as Democrats knew, and they smelled a trap.[6] Then nineteen GOP senators wrote a public letter to President Bush urging him to stick with his "no new taxes" pledge; deficit targets, they and other conservative House and Senate Republicans maintained, should be met by cutting federal expenditures.[7] Democratic wariness increased.

Nevertheless, preparations for the summit went forward. Twenty bipartisan House and Senate negotiators—principally the party leaders and the leaders of the taxing, spending, and budget committees—were named to meet with White House representatives, particularly Darman and Sununu, to try to work out a budget deal.

Leaders Meet in Budget Summit

The initial summit meetings on May 15 and May 17 produced little more than a lot of suspicious partisan circling. It was pretty clear to everyone that no real deficit reduction could be effectuated without a tax increase, but nobody wanted to be the first to advocate one. Democrats knew that John Sununu would not cease trying to place the onus for a tax increase on them. So they adopted a go-slow approach and made clear that the president had to be the first to propose new taxes, or the budget talks would stall or even collapse altogether. This Senate Democratic tactic worked.[8]

On June 26, to the shock and dismay of congressional Republicans, President Bush at last issued a statement (not in person, but by having it posted on the White House bulletin board) declaring that "tax revenue increases" would have to be a part of any budget-reduction plan. In that manner, his "read my lips" pledge was forsaken. With this astonishing breakthrough (which was to have momentous political repercussions), summit negotiations began in earnest, and an understanding was soon reached. Through some mix of tax increases and domestic and defense cuts, the summit would try for a $50-billion deficit reduction for fiscal 1991 and a total deficit cut of at least $500 billion over the five-year period 1991–1995.[9]

It became clear, though, that no budget agreement would be arrived at easily or soon when, in the July 10 and 11 summit sessions, White House negotiators continued to press for a reduction in the capital-gains tax rate, and Senate Majority Leader Mitchell, just as adamantly, reaffirmed his determination to block any budget proposal that contained such a cut.

Complicating matters further, rank-and-file House Republicans, outraged at Bush's abandonment of his no-tax pledge, met in conference. By a reported 2–1 vote they expressed their opposition to any new taxes and condemned the Democrats, especially Mitchell, for not negotiating in good faith and for blocking cuts in government spending.[10] On these discordant notes summit talks came to a halt at the end of July, so that members of Congress could go home for their regular month-long August recess.

When Congress returned to the Capitol in September, budget negotiators resumed their talks, but this time in even greater seclusion and secrecy, at Andrews Air Force Base just outside Washington. The pressure on the negotiators was growing, now. Under the Gramm–Rudman–Hollings law, unless a new budget could be adopted by October 1, across-the-board, draconian spending cuts (known as a "sequester") were supposed to go into effect automatically. Those cuts would total $84 to $100 billion, necessitating a 32-percent reduction in domestic programs (excepting Social Security, Medicaid, and food stamps) and a 35-percent reduction in defense spending for research, procurement, operations, and civilian employees.[11]

There were other serious complications, too. The American economy seemed to be headed into a recession. Was this a time to put on the economic brakes through tax increases and other efforts to reduce the deficit? Some economists said no.[12] And what about the new and heavy expense of maintaining the great numbers of troops and other forces President Bush had sent to the Persian Gulf after Iraq invaded Kuwait? Was this a time to talk of heavy cuts in defense spending? Not according to Senate Minority Leader Robert Dole.[13]

The Andrews Air Force Base sessions flared into angry partisanship when press leaks quoted a Joint Tax Committee analysis indicating how Republican tax proposals would benefit rich taxpayers and raise rates for wage earners, while Democratic proposals would have the reverse effect.[14] Dole declared that he was sick of such "Democratic propaganda," which he said had been deliberately leaked by the Democrats "to cast this as 'Republicans for the rich, and they are for everybody else.'"[15]

After ten tense days, the Andrews summit, which some of its participants had taken to calling "Camp Run-Amok," fell apart on September 17. The Democrats claimed that the sticking point continued to be the capital-gains tax cut, and they charged that the Republicans were still pushing this break for the rich while trying to increase the burden on low- and middle-income taxpayers.[16]

"Big Eight" Group Takes Over

The September 30 end of the fiscal year loomed, though. Absent a new budget agreement by then, or a "continuing resolution" to extend spending at existing levels, much of the federal government could not continue to operate. This and the dread prospect of an October 1 Gramm–Rudman–Hollings "sequester" obliged congressional leaders to keep trying for a budget deal.

The larger summit group was pared down to a kind of "big eight" team—for the president, Sununu, Darman, and Treasury Secretary Nicholas F. Brady; for the House, the Democratic speaker and majority leader, Thomas Foley of Washington and Richard Gephardt of Missouri, respectively, as well as the Republican minority leader, Robert

Michel of Illinois; and for the Senate, the Democratic and Republican leaders, Mitchell and Dole.

This hyper-summit began to meet in the speaker's office on September 18, just a week after President Bush had again insisted in a speech that a capital-gains tax cut had to be part of any budget deal.[17] There were new studies that showed that the richest taxpayers had benefited most from tax changes in the 1980s,[18] and Senate leader Mitchell declared that Democrats would accept a capital-gains tax cut only if it was coupled with new taxes on the wealthiest taxpayers.[19]

A break in the negotiations came at last, when first Dole and then House Republican leader Michel declared publicly what they had been hinting at privately—that the capital-gains question should be stripped from the budget package and handled separately.[20] The White House had no alternative but to cave in on this issue, too. A deal was struck and announced on September 30 that provided for a first-year deficit reduction of $40 billion and for $500 billion over five years.[21] The package called for $134 billion in new taxes over the period, including higher excise taxes on gasoline, tobacco, and alcohol (beer, wine, and spirits) and luxury taxes on furs, jewelry, and pleasure boats, as well as increases in Medicare taxes and the costs to Medicare recipients of premiums and deductibles.

Things moved rapidly after that. Pending passage of the budget package, Congress rapidly adopted a "continuing resolution" to keep government running at existing spending levels until October 5 and to suspend the automatic spending cuts of the Gramm–Rudman–Hollings law. Also passed was a measure to raise the federal debt limit until October 5. President Bush signed both measures.

In the Senate, Mitchell and Dole came away from their separate party caucuses with the feeling that they would each be able to rally a majority of their fellow partisans for Senate passage of the budget package.[22] But the first vote was set to be taken in the House, and fears began to rise there that House Republican leaders, particularly, would not be able to deliver a majority of their members in support of the deal—especially because of the opposition of the House Republican whip, Newt Gingrich of Georgia.

Worried, President Bush made a hastily arranged nationwide television address on October 2 to rally public support. He called the budget deficit a "cancer gnawing away at our nation's health," obliquely criticized the Reagan years, of which he had been a part, saying, "No family, no nation, can continue to do business the way the federal government has been operating and survive," and called the budget deal "the best agreement that can be legislated now."[23]

Bush's address was a failure. It moved neither the country nor the House of Representatives. The vote in that body came at about one o'clock in the morning on October 5. A majority of House Republicans, unyieldingly opposed to the tax increases contained in the budget, deserted the president and voted against the package. A sizable group of swing Democrats in the House held back their votes until the last to see what the Republicans were going to do and because they were worried about one aspect or another of the budget deal. When the outcome was clear, they too weighed in against the budget reconciliation resolution. It failed 254 to 179.[24]

Now what? That very midnight, the temporary debt-limit extension and the continuing resolution that had prevented the government from being shut down on October 1 were scheduled to expire. And the crippling bludgeon of the Gramm–Rudman–Hollings sequester was set to hammer down too at the same hour. On the afternoon of October 5, the House rushed through a second continuing resolution, this time extending all these deadlines until midnight October 12. The Senate rapidly followed suit. But now, the hardliners advising President Bush convinced him that he could recoup some of the public approval he had lost in violating his no-new-taxes pledge by vetoing the continuing resolution and blaming congressional Democrats for the budget mess. On Saturday, October 6, that is exactly what he did. Later that same day, the House failed by six votes to override the Bush veto.

The government thus prepared to close down, except for essential services and the military operations in the Persian Gulf. Little impact was expected from the shutdown right away because it came on a long weekend, the following Monday being the Columbus Day holiday.[25]

Leaders and Committees Pick Up the Pieces

Congress stayed in session over the holiday weekend. On Tuesday, October 9, the House adopted a stripped-down budget resolution that still called for a $500-billion reduction in the deficit over five years but more or less left out the spending-cut and tax-hike details for later decision. The House then passed a third continuing resolution to fund government at existing levels, extend the debt ceiling, and suspend the Gramm–Rudman–Hollings sequester until midnight October 19.

It was the Senate's turn next. Both party leaders made strong statesmanlike pleas for Senate approval of the budget measure. On the Senate floor, Minority Leader Dole said:

> Some of us want to reduce the deficit; but [say] do not touch Social Security, do not raise taxes, do not touch Medicare, do not touch Agriculture—but $500 billion is not enough, the same speaker will say. . . .
> The American people want leadership, not speeches. They may disagree with us. They may vote against us. But they have children, and they have grandchildren. And if we do not act now, when?[26]

Majority Leader Mitchell, in turn, made a similar appeal:

> Tonight, we reap the bitter harvest of a decade of national indulgence. For 10 years, the American people have been told that we can have it all—more

for defense, more for Medicare, less in taxes. Way deep down, in our national heart of hearts, we all knew it was not true. But it was easy to believe, easy to ignore the truth, easy for a nation to indulge itself. And our nation did.

From one side, we have been hit with the insistent demand that we reduce the deficit. The American people know, and we know, that we cannot go on spending hundreds of billions of dollars more than we take in. But with the other side has come with equal insistence fierce resistance to higher taxes and lower spending. . . . Let us try to begin now and make sure that one decade of national self-indulgence is enough.[27]

The Senate passed the stripped-down budget resolution and the continuing resolution. President Bush signed the stopgap measure, and the government was operating again, though the final vote on the detailed budget had not yet been taken.

There had always been an informal understanding among the budget negotiators that in each house, each party would deliver a majority of its members for the budget package agreed upon. With House Republicans hopelessly and bitterly divided and unable to do that, House Democratic leaders and the Democrats on the tax-writing Ways and Means Committee began work drafting their own budget plan. This course of action, without majority Republican support, both permitted and required House Democrats to produce a more liberal budget package. The package could, and would need to, recommend greater progressivity in new taxes—that is, a greater tax burden on the wealthy and a lesser burden for the poor and middle class than the defeated budget plan would have provided—as well as less of an impact on Medicare recipients. Only in that way could the votes of House Democratic liberals be counted on; without the Republicans, broad support from this group was essential. The pressure was on to raise income tax rates on the wealthy.

President Bush Equivocates

President Bush's response, or responses, came in a series of highly publicized "flip-flops"[28]—accompanied by further precipitous slides in his public approval ratings. In a morning news conference on October 9, the president stated that he could accept an increase in the top tax rate in exchange for a cut in the capital-gains tax, saying in his own unique way, "That's on the table. That's been talked about. If it can be worked out in the proper balance, the capital gains rate and the income tax changes, fine."

The Senate Finance Committee began working on a package that would raise the highest tax rate from 28 percent to 33 percent and lower the capital-gains rate to 23 percent. But that afternoon, a group of seventeen Republican senators held a meeting with President Bush, and afterward their spokesman, Bob Packwood of Oregon, said, "The President agreed, our unified position was we will not go up on the rate, not 1 percent, not 2 percent, not one penny. . . ." Packwood said further that, "We all put up our hands and said, 'No deal on rates at all.'"[29] Work in the Senate Finance Committee on such a package stopped.

The next day, October 10, the confusion mounted. First, Republican leader Robert Dole said in the Senate that Bush had not in fact expressed agreement with the group of Republican senators. "He listened to us," Dole said. "He did not announce his position. He did not acquiesce in what we said." On a campaign swing that day in North Carolina, when asked by the press to clarify his position, Bush said, "Let Congress clear it up." Worse, while he was jogging later that day in Florida, Bush was asked by a reporter if he was ready to drop his capital-gains tax cut, and his flippant answer was, "Read my hips." This frivolous variation on his "read my lips" pledge against new taxes was heavily featured, and widely criticized, in the media.[30]

Two days later, October 11, a group of House Republicans spent an hour and a half with President Bush at the White House. Emerging from this meeting, Representative Bill Archer of Texas, the ranking Republican on the House Ways and Means Committee, said, "He [Bush] . . . is willing to equalize or level the [top] rate at 31 percent . . . in exchange for a 15-percent capital gains rate. . . . I'm telling you that he told us today that he has been consistently for this all the way through." But before that very day was over, Marlin Fitzwater, the White House press spokesperson, issued a contrary statement, quoting the president as saying, "I do not believe now that such a compromise is possible. . . . The ball is now in Congress' court."

All this prompted one mockingly gleeful House Democrat to quote Bush as saying, successively: "Read my lips! Read my slips! Read my hips! Read my flips!"

The president's public standing suffered badly. His job approval rating, according to the Gallup poll, had been 65 percent in July, after he had abandoned his no-new-taxes pledge, but had gone back up to 76 percent in September, with the troop buildup in the Persian Gulf. Then, following his widely publicized flip-flops on the budget, Bush's approval rating slumped to 66 percent by early October and to 54 percent later that month.[31] On the campaign trail, Bush tried to toughen up his image. In a Republican rally in Dallas, speaking of the approaching October 19 budget deadline and threatening a macho veto of any new continuing resolution, he shook his fist and said, "Enough is enough. This Friday, time's up."[32]

Budget Work Progresses

Back in Washington, the members of the crucial budget and taxing committees of the House and Senate were scrambling to put a reconciliation bill together in time. Their task was made more complicated by the swarms of lobbyists that hovered around them, more agitated than usual because of this year's budget spectacle, which had lurched back and forth from impending disaster one minute to near miraculous escape the next.

Representatives of the Tobacco Institute, for example, grimly pressed key congressional members to hold any eventual tax on cigarettes to no more than eight cents a pack. The American Trucking Association put a one-hundred-member phone bank into operation, calling truckers all across the country fifteen hours a day and asking them to contact their members of Congress in a "damage-control" operation. Lobbyists for the thirty-million-member American Association of Retired Persons persisted in the fight that they had launched earlier against sharply increased costs to Medicare recipients.

On October 12, the Democrats on the House Ways and Means Committee acted. They adopted a tax plan that would, among other things, raise the tax rate from 28 to 31 percent on those with incomes over $200,000; enact a surtax of 10 percent on those with incomes above a million dollars; strictly limit any increased costs (or, put another way, reduced benefits) for Medicare recipients; provide no rise in federal gasoline taxes; and permit a restricted capital-gains tax cut only for those in middle-income brackets.[33] On the following day, the Senate Finance Committee, in a bipartisan vote, adopted its own tax package, which stayed pretty close to the summit budget package that the House had earlier rejected—raising gasoline taxes and cutting Medicare benefits (but not, in either case, quite so much as the rejected plan) and dividing the increased tax burden about equally between middle-income and rich taxpayers.

In a caucus on October 16, Democratic senators let their leaders know that most of them favored the Ways and Means plan over that of the Finance Committee; they wanted to keep the Democratic party on the march as the party of the people, branding the Republicans as the party of the fat cats. Majority Leader Mitchell and Finance chair Bentsen told the caucus that if Democratic senators would hold their noses, if necessary, and help pass the Finance Committee plan on the floor, the leaders would work in the conference with the House to make the Senate measure more like the more progressive Ways and Means Committee version.[34]

On October 15, the House passed a budget reconciliation resolution, after incorporating in it the Ways and Means tax plan. Only ten Republicans voted for this measure. The following day, debate began in the Senate on that body's own budget plan. On October 18, as Senate argument continued, the House passed a fourth stopgap continuing resolution, this time extending the debt limit, the Gramm–Rudman–Hollings sequester deadline, and existing spending approval until midnight October 24.

At 1:30 in the morning on October 19, the Senate passed its version of the budget reconciliation resolution. Republican leader Dole was able to hold enough of his fellow partisans in line to beat off a serious challenge from more conservative Senate Republicans who opposed any increase in taxes. Democratic leader Mitchell knew that a bipartisan vote would be required for Senate passage and that the adoption of a more progressive tax amendment would result in the loss of all Senate Republican support. He used a pragmatic argument to turn aside liberal attempts to tax the rich more: "Do you want to make a statement or do you want to make a law?"[35]

The budget reconciliation resolution was adopted without amendment, and the Senate, later that morning, also passed the fourth continuing resolution. Despite his earlier "enough is enough" macho, Bush said that sufficient progress was being made toward a budget so that he could now sign the continuing resolution, and he did. House and Senate conferees, joined by the usual White House representatives, immediately began their work toward a final budget reconciliation compromise.

Last Push Starts

The hard part was not yet over, and the compromise would not come easily. But the end was in sight. Again, the lobbyists swarmed. Again, Bush's representatives were maddeningly intransigent, and as before, as one observer put it, they "succeeded in dragging out the process and making it more partisan and ragged."[36] (At one point, Sununu, Brady, and Darman quit the budget talks in anger, claiming that the Democrats were unwilling to negotiate a fair compromise.[37]) Again, President Bush was seen to change his mind on the question of an increase in top rates; he finally committed himself, and so did the negotiators, to a rise in the peak income tax rate from 28 percent to 31 percent (although still taxing all capital-gains income at 28 percent).[38]

Conferees agreed: to make substantial cuts in agriculture subsidies (as had been planned all along); to raise gasoline taxes by five cents a gallon (compared with no increase in the House bill and a nine-cent increase in the Senate bill); to hold Medicare cuts to $43 billion (somewhat less than the Senate's $50 billion and considerably less than the original, rejected budget's $60 billion, with $30 billion of the cuts to come from lower payments to health-care providers, rather than higher costs to beneficiaries); to impose higher taxes on tobacco, alcoholic beverages, luxury cars and boats, furs, airplanes, and jewelry; and to raise the ceiling on wages taxed 1.45 percent for Medicare from $51,300 to $125,000.[39]

The main barrier to settlement, though, still remained: how to tax the richest taxpayers more. Democratic negotiators pressed even harder for approval of the House Democrats' surtax on incomes over a million dollars. Republicans continued to resist just as strongly.

But many Republicans began to see that they were losing the battle for public opinion. "We have lost this issue in the public's mind,"

Republican Senator Bob Packwood of Oregon, for example, declared at one point. "They are now convinced that we don't want to tax the rich."[40] Out trying to help Republican congressional candidates, President Bush found himself publicly scolded by a Vermont GOP incumbent whom he was campaigning for and snubbed by another Republican in New Hampshire who failed even to show up, citing Washington duties, at the very meeting in which Bush was scheduled to stump for him.[41]

Senator Packwood, who was opposed to the surtax or to raising rates higher, came up with another way to increase taxes on the rich. He suggested a phase-out of the benefit of deductions and the personal exemption for those with annual incomes in excess of $100,000.[42] Mitchell and Dole both immediately endorsed this approach, and the White House soon had to, too.

The income-tax changes finally agreed to meant that those with incomes of $10,000 or less would get a 2-percent reduction, and those with incomes between $10,000 and $20,000 a reduction of 3.1 percent; taxpayers with incomes between $20,000 and $200,000 would pay an average of 2 percent more, and those with incomes above $200,000, 6.3 percent more.[43]

With a final deal on reconciliation now probable, the House, on the evening of October 24, passed the fifth stopgap measure since the beginning of the fiscal year. It kept the government running until midnight October 27 by again extending existing spending levels, suspending Gramm–Rudman–Hollings, and permitting a temporary increase in the debt limit. The Senate quickly followed suit.[44] President Bush signed the temporary measure and then went to work to try to shore up his support among congressional Republicans, while honing a new political message for the country: the Democrats were responsible for raising taxes, instead of cutting spending, as the Republicans wanted to do.[45]

Three days later, just before dawn on October 27, the House passed the final budget reconciliation bill by a vote of 228 to 200, a majority of the Republicans voting in opposition. Later that evening, the Senate also approved it, by a vote of 54 to 45.[46] Nobody disagreed with Senate Budget chair Jim Sasser of Tennessee when he said, "I think I hear a loud collective sigh of relief coming from 100 offices in the near vicinity of this chamber."[47] Congress finally closed down for the year in the early morning hours of October 28—twenty-three days past its originally scheduled adjournment date.[48]

When the dust had cleared, it was said that Congress had reduced the deficit by $42.5 billion for fiscal 1991 and $496.2 billion during the five fiscal years 1991–1995.[49] But included in the deal were some important last-minute changes in the budget process for the future: a further relaxation of Gramm–Rudman–Hollings deficit limits; a shift of spending power from Congress to the executive department; a "pay-as-you-go" requirement, so that any tax cut, for example, would have to be offset by spending cuts; and caps on spending by categories, thus constructing a "budget wall" to prevent future cuts in military spending from being used to increase domestic expenditures.[50]

President Bush went back on the campaign circuit, at first trying a kind of "the Devil (meaning the Democrats) made me do it" argument.[51] (He was later, with some marginal success, to change campaign tactics and try to switch public attention from the budget fight altogether to the situation in the Persian Gulf.) But it was clear that Bush and his party had sustained a serious setback, and that the Democrats had come out ahead.[52]

There were those who felt, though, that government itself, and public confidence in it, had been the big casualty, that the political system was paralyzed.[53] By contrast, political scientist Nelson Polsby declared that the budget fight had been a "healthy duel," and he counseled the critics to "calm down," writing:

> There is nothing new, and nothing wrong, with disagreement between Congress and the President, between Democrats and Republicans or within parties on important matters of policy. So what if the first package of taxes and cuts failed to clear Congress? Trial and error isn't a bad way to go about constructing a complicated bargain. Indeed, if the bargain is complex enough, and requires the assent of a large enough number of officials, it may be the only way to go. . . . So Congress and the President bargained, according to the dictates of our constitutional design. This is the way the system is supposed to work.[54]

Was it true that the fiscal 1991 fight had been healthy? Was this really the way the system was supposed to work? Under our system, Congress has what is called "the power of the purse," but some think that the exercise of that power, as well as the purse itself, has become a little frayed around the edges.

The Power of the Purse

The fiscal power of Congress derives from Article I, Section 9 of the U.S. Constitution, which provides that "No Money shall be drawn from the Treasury, but in Consequence of Appropriations made by Law. . . ." And Section 8 of the same first article states that Congress shall have the power "to lay and collect taxes," as well as the power "to borrow money on the credit of the United States." Clearly, Congress controls the purse strings. Thus it is said correctly that, in regard to taxing, spending, and borrowing, the president proposes and Congress disposes.

These fiscal decisions all come together in the annual budget of the federal government, the numerical expression of federal policy. Political scientist Harold Lasswell defined politics as: "Who gets what, when, how?"[55] In that sense, the federal budget is the nation's most important annual political manifesto. It determines, as one modern authority has written, "How large government will be, the part it will play in our lives, whether more or less will be done for defense or welfare, how much and what sort of people will pay for services, what kind of society, in sum, we Americans want to have. . . ."[56]

What Congress does in the exercise of its fiscal power—the power to tax, spend, and borrow—has regulatory, distributive, and fiscal (economic) effect. Through tax measures and appropriations bills, aside from just raising and spending money, Congress regulates behavior, or attempts to do so—providing subsidies to those who will agree to grow tobacco, for example, or tax advantages for those who will agree to invest their money in municipal bonds. Congress redistributes money— taking from some (not always those most able to pay) and giving to others (not always those who are most deserving). Congress makes fiscal policy (though not always explicitly and consciously, and often not fairly), and this affects the economy—the rate of inflation, the rate of unemployment, and sometimes more indirectly, the rate of interest on borrowed money.

Until the end of the First World War, though, there was no attempt on either end of Washington's Pennsylvania Avenue to deal with federal budget policy in any unified, rational way.[57] Each agency of the executive department simply submitted its proposed budget directly to Congress. Separate congressional committees then reacted, and Congress finally sanctioned these separate judgments. The budget that came out at the end of this process was the sum of many distinct and largely unrelated decisions.

To make the federal government more efficient, more businesslike, but especially to hold down overall spending, which legislators felt had increased too much during World War I, Congress passed the Budget and Accounting Act of 1921. The new law created the Bureau of the Budget (now the Office of Management and Budget [OMB]) and provided for an annual presidential budget to be proposed to Congress. For the first time, the executive branch was expected to produce a consolidated budget, to recommend concerning federal taxation, spending, and borrowing in one, universal document.

Fragmented Congressional Budgeting

The congressional approach to budget-making remained a study in fragmentation, although the 1921 Act did create a congressional agency, the General Accounting Office (GAO), to work with congressional committees and subcommittees in the oversight function, as a "watchdog" over executive expenditures. The purpose was to give Congress more "postappropriations budgetary control," as it was called. But what about more *pre*appropriations budgetary control, some central way for Congress to deal with the budget in advance? This was still largely outside the capability of the legislative branch (although the House and

Senate did at this time centralize spending jurisdiction in their two appropriations committees).

The Senate Finance Committee (dating from 1816) and the House Ways and Means Committee (from earliest days) once exercised authority over *both* taxation and appropriations measures.[58] But at the time of the Civil War, each committee was stripped of its jurisdiction over spending. The taxing and spending functions were separated from each other like the two halves of a dividing cell.

Most peculiarly, the spending half was already, itself, a segmented power, split into two legislative stages, each with separate committees in charge—the *authorization* stage and the *appropriations* stage. Very early in Congress (1837 in the House and 1850 in the Senate), members had grown tired of "riders," or extraneous legislative amendments, being saddled onto appropriations bills, burdening and slowing the process. So each house had adopted a rule that provided for a two-phase spending procedure. The basic idea was that before an appropriations bill could be passed to provide spending or budget authority for a federal agency or program, it had to be preceded by an authorization bill, particular legislation creating or continuing a federal program or agency and setting an upper limit on what could be spent for it.[59]

First, the authorization bill was required to go through all the regular legislative steps in both houses—introductory, committee (and subcommittee), floor, and, if dictated by differing versions, conference—to become law. Then the appropriations bill, dealing with money only, had to go through all the same steps. An appropriations measure could approve of the expenditure of a lesser, but not a greater, sum of money than that previously set forth in the authorization bill. Until the end of World War II, most agencies and programs were authorized on a permanent basis. After that, though, it came to be the practice in Congress to vote authorizations for increasingly shorter periods, finally only for a year or two at a time.

In that way the authorizing committees—in the Senate the present-day Energy and Natural Resources Committee, the Labor and Human Resources Committee, or the Environment and Public Works Committee, for example—were able to retain some authority and control over programs and agencies within their jurisdictions, not giving up so much power to the appropriations committees.

The two-phase procedure—first authorizations and then appropriations—is still the rule in today's Congress, although there are some ways around it. Congress, with the power to make the laws, also has the power to change them, including the power to allow some legislation on appropriations bills—especially, in the Senate, for example, when the Appropriations Committee itself recommends such a rider. Also, authorization bills may provide so-called "back door" funding by giving an agency "borrowing authority" or "contract authority"—authority to borrow money or to enter into a contract obligation to pay money, legal commitments that later must be covered by appropriations.

When the Senate Finance Committee and the House Ways and Means Committee lost their power over spending during the Civil War period (to the authorizing and appropriating committees), the appropriations part of the spending function was itself at first divided among several committees in each house, then concentrated in the House and Senate appropriations committees, then split up again among about nine committees in each body, and finally, by the 1921 Act, reconcentrated once more in the two appropriations committees. But this last centralization of the appropriations power was only momentary: each appropriations committee soon fragmented into numerous and permanent subcommittees, each with enough autonomy and independence that its decisions about appropriations within its own jurisdiction were largely allowed to stand.[60]

In today's Congress, each appropriations committee is divided into thirteen subcommittees—including, for example, the Senate Subcommittee on Defense Appropriations, the Senate Subcommittee on Energy and Water Development Appropriations, and the Senate Subcommittee on HUD-Independent Agencies Appropriations. Each subcommittee holds its own separate public hearings and then its own separate "markup" session, to decide on the final form of the measure it will recommend to the full committee. A 1980s study of the Senate found that the Senate Appropriations Committee and its subcommittees met more than three hundred times a year, far more than any other committee, and that more than 90 percent of these meetings were subcommittee meetings.[61]

The result is incredible fragmentation:

- Two houses of Congress
- Within each house, taxing power separated from spending power
- Spending power divided into authorization stage and appropriations stage
- Authorization jurisdictions spread among numerous committees
- Appropriations power decentralized to numerous subcommittees

And then, in a kind of doubling back that makes the budget process even more complicated, it turns out that the taxing committees also handle a great deal of the spending, through so-called "tax expenditures" and "entitlements."

Tax Expenditures and Entitlements

Let's start with tax expenditures. If you wanted people to invest in rental housing, you could simply pay them to do so with a direct federal subsidy. That would be a straightfoward expenditure; in the Congress it would require recurrent authorizations and appropriations. Or you could use a *tax expenditure*—reduce the taxes that those who invest in rental housing would otherwise have to pay (allowing them favorable tax write-offs for depreciation, for example). With either choice the Treasury would lose the same amount of money. But the choice of a tax expenditure rather than a regular expenditure avoids having to go through the annual authorization and appropriations process. (This is a major reason why those who enjoy such privileges prefer to have them as tax expenditures.)

Tax expenditures fall within the jurisdiction of the Senate Finance Committee and the House Ways and Means Committee (not that of the authorizing and appropriating committees). The costs of tax expenditures run into the billions of dollars, and the benefits mostly go to people in upper brackets so that, as Howard Shuman, an authority on the federal budget process, has written:

> There is a political schizophrenia between those who favor tax expenditures but oppose spending expenditures and those who defend spending expenditures but oppose tax expenditures. The reason why this division exists is relatively simple, if unstated. It is partisan and political. Tax [expenditures] . . . generally benefit those with higher incomes. The poor, who pay no income taxes, only casually or by serendipity receive tax expenditures.[62]

Republicans in Congress have been the strongest supporters of tax expenditures. Democrats have been the strongest backers of *entitlements*. If you wanted to issue subsistence checks to old people, or help them with their medical bills, you could do so by regularly approving direct expenditures for this purpose from the general fund of the federal government. Again, this would require recurrent authorizations and appropriations. Or you could accomplish the same objective by earmarking tax funds for the purposes and obligating yourself by law to pay as many people as qualify—as many as turn out to be *entitled* to the benefits each month—no matter the overall cost.

It is the second of these alternatives that the U.S. government has chosen, and the tax aspect of such programs—Social Security and Medicare, most notably—means that jurisdiction over them is lodged solely in the Senate Finance Committee and the House Ways and Means Committee (not the authorizing and appropriating committees). Medicaid (medical assistance for poor people), though it is not funded by an earmarked tax, is also treated as such an entitlement program. By 1974, when Congress reformed the budget process, entitlement spending amounted to 53.6 percent of federal outlays (and it was to go higher, after the reform).[63]

Howard Shuman, again, has pointed out the political problems that tax expenditures and entitlements, each with its backers from opposite ends of the political spectrum, present in the budget-making process:

Tax expenditures and spending entitlements are a common breed. They are like the animals in Noah's Ark which marched aboard side by side. Both are automatic and paid out by law and formula. Neither is regularly reviewed. Both are uncontrollable without a change in the law. Once legislated, they create powerful interest groups that are dependent on their benefits, deeply entrenched, and difficult, if not impossible, to oppose.[64]

Congressional Budget Reform

As far back as 1921, when procedure for a national executive budget was first put in place, there was a parallel effort in Congress to provide for a unified congressional budget. Democratic Representative Carl Hayden of Arizona (later a senator from the same state and chair of the Senate Appropriations Committee) led that effort, declaring: "Let us have a legislative budget. Let there be a committee on the budget, a joint committee of the House and Senate, and let its findings be binding upon both bodies. It is only by the creation of such a committee that there can be any real reduction in governmental expenditures."[65] When the effort failed, Congress did not try again for more than forty years.

The 1921 Budget Act, with its centralized executive budget, enhanced the president's influence in the budgetary process. Presidential budget-making leverage grew even more with passage of the Employment Act of 1946, which called for an annual economic report by the president together with presidential fiscal-policy and programmatic recommendations for the achievement and maintenance of full employment. This Act institutionalized the management of fiscal policy in the office of the president.[66]

Passage of the Employment Act amounted to a kind of official endorsement of "Keynesian economics." British economist John Maynard Keynes, in his classic 1936 work *The General Theory of Employment, Interest, and Money*,[67] had asserted that in times of depression or recession, governments should stimulate aggregate demand through direct spending, or through tax cuts to leave more of people's own money in their own hands, or both, even if this produced government deficits. (Keynes had also argued that when the economy is overheated, governments should reduce demand by cutting spending, raising taxes, or both; this side of Keynesian economics is, of course, more difficult for politicians to put into practice.)

The Employment Act of 1946 placed the president squarely in the center of the budget-making spotlight. Partly to offset this growing presidential dominance in budget-making but also to redress the budget problems caused by congressional fragmentation and, it was hoped, to reverse the growing trend toward deficit financing of the federal government, Congress enacted a new legislative budget system as a part of its 1946 Legislative Reorganization Act.[68] This attempt, however, was no more successful than the failed 1921 effort.

The one-hundred-member Joint Committee on the Budget that the 1946 measure created, made up of the membership of the four taxing and appropriating committees, was much too unwieldy. The government was racked by acrid partisanship at this time, too, with a Democrat, Harry S. Truman, in the White House and the Republicans in control of Congress. As a result, the new budget committee was unable to agree on a budget ceiling as it was supposed to do; the system was soon dismantled and the committee dissolved.[69]

The congressional budget process remained highly fragmented. Contention—between the president and Congress, between the House and Senate, and between the taxing and appropriating committees— grew more and more intense, especially in the years just before budget law received major revision in 1974. Expert Allen Schick has called this period "The Seven-Year Budget War: 1966–1974."[70]

Republican Richard Nixon was president when reform finally came. In fact, reaction to him and his policies by a Democratic Congress was a big factor in bringing on budget reform.[71] Nixon and congressional Democrats were in harsh disagreement about priorities. The majority in Congress by then thought that the Vietnam War should be wound down more rapidly and that larger sums should be spent on domestic programs. Nixon brazenly "impounded" some of the funds Congress appropriated for social programs—that is, he refused to spend the total amounts that Congress approved. Democrats were outraged by these impoundments.

They were worried, too, about the fact that Congress had no mechanism of its own for considering and deciding on national spending priorities. In those days, when the Senate considered an appropriations bill—for the Department of Interior, say—the bill was compared only with itself. That is to say, the chair of the Senate Subcommittee on Appropriations for the Interior Department would rise in the Senate when the money bill for that federal agency was called up and make a beginning declaration something like this:

> Mr. President and members of the Senate, this bill appropriates $9 billion for the Department of the Interior. That amount is $1 billion more than was appropriated last year. It is $500 million less than recommended in the President's Budget and $250 million more than the amount approved by the House of Representatives.[72]

Notice that there is no comparison of this measure with any other appropriation; no attempt to argue, for example, that funding for Indian education (the Bureau of Indian Affairs being in the Interior Department) should have priority over the construction of a new main battle tank for the Defense Department, no admission, reluctant or otherwise, that the costs of new fencing on Bureau of Reclamation lands (also within the jurisdiction of the Interior Department) should receive a lower priority than Department of Health and Human Services needs for increased medical-care funds for poor people.

Not only was Congress disturbed by Nixon's impoundments and concerned that it had no good way to determine national spending priorities. Many members were also troubled by the lack of a congressional mechanism, corresponding to the president's Office of Management and Budget, for deciding on overall national fiscal, or economic, policy—to stimulate the economy when it was in recession, brake it when it was overheated.

President Nixon, as a candidate for president in 1968, had run against Keynesian economics so hard that some voters might have been pardoned for believing that John Maynard Keynes, rather than Hubert H. Humphrey, was Nixon's Democratic opponent. But after he was in office and the economy had slumped into recession, largely brought on by the Federal Reserve Bank's tight-money fight against inflation, President Nixon embraced Lord Keynes as though the two were going steady. "We are all Keynesians now," Nixon declared, while moving to stimulate demand in the very way that Keynes had advocated, running up the deficit, in his reelection year of 1972.[73] Republicans in Congress, and a lot of Democrats, too, were worried about Nixon's deficits and about rising federal deficits in general over the years—some planned as a way of getting out of recessions, some unplanned, the result of under-collections in times of economic downturn.[74]

During the last six fiscal years of the Truman administration (1948–1953), there had been budget surpluses in three years and relatively small deficits in the other three, so that, overall, the period had produced an average fiscal-year *surplus* of $1.2 billion. Dwight Eisenhower's eight fiscal years (1954–1961), three of them with surpluses, averaged an annual deficit of $1.9 billion. The Kennedy–Johnson fiscal years (1962–1969), with only one year out of the eight, fiscal 1969, in surplus (this being the last year, to date, that the federal budget has *ever* been in surplus), saw annual federal deficits average $6.7 billion.

In Richard Nixon's first four fiscal years (1970–1973), the average annual deficit increased to $11.7 billion, and in the 1974 session Congress was just beginning to be aware that the average annual deficit for the following two years would leap to $63.5 billion. Even allowing for inflation, which began to creep up alarmingly during the later Nixon years, these deficits looked big (although, of course, at their highest annual average they were only one-fourth the average deficits that would be produced during the coming Reagan administration). Campaigning for reelection in 1972, Nixon had tried to put the blame for deficits on Congress, declaring in one campaign radio speech:

But let's face it, Congress suffers from institutional faults when it comes to federal spending. In our economy, the President is required to operate within the discipline of his budget, just as most American families must operate within the discipline of their budget.

In the Congress, however, it is vastly different. Congress does not consider the total financial picture when it votes on a particular spending bill; it does not even contain a mechanism to do so if it wished.[75]

Yet another distressing complication in the budget process was that, by 1974, Congress had grown increasingly unable to get its budget work done on time, to pass appropriations bills by the start of the new fiscal year (then July 1). The passage of continuing resolutions to allow federal departments and agencies to operate at the previous year's spending levels until the new appropriations bills could be adopted, sometimes as much as six months late, had become a worrisome and blameworthy commonplace.[76]

Therefore, to stop presidential impoundments and give itself a way to set national priorities; to counter presidential dominance in budget-making and provide a congressional mechanism for making responsible fiscal policy; to control government spending and cut the deficits; and to facilitate timely budget action—Congress passed the Budget and Impoundment Control Act of 1974.

Budget and Impoundment Control Act of 1974

The 1974 Act adopted by Congress was the result of great study by a number of committees and leaders. It was a consensus measure. Though members had various motives for their support of the proposal, they were not, at the end, in disagreement about what ought to be done. In the Senate, for example, the 1974 act was unanimously adopted by an 80–0 vote, and the conference report on it was later similarly agreed to by a vote of 75–0.[77]

The Act curtailed the power of the president to impound funds appropriated by Congress. It changed the fiscal year to October 1–September 30 (in place of July 1–June 30). Apparently this was done in the hope (a vain one, as it turned out) that the later date would afford more time for Congress to get its budget work done prior to the beginning of each new fiscal year.

The 1974 Act created a special joint staff for Congress, called the Congressional Budget Office, to parallel the Office of Management and Budget in the White House and to provide Congress with its own, independent source of information on the economy and on federal finances.

Nothing was done in the new budget law, though, to collapse the bifurcated authorization and appropriation functions into one, nor to fuse the splintered taxing and appropriating jurisdictions. Instead, a new and *extra* committee layer was added in each house—the Senate Budget Committee and the House Budget Committee—as a kind of superstructure to bridge the different functions and jurisdictions and impose the needed overall, macrolevel budgeting. Neither was there any deficit-cutting requirement included in the new law. The 1974 Act simply declared that Congress should agree on that budget deficit, or surplus, "which is appropriate in the light of economic conditions and all other relevant factors."[78]

Congress did seek to discipline itself in regard to time by writing a strict budget-action schedule into the Act. After the president presented his budget, as required, within fifteen days after the convening of Congress in January of each year, there were to be certain self-imposed congressional deadlines. One was May 15. By that date the first tentative congressional budget, reported by the two budget committees—after hearings, and in consultation with the authorizing, appropriating, and taxing committees—was supposed to be approved in the form of a first concurrent resolution adopted by both houses. Another deadline was September 15, the date by which both houses were supposed to have adopted a second, and "binding," concurrent resolution containing the final budget reported by the two budget committees.

Results of the 1974 Budget Reform

The impoundment-control part of the new budget law worked well.[79] The Act had provided that a president could not permanently halt the expenditure of appropriated funds unless both houses of Congress thereafter, within forty-five days, approved such a "rescission," as the law called it. President Gerald Ford tried to carry on Nixon's practice of making policy-based rescissions as a kind of line-item veto, but Congress refused to approve about 90 percent of the Ford rescissions. During the Carter and Reagan administrations that followed, Congress likewise refused to weaken the 1974 impoundment controls.[80] After passage of the Act, then, presidents were no longer able to use impoundments as a way of frustrating congressional spending priorities.

The Congressional Budget Office (CBO) established by the Act worked well too. Dr. Alice Rivlin, a respected economist, was appointed as the first CBO director, and she carried out her duties in a highly professional and praiseworthy manner. Through the CBO and the staffs of the two budget committees, Congress was able for the first time to produce its own independent and reliable information on the economy and the budget.[81]

In earlier times, Congress had to rely solely on the president's Office of Management and Budget (OMB) for the economic projections on which revenue estimates and spending estimates were based. The president's budget could be made to look good, closer to being in balance, when OMB overestimated future economic growth and thus projected higher tax revenues and lower welfare-type costs.

OMB could also make the president's budget look better by underestimating spending (as, for example, the spending required for cost-plus defense contracts) or by underestimating interest rates for the coming year (thereby reducing the large item in the budget that OMB said would be necessary for payment as interest on the national debt).

David Stockman, President Ronald Reagan's first OMB director, once boasted, in effect, that "I've just lowered the temperature from 110 to 78." To achieve the first big budget cuts of the Reagan administration, he had purposely made an economic forecast that was too optimistic, and many of his "cuts" were actually only reductions in earlier, inflated projections of spending growth.[82]

Stockman and other OMB directors could, indeed, raise or lower the mercury in the thermometer, and they did. Although Congress could make its own economic forecasts and revenue and spending projections now, from time to time, as in 1981 and at first in 1990, too, it chose the easier path of accepting OMB's rosier projections in place of the gloomier ones of its own Congressional Budget Office.[83]

In the Senate, the 1974 Act served to refocus decision-making more centrally, initially in the Senate Budget Committee and then on the Senate floor.[84] The fact that the Budget Committee's membership (initially numbering sixteen, now twenty-two) was permanent, like any other standing committee in that body, helped make it a powerful committee from the first. (In contrast, no House member could serve on the House Budget Committee, whether regularly appointed or drawn from the Appropriations Committee or the Ways and Means Committee, for more than four [now six] years in any ten.)[85]

During the first six years under the Budget Act (1975 through 1980), bipartisan coalition-making within the Senate Budget Committee was normally necessary for the adoption of a budget there. The Senate committee's first chair, Democrat Edmund Muskie of Maine (who could usually expect three of his more conservative fellow Democrats to vote like the Republicans), and the ranking Republican, Henry Bellmon of Oklahoma, worked fairly cooperatively together.

Budget partisanship in the full Senate was at first restrained, too. The Democrats were in control there, but because their ranks included a number of conservative Democrats, their margin was not great enough to permit the adoption of budgets without some Republican support. Budget resolutions regularly passed the full Senate during this period by 2–1 margins.[86]

The Senate budget process began to grow more and more partisan, though, soon after the inauguration of Republican Ronald Reagan as president and the concurrent Republican takeover of the Senate the same year as a result of a net gain of twelve seats for the GOP (given, also, the continued disappearance of conservatives from Senate Democratic ranks). Reagan was elected on a platform that promised severe tax cuts, sharp increases in military spending, and elimination (simultaneously!) of federal deficit financing altogether by 1982.[87] It was a tall order—too tall, as it turned out. And the push to accomplish it polarized the Senate (and the House, too).

Pressed by OMB Director David Stockman, the Republicans short-circuited the budget process in 1981 and put reconciliation, usually associated with the second budget resolution, first (something that had been done the preceding year also, but on a much smaller and less far-reaching basis). Included in reconciliation, which was binding on the authorizing as well as the other committees, ordering them to reduce established programs, were all the Reagan-backed cuts—nearly $40 billion worth—in domestic spending and a mammoth shift of funds to the military. Next came a giant tax cut. With the 1981 changeover in the Senate, only three Democrats and one Republican remained from the original membership of the Budget Committee, and the public bipartisanship of the Muskie–Bellmon era went rapidly into decline.[88]

The following year, 1982, Republican Majority Leader Robert Dole of Kansas adopted a kind of Republicans-only, King Caucus strategy in the Senate. He produced a budget resolution and related tax-increase legislation (to make up some for the deficit caused by the preceding year's budget decisions) that were largely drafted behind closed doors in Republican caucuses. He was able to get his way in the full Senate because the 1974 Budget Act limited debate and amendments on budget reconciliation measures in the Senate without the need to get sixty votes to do so as under the filibuster rule.[89]

The provisions of this first 1982 budget resolution (for fiscal 1983, of course) made it binding and final if a second resolution was not passed by October 1. Senate debate on it was rancorous and partisan and featured almost straight party-line voting:[90] forty-six of forty-eight Republicans voted for the budget resolution and forty-one of forty-four Democrats voted against it. That set the pattern for continued party-unity voting (a majority of one party voting against a majority of the other party) on budget resolutions in the Senate in the years that followed, with the exception of the vote in 1986.[91]

Thus the 1974 budget process did not lessen Senate partisanship; on the contrary it seemed to heighten it.[92] Moreover, it was the party leadership in the Senate (and House) that was increasingly called on to "shape the budget panel's membership, monitor their activities, and mobilize the winning coalitions needed to pass the often controversial fiscal package."[93]

Not only were budget votes on the Senate floor increasingly partisan, but there were a lot more of them. Under the budget law, "multi-layered and overlapping" congressional actions are required, as one authority has put it, on budget resolutions, authorizations, appropriations, and continuing resolutions (as well as tax measures). Thus, on the Senate floor itself, "losing one key budget vote, opponents can simply take aim at the next stage, or the one after that."[94] In the Senate, of course, first budget resolutions, reconciliation measures, and continuing resolutions were (and are) all subject to amendments proposed on the floor, and plenty were proposed and voted on. In the years 1981 through 1987, there were an average of twenty-three Senate roll-call votes a year on amendments offered to first budget resolutions, an average of eleven or twelve Senate roll calls a year on amendments offered to reconciliation measures, and an average of fourteen Senate roll-call votes a year on amendments offered to continuing resolutions.[95] No wonder that twelve years after the 1974 Budget Act was adopted, Arkansas Democratic Senator David Pryor decried what he saw as increasing Senate stalemate, declaring that the Budget Act itself was a big part of the problem because it had made the Senate into "a giant committee on the budget," and reported:

> In 1970, before the Budget Act was implemented, that Senate that year had 170 roll-call votes on budget related items. By 1975, 246 budget related votes in the Senate were cast. In 1981, 344 budget related votes were cast. In 1985, 67 percent of all of the roll calls in the Senate were on budget related items.[96]

Thus the 1974 Act added to the trends toward individualization of power and nationalization of outlook in the Senate. With more budget-related issues being taken to the floor for majoritarian decisions, individual senators, as opposed to committees and subcommittees, gained power in the process.[97] Senators could not focus alone on locally important programs; they were forced to become more national in their viewpoint because of the great national attention on deficit reduction.[98] They were forced to become more national in their outlook, also, because of the new openness of the system. As James Thurber has written, what was once "a relatively closed process for financing government agencies . . . has been transformed into an open process for providing benefits and contracts to Americans"[99]—especially those Americans, one might add, represented by national interest groups.

The 1974 Act was supposed to expedite the budget work of the Senate (and House), so that all the appropriations bills could be passed before the end of one fiscal year and the beginning of the next. But the reality turned out to be just the opposite: congressional performance got worse, not better.[100] Appropriations bills were enacted later and later, and the government was forced to operate for longer and longer periods under continuing resolutions.[101] In the 1970s, Congress had almost always passed all but one or two of the appropriations bills before it adjourned for the year. The 1980s were different. In 1982, 1984, and 1985, when Congress adjourned over half the appropriations bills had still not been enacted. In 1986 and 1987, *none* of these regular bills had been passed when Congress quit for the year.[102]

Liberals, particularly, had supported the Budget Act of 1974 because they thought it would help them get spending priorities shifted from defense to social programs. That did not happen. Prior to Reagan, one authority said that Congress had done more of that kind of liberal adjustment of priorities "before it had a budget process than it has since."[103] After Reagan, priorities were changed—but not in the direction that liberals had in mind when they voted for the 1974 Act.

The record was no better on the supposedly enhanced ability of Congress to make fiscal policy—to coordinate and consciously decide on overall spending and tax decisions so as to stimulate the economy in sluggish times, chill it in overheated periods.[104] One problem was that the economic advice Congress received, or could draw on, was not always reliable. In 1986, for example, the economic forecasts for the coming year by five leading U.S. economists, two in government and three in the private sector, were widely off the mark: by an average of 25 percent in regard to economic growth (that is, growth in the gross national product); by an average of 99 percent in regard to the expected rate of inflation; and by an average of 19 percent in regard to what interest rates would be.[105] That same year both OMB and CBO were off an average of 44 percent on estimated economic growth.[106]

Another problem was that neither Congress nor the president had (or has) direct control over the Federal Reserve System (the "Fed"), America's central bank. It is the Fed that manages monetary (as contrasted with fiscal) policy, and often the two policies are applied at cross purposes, one attempting to brake the economy, the other attempting to speed it up. The Fed can affect the money supply (and interest rates) in three major ways: (1) Through its "open market committee," the Fed can sell government securities that it holds in its own portfolio and thus take money out of the economy, dampening it (and pressing interest rates upward); by buying such bonds the Fed can put more money in circulation, stimulating the economy (tending to lower interest rates). (2) By raising or lowering the "discount rate" it charges member banks for loans, the Fed can encourage or discourage such borrowing, thus increasing or decreasing the money supply. (3) By changing the "reserve requirements" for member banks (the amount of cash, deposits in other banks, and government bonds that they are required to keep on hand), the Fed can raise or lower the amount of money the banks are allowed to loan out—with more loans, again, meaning greater growth in the money supply, and fewer loans meaning less growth in the money supply.

In making monetary policy (and affecting interest rates in the country), the Fed represents the interests of the banks, lenders, and investors (creditors, not debtors), and the Fed is always much more worried about the consequences of inflation than those of unemployment, if a choice has to be made.[107] The Congress, on the other hand, especially when under the control of the Democrats, is usually more concerned about unemployment than about inflation.

Aside from policy differences, though, there is a serious problem here in regard to economic-policy management, in that no institutional channel exists for coordinating monetary and fiscal policy. Democratic representatives Byron Dorgan of North Dakota and Lee Hamilton of Indiana introduced a bill to correct this. Their proposal would have: required the secretary of the treasury, the chair of the Council of Economic Advisers, and the director of OMB to meet three times a year with the Fed's Open Market Committee; permitted a new president to name the chair of the Fed soon after taking office; and required the Fed to announce immediately any changes in its monetary-policy targets.[108] Enactment of this measure would make it less likely that fiscal policy-making by Congress would be thwarted by the Fed and its monetary policy.

One more problem is that just what kind of fiscal policy Congress *should* make has not always been clear—as, for example, during the last part of the administration of President Jimmy Carter. As had been true for a time under President Nixon, too, the economy confusingly suffered from *both* unacceptably high unemployment and unacceptably high inflation (as well as, as a result of the inflation-combating tight-money policy of the Fed, unacceptably high interest rates).[109]

In large measure, though, the poor record of Congress in making fiscal policy resulted from its lack of serious attention to that budget-policy aim, compared with others. One of the principal purposes of the Budget Reform Act of 1974 had been, of course, to get control of federal spending and bring down federal deficits. Especially with the commencement of the Reagan administration, this became the central focus of congressional budgetary policy, virtually heedless of fiscal effect. As one authority on budget-making during the Reagan era has written: "For Republicans and Democrats alike, in 1981 the balanced budget quickly became an end in itself, regardless of what it might mean for an economy that rapidly headed into a deep recession."[110]

But it seemed that the more Congress concentrated on deficit cutting, the more the deficit grew. And the more the deficit grew, the less Congress was at liberty to try to manage the economy through fiscal policy. Here—concerning the goal of holding down federal spending and eliminating deficit-financing, while developing a tunnel-vision focus on this goal—was the most spectacular failure in the implementation of the 1974 Act.

A Concentration on Deficits

It took more than 190 years—from President George Washington until the first months of Ronald Reagan's presidency—for the U.S. national debt to reach $1 trillion.[111] When that total was reached in 1981, Reagan called the figure "incomprehensible," but nevertheless tried to explain it graphically by saying that it was the same as a stack of $1,000 bills sixty-seven miles high. About five years later, in 1985, Reagan's stack would have had to be 134 miles high. The national debt had doubled, had gone to $2 trillion, during the first five years of his term. It was to almost triple by the end of his second four years in the White House. How had the deficits and the national debt grown so unbelievably rapidly?

Riding a crest of popularity and, some said, a popular mandate to carry out his campaign promises, President Reagan pretty much had his way with Congress after taking office in 1981, and he accomplished a "federal budget revolution."[112] He had promised a huge increase in military spending, and he got it. A 7-percent rate of annual increase (a rate more or less pulled out of the hat by Reagan's OMB Director David Stockman) actually worked out to be a 10-percent annual increase because of a Stockman miscalculation.[113] Military authorizations nearly doubled, from $141 billion to $283 billion, between 1980 and 1985, rising to 28 percent of the total federal budget. This was the highest peacetime level ever, according to the CBO, even after adjusting for inflation.

President Reagan got a decline in social spending as a percentage of gross national product. He did not push for a decrease in the cost of "entitlements," such as Social Security, Medicare, and veterans' pensions. He also got an unprecedented tax cut, the largest in history. The so-called "supply siders" in his administration sold it as a measure that would increase, not decrease, federal revenues because it would stimulate investment and production.[114]

They were wrong. By the time the president had signed the 1981 budget reconciliation and tax-cut measures, his advisers were already privately in a "deficit panic" (and some of them, OMB Director Stockman included, had known all along that their projections were too optimistic).[115] The ranking Democrat on the Senate Budget Committee, Ernest Hollings, Jr., of South Carolina, called the massive tax cut, which was tilted toward wealthy taxpayers and business and which was to help cause unprecedented deficits, "revenue hemorrhage legislation."[116] Hollings knew what he was talking about. The tax cut was projected to reduce federal revenues by $787 billion over just the following three years (although a third of this was later recouped through a tax increase that had to be enacted in 1982).[117]

Before the 1981 tax cut was passed, the Fed, operating independently but with the approval of the Reagan administration,[118] had already put a "choke hold" on the economy, instituting a severe tight-money policy to fight what it saw as intolerable inflation.[119] As a result, the country plunged into a recession in 1981 and 1982, the worst since the Great Depression of the 1930s. This, of course, worsened the deficit (by decreasing revenues collected and increasing welfare-related expenditures). High inflation that had begun in the last years of the administration of Jimmy Carter caused entitlement payments, indexed to the cost of living, to mushroom. At the same time, inflation helped to increase the deficit. Ironically, when the Fed later put on the brakes, raising interest rates and cutting the rate of inflation, the fall in the inflation rate itself reduced total federal revenues, while at the same time higher interest rates meant that the government had to pay out much more to finance the national debt.[120]

Thus a number of candidate Reagan's campaign promises came true. He got a colossal increase in military spending. He got a cut in the rate of growth in social programs. He got his monumental tax cut. What he did not get, of course, was a balanced budget. In his first full fiscal year in office (fiscal 1982), the deficit nearly doubled (to $128 billion) what it had been the year before. The next year it soared even higher, above $200 billion,[121] and it stayed in or near that high-altitude range for a large part of the rest of the Reagan presidency. As the deficit grew, of course, so did the amount of money the federal budget had to allocate for paying interest on the national debt. This interest figure rose sharply from not quite $51 billion in fiscal 1980 to nearly $110 billion in fiscal 1984 (and on up by big jumps each year to $169 billion in fiscal 1989, $14 billion more than the deficit itself that year).[122]

By the end of his first year in office President Reagan was already forced to admit what his advisers had known for some time: horrendous federal deficits were in the offing. In a December press conference, Reagan limply claimed that his campaign pledge to balance the budget "was our goal, not a promise."[123] Earlier Reagan had claimed that "inflation results from all that deficit spending."[124] Now, Chairman Murray Weidenbaum and William Niskansen of the President's Council of Economic Advisers began pronouncing, to the contrary, that federal deficits were not, after all, the cause of inflation, nor of high interest rates either.[125] Before long President Reagan himself was saying that deficits were not always bad, sounding almost Keynesian when he declared in November of 1982, for example, that the budget deficit could not be eliminated by budget cuts alone, nor by tax increases, but that the answer was in "stimulating the economy and productivity."[126]

Reagan and these advisers were, in fact, correct. The big deficits did help, beginning in 1983, to bring the nation out of a recession; the rate of unemployment began to come down.[127] So did nominal interest rates and inflation. Keynesians, or more properly neo-Keynesians, were able to show that the Reagan deficits of 1981–1982 produced the economic recovery of 1983–1984, though they argued that the deficits should have been directed more toward the financing of human capital and public infrastructure investment, instead of toward underwriting the escalation in the costs of military operations and procurement.[128]

Reagan's treasury secretary, Donald Regan, at one point in 1983 issued a general challenge, saying that "I will offer a prize to anyone who can show me the connection between high rates of interest and high deficits." He also declared defiantly to the U.S. Chamber of Commerce that same year (indirectly disagreeing with Martin Feldstein, by then chairman of the President's Council of Economic Advisers, who believed a tax increase was needed), "Economists who continue to claim that deficits make for higher interest rates should climb down from their celestial observatories and acquaint themselves with terrestrial facts."[129] Regan's views, as well as the earlier statements of Weidenbaum and Niskansen contradicting a presumed relationship between deficits and inflation, were borne out by both government and private studies.[130] In the 1980s, interest rates and inflation dropped as rapidly as federal deficits rose.[131] Nor, a number of economists found, had high deficits soaked up private capital and "crowded out" private investment to any significant, unmanageable degree; if anything, it appeared that the contrary had taken place, that stimulative deficits might have increased private investment, in effect a kind of "crowding in."[132]

Deficits: How Serious a Problem?

Was the size of federal deficits, then, a problem worth anguishing over? Answering that question requires, first, an acceptance that merely comparing the dollar amount of the deficit and the national debt with similar dollar amounts in 1945, say, is not very helpful. A dollar owed today, because of inflation, is much less than a dollar owed then. Economist Robert Eisner of Northwestern University, former president of the American Economics Association, has shown that the net federal debt, uncorrected for the value of the dollar, rose from $250 billion to $600 billion between 1945 and 1980, but the real value of the debt, taking inflation into account, actually fell during that period by almost 60 percent. Eisner therefore concludes: "With inflation, we can have substantial budget deficits by the official measure while the real value of the net debt goes down."[133] Similarly, the fiscal 1983 federal deficit of $195.4 billion, for example, amounted to only $89.8 billion in 1972 dollars.[134] Dollar amounts just do not explain the debt and the deficits.

Neither can we understand the debt by looking only at what we owe—something over $4 trillion in late 1992. It is a peculiarity of government accounting that, unlike a private business, the federal government lists on its balance sheet (as we might call it) only its liabilities and not its assets. And we have a lot of assets.

Though there is disagreement about how to value certain holdings and possessions, indications are that the government, any time it wanted to, could almost certainly sell out for enough to pay off the national debt. In 1986, for example, when all U.S. Treasury obligations amounted to $2.2 trillion, federal assets—including "structures such as federal office buildings, post offices, NASA equipment, the value of military hardware, gold, the national parks," and myriad others (many, such as land, gold, and reserves of oil, gas, and coal, valued much below their market value)—totaled $3.9 trillion.[135]

Of course the government is not about to hold a garage sale

(although President Reagan once proposed a one-shot sale of some assets, housing loans for example, as a way of making the budget look better that year). Still, remembering that the government has assets to offset its liabilities is useful in at least one way: it helps us see that we would not serve the best interests of our children and grandchildren, whose names are often invoked when the dread of the federal debt is discussed, if we were to pay off the debt by selling out, leaving them no assets. Would it not be of more value to leave them with good schools and an educated work force, good highways and water systems, and a bustling economy?

Related to that point is another curious fact, and that is that the federal government, when it spends money, does not distinguish between operating costs and capital investments. If it did, as all businesses do, it would be clear that, like businesses, the federal government will always be in debt and that the total debt will surely continue to grow, the same as when the local trucking company, as it wears out one truck, borrows the money to buy another, probably more costly, one. If ordinary federal expenditures were separated from investments, and especially if operating costs were kept within tax revenues, the "deficit" would disappear, becoming simply an investment made with borrowed money. This is important because, as Eisner has written:

> What we really bequeath to the future, however, is our physical and human capital. A "deficit" which finances construction and maintenance of our roads, bridges, harbors, and airports is an investment in the future. So are expenditures to preserve and enhance our natural resources, or to educate our people and keep them healthy. Federal budgets which are balanced by running down our country's capital or mindlessly selling public assets to private exploiters are real national deficits.[136]

According to Eisner, the 1988 deficit of $155 billion would have been reduced to $85 billion by deducting the $70 billion the federal government invested that year in schools, training programs, and roads.[137]

The only meaningful way, of course, to evaluate the federal deficits and debt is on the basis of "ability to pay," the principal factor a private lender looks at when making a loan. For the federal government, that measure is gross national product (GNP), the total value of all goods and services produced in the United States annually. It is GNP that generates the federal government's tax revenues, our ability to pay.

Back in 1946 the national debt was over 134 percent of GNP. Then the percentage began to drop—to around 82 percent in 1951, about 66 percent in 1956, about 40 percent in 1971, and around 35 percent in 1981, when President Reagan took office. Since 1981 the national debt has not only climbed in total dollar amount, it has also climbed as a percentage of GNP—to 49.4 percent in 1986 and 56.7 percent in 1990. But these percentages are not as high as they were in 1961 and earlier.[138] Conservative economist Milton Friedman said in 1988 that "the size of the deficit has been exaggerated by use of such adjectives as 'tremendous,' 'gigantic,' 'obscene.' As a percentage of the national income, the deficit is not out of line with levels reached in the past."[139]

Considering only the publicly held federal debt (excluding the 20 percent or so that is owed to the Federal Reserve System or another agency of the federal government), the conclusions are the same: the federal debt of $2.05 trillion held by the public in 1988, for example, was 43 percent of that year's GNP, compared with the public debt's 63 percent of GNP in 1952 and 45 percent in 1962.[140]

Annual federal budget deficits, which averaged around 2 percent of GNP during most of the 1970s and the first part of the 1980s, hit 5.7 percent in 1983, but the deficit of 1988 was back down to 3 percent.[141] The fiscal 1991 deficit, $268.7 billion, pushed up by a continuing recession and the cost of the savings-and-loan bailout, was 4.8 percent of GNP (compared to the 4.1-percent average for the decade of the 1980s), and the unprecedentedly high 1992 fiscal-year deficit of $333.5 billion was 5.7 percent of Gross Domestic Product (the new way of reporting, leaving out overseas operations).[142]

"Much of the economic debate over the public debt," as one expert, James Savage, has put it, "is philosophical in nature rather than economic, which partially explains why the debate continues."[143] That is the view of experts Joseph White and Aaron Wildavsky, too. Before the 1990 budget-summit deal, they advocated about the same amount and type of budget reduction that effort eventually accomplished—because it represented a reachable goal, would limit the nation's risk if bad economic times were coming, and would "placate" people who believed deficit reduction to be vital.[144] After that, White and Wildavsky said, their advice to Congress would be: "then you can get on with the rest of the task of governing."[145] Other important questions need to be addressed.

Should we put more money in human and other investment, rather than in military operations and procurement? Do big deficits make it politically infeasible to increase funds, now, for social programs? Some, including Democratic senators Ernest Hollings, Jr., of South Carolina and Daniel Patrick Moynihan of New York, have said that was Reagan's intention all along.[146] Is the borrowing we are doing now, the debt we are leaving for our children and grandchildren, justified by the way we are spending it—for investment in human and physical capital and to leave them a healthy and growing economy? Is growth in the economy keeping up with the deficits and debt?

The deficit is affected as much by the economy as the economy is affected by the deficit, and maybe more.[147] Publicly held federal debt went down as a percentage of GNP from the end of World War II through the 1960s and into the early 1970s, but climbed again during the recessions of 1974–1975 and 1981–1982. As the American economy slipped into recession in 1990, deficits and debt began to mount again and continued to mount as the recession persisted through 1991 and 1992, the new estimate projecting a fiscal 1992 deficit of $333.5

billion.[148] How can fiscal and monetary policy be used to keep the economy strong and growing?

These are the philosophical and political questions that need asking. But they were neither the focus of the Congress nor of the country as the first shocking Reagan deficits began to appear.

Obsession with Deficit Reduction

The overpowering conventional wisdom during the early Reagan years was that nothing—nothing whatsoever—mattered so much as getting the deficits down. President Reagan spoke passionately against increasing taxes, if that was the price for deficit reduction, but he still made strong attacks against deficit financing. Even while proposing unbalanced budgets himself (and blaming Congress for too much domestic spending), Reagan made political speeches in favor of a budget-balancing amendment to the U.S. Constitution that would force him and Congress to do what they would not do voluntarily.[149]

Officially, and politically, deficits were still anathema to the Republicans (even though the biggest deficits had accumulated under Republican presidents). And the Democrats, to hear them tell it, had become even more fierce deficit fighters than the Republicans. Democratic rhetoric helped fuel the country's deficit panic. House Democrats, for example, put out a manifesto that declared: "We reject any economic program that projects annual federal budget deficits of $100 billion or more well into the foreseeable future. Such deficits will keep interest rates high, choke off investment and, over time, prove inflationary."[150]

A Democratic party radio commercial featured the voice of an announcer saying, as the sound of dripping water in the background grew into a cascade, that huge Republican cuts in taxes and increases in military spending had produced horrendous red-ink federal deficits.[151] Jesse Jackson, campaigning unsuccessfully for the 1984 Democratic presidential nomination, called for a cut of $70 billion in the federal deficit, but he gave a populist reason for his position. Jackson asserted that "chronic and expanding budget deficits, combined with bloated interest rates" had become "a near-permanent engine for the redistribution of income from the general public to leaders."[152]

The 1984 platform of the Democratic party took a hard line on deficits, proclaiming that "The Democratic Party is pledged to reducing these intolerable deficits."[153] And Walter Mondale, chosen as the party's presidential candidate, went even further in his speech accepting the nomination:

> Here is the truth about the future: We are living on borrowed money and borrowed time. These deficits hike interest rates, clobber exports, stunt investment, kill jobs, undermine growth, cheat our kids, and shrink our future. . . .
>
> I mean business. By the end of my first term, I will cut the deficit by two-thirds.

Let's tell the truth. Mr. Reagan will raise your taxes, and so will I. He won't tell you. I just did.[154]

Polls showed that the American people agreed with Mondale on the overriding importance of the deficit problem, but as the general election vote soon dramatically demonstrated, they agreed with Reagan that increasing taxes was not the way to solve the problem.[155] Reagan won reelection in a landslide, and he did it while taking strong positions against a tax increase of any kind and against any delay in regular cost-of-living adjustments (COLAs) of Social Security and other entitlement benefits.

The people wanted the deficits cut. That was clear. But how? That was not so clear. The job could be done by raising taxes, cutting spending, or both. Earlier polls had indicated that Americans favored reducing defense expenditures and eliminating the third year of the 1982 tax cut—and not cutting programs for poor people—as the best way to pare the deficit, and a more recent Gallup poll reported the same public-opinion result.[156]

Instead of a serious deficit-cutting proposal, though, President Reagan sent Congress, when it reconvened in 1985, a budget that was "dead on arrival"—a budget based on unrealistic economic projections and lower interest rates, calling for cuts in controllable domestic programs that Congress felt had already been cut to the bone and increasing defense spending.[157] Indications were that the people, and Congress, might support a moderate tax increase as a way of getting the deficits down, and Senate Republican leader Robert Dole of Kansas backed a proposal for a modest $59-billion tax hike over three years. But Reagan remained adamantly opposed, having said just days earlier, "I'll repeat it until I'm blue in the face: I will veto any tax increase the Congress sends me."[158]

Dole and the chair of the Senate Budget Committee, Republican Pete Domenici of New Mexico, pushed a budget measure through the Senate that would have frozen Social Security and other COLAs for one year. They made GOP senators "walk the plank" for this politically dangerous proposal, only to have Reagan shoot this plan down, too, by the time it got to the House. "People feel they flew a kamikaze mission and ended up in flames and got nothing for it," Republican Senator Warren Rudman of New Hampshire said bitterly.[159] House Democrats wanted to freeze military spending, but Reagan's steadfast opposition made this impossible.

So now what? Congress thought controllable domestic spending had already been cut enough. Money required to pay interest on the national debt could not be shaved. As a practical matter, Congress could not reduce defense spending without Reagan's approval. Nor could they cut Social Security COLAs or raise taxes over Reagan's opposition. Still, every member of Congress was under enormous public pressure to reduce federal deficits. Something had to give.

Gramm–Rudman–Hollings

The Gramm–Rudman–Hollings Balanced Budget and Emergency Deficit Control Act of 1985 came up in the Senate as a floor amendment to the 1985 debt-limit extension resolution. A good many of those who voted for the measure had serious misgivings about it. But, as co-sponsor Rudman himself said, it was a "bad bill whose time had come."[160] An overwhelming bipartisan Senate majority, by an initial vote of 75 to 24, approved Gramm–Rudman–Hollings and attached it to the debt-limit extension bill.[161]

Congress, if it were to decide to, could simply delegate to the secretary of the treasury the general authority to borrow annually whatever amount is necessary to cover that year's federal deficit. Congress, after all, makes borrowing necessary in the first place, by approving the spending of more money than the government collects in tax revenues. But the national legislature has jealously guarded its constitutional power over federal borrowing, preferring to dole out borrowing authority only in limited increments. So Congress periodically passes a legally binding joint resolution (separate from taxing, spending, budget, and other legislation) to boost the debt limit up another rung—by something like a couple of hundred billion dollars at a time!

With persistent and growing deficits during the 1970s and 1980s, legislation to raise the debt limit came before Congress with increasing frequency. A good many members of Congress viewed each such occasion as an opportunity for political posturing and speech-making, much of it highly partisan, about fiscal responsibility. During the 1960s, with Democrats in the White House, it was Republican members of Congress who were most heard in opposition to deficit financing. By the time of the monumental Reagan deficits of the 1980s, increasing numbers of Democrats were also delivering opposition orations.

In addition to providing a good excuse for partisan speech-making, the recurring debt-limit resolutions are also seen by many senators and representatives as likely legislative vehicles to which their favorite proposals can be offered as amendments. Everybody knows that a debt-limit measure has to pass and has to be signed into law by the president; otherwise, since the money, above and beyond revenues, has already been spent or obligated, the government would have to shut down. Thus in the Senate especially, where floor amendments are always in order and usually need not even be germane to the legislative measure under consideration, the must-pass, veto-proof debt-limit resolution, each time it comes up, is a superior horse to try to saddle a rider on. It is a horse that is sure to go the distance.

A balanced-budget rider was what freshman Republican Senator Phil Gramm of Texas had in mind when, in September 1985, the House-passed joint resolution to raise the limit on the national debt from $1.824 trillion to $2.078 trillion came over to the Senate. Gramm was not only new to the Senate, having just been elected; he was also new to the Republican party. He had served briefly in the House as a Democrat. There, almost immediately on taking office in 1979, he had tried to attach a balanced-budget amendment to an earlier debt-limit extension bill and, as a conservative "boll weevil" Democrat, had regularly joined House Republicans in support of President Reagan's domestic budget cuts. Expelled from the House Democratic Caucus for his apostasy, Gramm had defiantly resigned from the House, changed parties, and gotten himself reelected as a Republican. Soon after, he won election to the U.S. Senate, and he came to that body still determined, among other things, to pass a balanced-budget measure. His best chance came with Senate consideration of the 1985 debt-limit resolution.

Do *something* about the deficit! That was the emphatic message senators had received from their constituents during the August recess of that year. But what? They could not pass a tax increase. They could not cut defense or entitlement spending. They could pare little more from domestic programs. They could not cut interest payments on the national debt. But do *something!*

The most politically practical something turned out to be the measure that Senator Gramm had been perfecting as a bill and which he had persuaded fellow Republican Rudman of New Hampshire and Democrat Hollings, a fiscal conservative from South Carolina, to co-sponsor. Their plan would attempt to put the federal budget on automatic pilot (as then Democratic Senator Gary Hart of Colorado was to charge).

In its final form, Gramm–Rudman–Hollings had four main provisions: statutory deficit limits; automatic cuts or sequesters; a faster budget timetable; and tighter enforcement procedures. First, the Act starkly mandated (with exceptions for war and recession) that the federal deficit be "zeroed out" in equal steps—that is, that it be pruned in five annual cuts of $36 billion each until, by fiscal 1991, there would be no annual deficit at all, and the federal budget would be balanced.

Simple enough. But how to enforce compliance? A threatened "train wreck," as some called it, was the answer. If Congress did not meet the deficit targets each year, the Act provided for automatic, across-the-board budget cuts—sequestration.

The Office of Management and Budget and the Congressional Budget Office, together, would agree on a deficit estimate each year. (Congress did not trust OMB to make an honest and correct estimate by itself; an average would be used, if the two agencies could not agree.) Should congressional budget actions not meet the required deficit targets, the comptroller general, head of the congressional General Accounting Office (GAO), would draft, and the president would issue, a sequestration order, finally effective October 15, for automatic spending cuts—half to come from defense (with exceptions, including multiyear procurement contracts and, as amended later, personnel and certain other costs if the president so decided) and half from domestic spending (with exceptions, including interest on the national debt, Social Security, Medicare, veterans' pensions, Aid to Families with Dependent Children, Medicaid, and certain other programs for the poor).

Senate Democrats Robert Byrd of West Virginia and Lawton Chiles of Florida tried to amend Gramm–Rudman–Hollings to provide that, in the event of sequester, two-thirds of the required deficit-cutting would come from reductions in defense and domestic spending and one-third would come from taxes. This proposal was defeated. Hence, spending reductions alone were made to bear the whole sequestration burden. Senate Republican leaders Robert Dole of Kansas and Pete Domenici of New Mexico wanted to make sequestration apply to Social Security and other entitlements, but they had to give up on this. Thus, with the various defense- and domestic-spending exceptions, less than half of the federal budget wound up subject to the automatic cuts of Gramm–Rudman–Hollings. Just as a majority of the members of Congress had been reluctant to vote tax increases or certain types of spending reductions directly, they did not want to make them the subject of automatic action under Gramm–Rudman–Hollings, either.

Gramm–Rudman–Hollings changed the budget law to speed up the process, ordering a month-earlier timetable. The measure also changed House and Senate rules to enhance enforcement and to expedite action. In the Senate, any budget amendment that would increase the deficit was made subject to a point of order that could only be overruled or waived by sixty votes (three-fifths of Senate membership, the same as for cutting off a filibuster). Anyone offering an amendment to increase spending, then—for a pet project at home, say—had to propose a corresponding tax increase or a cut in some other program to offset it. The Senate also adopted the so-called Byrd Rule, providing that the same sixty votes would be required to overturn a ruling from the chair that an amendment to a budget resolution was not germane.

When the Gramm–Rudman–Hollings amendment came up for a vote in the Senate, there was a confusion of motives on both sides of the issue. Democrat Moynihan of New York opposed it because he feared that it would result in too-severe cuts in military funds. But Georgia's Sam Nunn, also a Democrat and a consistent advocate of a strong U.S. defense, supported the measure. Republican Gramm and others wanted to cut domestic social programs, but Democrat Edward Kennedy of Massachusetts voted for Gramm–Rudman–Hollings as a way of saving social programs. Kennedy hoped, and expected, that the threat of sequestration's across-the-board cuts in defense would at last force President Reagan to ask for a tax increase.

Democrat Gary Hart blasted Gramm–Rudman–Hollings as a "fraud" and a "mindless" surrender of the constitutional responsibility of Congress. "We are lacing ourselves into a fiscal straightjacket—as if to say, 'stop me before I kill again,'" Hart said.[162] Another opponent, Republican Lowell Weicker of Connecticut (later governor of that state) called the measure a "substitute for the guts we don't have to do what needs to be done."[163]

Privately, a good many senators agreed with Hart and Weicker. But in the end, most of them decided to support Gramm–Rudman–Hollings, despite misgivings. Voting against it would have been like voting against the Ten Commandments, while a vote for the measure gave them something to point to when they were asked what they were doing to cut the deficits. Many senators probably also thought that Gramm–Rudman–Hollings would be killed in the House. They were wrong in this, because the members of that body soon lined up to back Gramm–Rudman–Hollings too, just as President Reagan, though initially surprised by its appearance and strong backing, had earlier done.

So with its heavy balanced-budget rider hanging on, the debt-limit horse raced all the way to the finish line and into the White House winner's circle. Reagan's December 1985 signature on the debt-limit extension resolution made both it and Gramm–Rudman–Hollings the law.

Gramm–Rudman–Hollings in Action

The U.S. Supreme Court soon held that the attempt to give to GAO's comptroller general the power to draft the sequester order was unconstitutional, a violation of the principle of separation of powers.[164] Thus Congress, for the first year under Gramm–Rudman–Hollings, simply enacted the relatively small cuts that would have been made automatically by the sequester. After that, in 1987 (with what was called Gramm–Rudman–Hollings II), they transferred the sequester-drafting authority to OMB, an executive agency. In the same measure, incidentally, Congress also relaxed and extended the Gramm–Rudman–Hollings deficit targets and deadlines, as they were regularly to do from then on.

Did Gramm–Rudman–Hollings work? In general, no. "Gramm–Rudman is like the cross-eyed archer," Democratic Senator Wyche Fowler of Georgia has said. "He doesn't hit anything, but he scares hell out of everyone."[165] If the question is whether the Act worked as an economic-policy tool, the answer is still negative. By making deficits the number-one priority, the Act reduced the ability of Congress to use fiscal policy in a countercyclical way to manage the economy.

Nor did Gramm–Rudman–Hollings cause Congress to get its budget work done on time. In the very first year of the new law's operation, Congress missed the new budget deadlines and was unable to pass a single one of its appropriations bills before the beginning of the new fiscal year. That became almost a pattern for what would happen in the future.[166]

As in the great 1990 budget struggle, final decisions continued, almost always, to be made at the last minute and under a threat that the government would have to be shut down. A key reason for the difficulties Congress experienced in meeting its budget timetable may have been that Gramm–Rudman–Hollings, itself, instead of diminishing some of the overlapping budget-process layers, added to them. The same decisions wound up being made many times.

Gramm–Rudman–Hollings encouraged the use of "blue smoke and mirrors," fiscal legerdemain, to meet deficit targets, or, as former CBO director Alice Rivlin has put it, actually "created incentives for sleight of hand and budgetary gimmickry."[167] OMB and the president regularly continued to underestimate spending totals and to rely on gimmicks to reduce them, while at the same time overestimating revenues, particularly by projecting a higher rate of growth in the economy than was realistic. And, more often than not, Congress continued expediently to adopt OMB's rosy view in place of the more pessimistic and restricting estimates and projections of its own CBO. Senator Hollings has since charged that the intent of the law has been subverted by "lies, deceit . . . sleight of hand, changing numbers, changing targets."[168]

Gramm–Rudman–Hollings did not eliminate the deficits and stop the growth in the national debt. Why were the deficits so hard to cut under the new law? For the same reasons that they had been produced in the first place. For the same reasons that they had been so hard to cut before Gramm–Rudman–Hollings. Nobody wanted a tax increase, and everybody wanted spending, at least spending that benefited them. The deficits hurt no one enough directly. As John Gilmour has written:

> There is specific opposition to almost every possible budget cut because all such budget cuts directly and significantly hurt some probably well-organized group. A budget deficit, however, hurts no one directly. Its cost is difficult to perceive because it is so widely distributed, both across the population and over time. If the $200 billion deficits of the Reagan era had plunged the nation into a depression, caused widespread unemployment or inflation, then the political support for deficit reduction would be stronger. But because the deficits coincided with a period of strong economic growth and low inflation, they appeared virtually costless.[169]

Another reason the deficits were so hard to reduce was that the numbers had ballooned hugely; no marginal cuts or easy "freezes" would do the job. For example, even if, with "rosy scenario" estimates and projections, the annual deficit could be pegged at $100 billion, that would still equal roughly all appropriations for the entire U.S. Navy, as two budget experts, Joseph White and Aaron Wildavsky, have pointed out, or would exceed combined federal appropriations for Medicaid, the Department of Education, the National Institutes of Health (including cancer and AIDS research), the Department of Justice, the Department of State, and the Federal Highway Administration.[170] Or, if the hypothetical $100-billion deficit had to be eliminated by tax increases alone, that would mean a steep hike in individual income taxes of 10 percent, plus a 20-percent increase in corporate taxes and a 50-percent increase in excise taxes.[171]

The enactment in a one-year chunk of such draconian spending cuts or tax increases, or some combination of both, not only would be politically unimaginable but would deal an unacceptably recessionary blow to the economy. And to get a clearer understanding of just how difficult cutting the deficit really is, dwell a moment on the fact that the projected fiscal 1992 deficit was more than three times greater than this hypothetical $100-billion, rosy-scenario deficit.[172]

Still, as White and Wildavsky have concluded, as a result of tax increases and cuts in defense and nondefense discretionary spending as a percentage of GNP, "politicians have done a great deal about the deficit," reducing the red ink substantially from what it would have been without such efforts.[173] The two Republican authors of Gramm–Rudman–Hollings, senators Gramm and Rudman, agree. They released figures in 1990 to show that the growth rate in federal spending had been cut in half and that spending was declining as a percentage of GNP.[174] There is reason to believe that the offset provisions of Gramm–Rudman–Hollings, making the budget process a kind of revenue-neutral, zero-sum game, have operated as a real restraint. "I see less demand by special-interest groups for increases than I have ever seen in my five years of being Budget chairman and in my fourteen years of serving as a senator," Republican Domenici said soon after Gramm–Rudman–Hollings became law.[175] The same restrictive effect was felt when Congress considered, and passed, the major tax reform law of 1986; Gramm–Rudman–Hollings offset provisions helped prevent new, revenue-losing tax loopholes or giveaways from being agreed to.[176]

Still, given the great and growing federal deficits and the inability of Congress to finish its budget work on time, set priorities, and make conscious fiscal policy, as well as the dominant, even overwhelming, effect of budget matters on the congressional agenda, nobody is satisfied with the present budget process. Additional reforms are essential.

Needed Budget-Process Reforms

With America's bicameral national legislature and separation of powers system, together with frequent split-party control of the presidency and Congress in recent times, there was bound to be a great deal of internal governmental conflict over serious matters, and nothing is more serious than budget policy.

Too, when there is plenty of money almost any budget process will work, but when money is tight, as in recent times, the opposite is true. There is some truth, then, in what one U.S. senator has said in discussing congressional budget problems: "The issue is lack of money, not the budget process."[177]

Lack of will is also a problem, according to some senators, including the former chair of the Senate Budget Committee, Democrat Lawton Chiles. Changes in process, Chiles has said, "won't work without the joint will of the executive and Congress." [178] Former director of the Congressional Budget Office Alice Rivlin has made the same point: "No set of procedures, however, can force participants to make choices that they do not want to make or do not regard as necessary. Reforms to the process cannot substitute for political will or for the exercise of leadership in working out compromises among warring parties."[179]

Still, Rivlin has argued, certain reforms are needed to make congressional budget-making more efficient and workable. A well-designed budget process, she has written, should do three things: give decisionmakers all the information they need; put the sequence of decision-making in logical order; and conserve time and keep the budget process from overwhelming other governmental activities. The present congressional budget process fails in regard to the third criterion, at least, and here is where reform should come, according to Rivlin.[180]

There is a widespread feeling among senators, too, that general improvement of congressional budgeting is overdue:[181]

> The budget process has failed miserably. (Dennis DeConcini [D., Ariz.], member of the Senate Appropriations Committee)

> The process is not working; it has broken down. (Don Nickles [R., Okla.], member of the Appropriations Committee)

> I doubt that anyone could be found in the Senate who thinks the budget process works. It doesn't work. Fix it or get rid of it. (Harry Reid [D., Nev.], member of the Appropriations Committee)

> There's more complexity now. The budget process is going to be changed in an orderly manner, or there will be a gradual breakdown until Congress abolishes it all of a sudden. (Pete Domenici [R., N.M.], ranking minority member of the Budget Committee)

Balanced-Budget Amendment, Item Veto No Answer

Some critics of the present national budget system look for structural changes outside Congress. One such proposal calls for amending the Constitution to require a balanced budget.[182] Another suggestion is that the president should be given the power to veto individual items in congressional appropriations bills. Both suggestions are objectionable.

Throughout his eight years in office, President Reagan supported a proposal for a balanced-budget amendment to the U.S. Constitution—even as he himself proposed unbalanced budgets. The same was true of President Bush. The movement for such an amendment predated Reagan's presidency and had long been a favorite cause of conservative groups. By 1985 thirty-two of the required thirty-four states had voted to call a national constitutional convention to propose a balanced-budget amendment.[183] That was the apogee of public support for the idea. In 1982 the U.S. Senate passed such a proposal only to see it die in the House. A similar proposal failed in the Senate in 1986, falling one vote short of the two-thirds required.

When recession-fueled deficits forecast for fiscal 1992 and later began to soar, the balanced-budget amendment drive gained renewed congressional steam.[184] But critics of the idea called it a futile effort. They argued that if the rigid statutory deficit targets under Gramm–Rudman–Hollings could be avoided by "creative accounting" and gimmicks, the same could surely be done in regard to constitutional restrictions. One opponent said that "a Congress so profligate as to require a constitutional amendment to limit its spending propensities will be ingenious enough to figure ways to avoid it."[185] Which federal official or agency would make economic projections and revenue and spending estimates—and who would keep them honest, the courts? Would all the definitions and restrictions, such as those relating to "off-budget" items, be written lengthily into a presently brief Constitution that has only been amended seventeen times in 200 years?

It is true that all of the state constitutions but one (Vermont's) require a balanced budget. But states do not have to fight wars. Nor can a state's fiscal policy be used to stabilize or stimulate the national economy. And state budgets, unlike those of the federal government, separate capital spending from operating expenditures, only the latter having to be equaled by revenues. A balanced-budget amendment to the U.S. Constitution would have to be so riddled with loopholes—for national emergencies or economic slumps, say—as to be meaningless. If really made restrictive, it could require Congress to adopt exactly the wrong fiscal policy, a big tax increase, for example, during a recession.

The proposal for a presidential line-item veto, also backed by President Reagan (and later by President Bush), is as unworkable as the constitutional requirement for a balanced budget.[186] In the first place, if a big spender rather than a big cutter were in the White House, this enhanced presidential power could increase spending. The president could use the threat of line-item vetoes against favorite programs and projects of members of Congress to force them to add the president's own favorite spending items to appropriations bills. As one Democratic congressional aide said in 1990: "The real head scratcher in this is that the Republicans must believe that no Democrat will ever be elected President, or that no President will ever stray from the conservative fold."[187]

The president also could use this added leverage to move Congress on nonspending matters as well, profoundly changing the delicate presidential-congressional balance established in the Constitution and centralizing too much power in the presidency. This is why some conservatives oppose the line-item veto idea—and others should.

In any event, the line-item veto would probably not cut spending much, one study estimating only a 2-percent budget reduction by this method.[188] Even a president like Reagan or Bush would be unlikely to strike defense items. Spending for interest payments could not be vetoed. And entitlement spending would not come to the president in a form subject to a line-item veto. When President Reagan once again told Congress in January 1988 that he wanted line-item veto authority, he said that to demonstrate what could and should be done with this power he would soon send them a list of items, totaling $600 billion, that he would have used the item veto to strike from a recently passed continuing resolution. But when the list finally came, the 107 items it proposed to cancel amounted to only $1.15 billion (out of a federal budget that was then about $150 billion in the red).[189]

Experience in the states, where forty-three of the governors have line-item veto power, shows that, if presidents had this prerogative, Congress would probably shift more of its spending to nonvetoable forms, such as entitlements. Moreover, it would add one more step and one more presidential-congressional confrontation stage (a postenactment and override confrontation, on top of the present preenactment confrontation) to a budget process that already has too many steps and stages. The presidential line-item veto is not an idea whose time has, or should, come.

However, three meritorious proposals have been made for improving the budget process inside Congress. They include: institutionalization of the summit device; multiyear budgeting; and elimination of the authorization-appropriations dichotomy. All these would improve the process considerably and should be adopted.

Reform of the Senate Budget Process

Institutionalizing Summits

Gramm–Rudman–Hollings, especially by eliminating the second budget resolution and making the first budget resolution binding (standardizing the actual practice that had already developed), put a premium on early, high-level congressional-presidential budget negotiations. As a result, party leaders and party bargaining became much more important in congressional budget-making.

After the New York Stock Exchange's "Black Monday" (October 19, 1987), the day when the Dow Jones average fell 508 points, the conventional wisdom was that if the federal government did not trim the deficits, worse financial disaster would occur.[190] That event pushed President Reagan into a budget summit with Congress in 1987. Although this first summit's achievements proved, in reality, to be less than spectacular, it did seem to expedite congressional budget action.

The Reagan summit was the pattern for the 1990 summit in the second year of the Bush administration. As Democrat J. Bennett Johnston of Louisiana, a member of both the Appropriations Committee and the Budget Committee in the Senate, said after the first summit:

The ad hoc budget summit of last year [1987] had all the principals, plus the White House, and came up with the parameters and exact numbers. Consequently, we'll have [all] thirteen appropriations bills on the President's desk in October of this year [1988], the first time since 1948. This is the prototype, with the right people.[191]

In addition to the other forces that have caused a simultaneous nationalization and individualization of the U.S. Senate, the changed budget process since the mid-1970s has been a factor in these developments, both affecting them and being affected by them. Many more decisions have been moved by the new process to the Senate floor, where senators use their increased individual power. The summit device, involving the party leaders and the principal committee leaders of both parties, offers a needed way of arching over Senate fragmentation and decentralization to foster action (though not always neatly or quickly, as we saw in 1990).

Multiyear Budgeting

Senate Republican Budget Committee leader Pete Domenici has made the case for institutionalization of the summit, tying the idea, as many do, to a multiyear budget. "We need a two-year budget—with a joint committee, leadership-appointed, in the first ninety days of the first year," Domenici said in 1988. "This would work very much like the recent [1987] summit, but with more meat on the bones."[192] The "meat on the bones" that Domenici supports would replace the ad hoc (now and then, when leaders feel like it) summit with a legally established one. The two congressional budget committees would be "folded into" a single joint one. To this group would be added from both parties, as was true of recent summits, the House and Senate party leaders and the chair and ranking member of the two appropriations committees and the two taxing committees (Senate Finance and House Ways and Means).[193] It is a good basic design for a structure; with the addition of White House negotiators it could help bridge the federal government's inevitable branch, house, and party fissures.

It is no accident that Senator Domenici mentioned the two suggestions—summit and multiyear budgeting—together. There seems to be overwhelming support for the idea of a multiyear budget. That is, in fact, what happened in the two great budget summits of 1987 and 1990: multiyear budgets were agreed to. Former Budget Director Alice Rivlin, as well as former Senate Budget Committee chair Lawton Chiles, have joined Senator Domenici in support of a two-year budget.[194] If Congress were thereby relieved of an annual budget fight, Rivlin has written, it could spend more time on long-range issues and on oversight of the executive department, particularly in alternate, nonbudget years.[195]

In addition to Domenici, a good many other Senate influentials have given their outspoken backing to the two-year budget idea. These include Majority Leader George Mitchell of Maine (who is also a member of the Senate Finance Committee), Budget Committee chair James Sasser of Tennessee (who is also a member of the Senate Appropriations Committee); other Budget Committee members including Wyche Fowler (D., Ga.) and Nancy Landon Kassebaum (R., Kan.); a number of other Appropriations Committee members including Patrick Leahy (D., Vt.), Don Nickles (R., Okla.), and J. Bennett Johnston (D., La.) (with Johnston being a member of the Senate Budget Committee, also), and David Boren (D., Okla.), a member of the Senate Finance Committee.[196]

With all that support for the concept, it is no surprise that multiyear budgeting was at the heart of the 1990 budget-summit deal.[197] In that agreement, Congress increased taxes and cut certain spending over a five-year period. It also relaxed and stretched out the Gramm–Rudman–Hollings deficit targets for the ensuing five years (with adjustments for changed economic conditions after the first three). And it went even further toward pulling the teeth of the basic law by taking away the threat, for the time being, of the overall sequester if the deficit targets were not met.

Instead of mandatory deficit targets, the new law (which has been called Gramm–Rudman–Hollings III) provided for a three-year budget and upward limits on spending for fiscal 1991, 1992, and 1993. Congressional taxing, entitlement, and spending decisions during that period were required to be deficit-neutral. That is, actions of Congress should not themselves increase the deficits, although it was assumed that other causes, such as recession and troop deployment in the Persian Gulf, would do so.

Under the 1990 law, the principal focus no longer was on the size of the deficits, but on whether discretionary *spending* stayed within annual caps set in three distinct categories: defense, domestic discretionary (not entitlements), and foreign aid. New enforcement sanctions were provided: fifteen days after congressional adjournment, OMB could institute a mini-sequester, or automatic, across-the-board cut within any of the three separate spending categories, if it was found that Congress had overshot that category's separate spending limit. (Congressional critics soon charged that this procedure gave too much power to OMB and complained that congressional savings in one category could not be used to increase spending in another—that is, that reductions in defense spending, for example, could not be used to increase spending for domestic programs.)

On an ad hoc basis, then, Congress has twice in recent years turned to summits and multiyear budgets. It should now institutionalize these concepts by formally enacting them into law. In both cases this would require little more than that practice be made requirement.

Committee Reform

There is strong sentiment in the Senate for some kind of budget-oriented restructuring of committees. There is support for expanding the Budget Committee membership to make it more inclusive or representative.[198] There is support for simply abolishing the Budget Committee altogether.[199] There is support for abolishing the Budget Committee and giving its functions to the Appropriations Committee.[200] There is support for abolishing the Appropriations Committee and giving its functions to the Budget Committee.[201] There is support for abolishing the Budget Committee and letting the authorizing committees set the spending caps.[202] And there is, in regard to one proposal or another, considerable support for combining the authorizing and appropriating functions.[203] Many senators feel that there are just too many steps and layers in the present budget process.

The best proposal for change is one authored principally by Senator Nancy Landon Kassebaum and co-authored by Hawaii Democrat Daniel Inouye,[204] which we discussed in more detail in Chapter 5. Under their plan, present Senate authorization and appropriations committees would be replaced by new legislative policy committees, including, for example, a Committee on Economic Policy (combining the jurisdiction of the present Finance Committee and the Banking, Housing, and Urban Affairs Committee), a Committee on Agriculture Policy, a Committee on Defense Policy, a Committee on Commercial Policy, and so on. Each of these committees would have both authorizing and appropriating power.

The present Budget Committee would be replaced by a Committee on National Priorities, which would have representation from the other committees (and which, though the Kassebaum–Inouye measure does not deal with this, could furnish the Senate's delegates to an institutionalized budget summit).

The Kassebaum–Inouye proposal makes sense. It, or something like it, should be adopted in the Senate (and in the House, too, for that matter). The Senate's budget-process system could thereby be rationalized and streamlined. Action on the budget, as well as on other matters, would be facilitated.

The congressional power of the purse—budget power: the authority over federal taxing, spending, and borrowing—is the central and most important power of Congress. Both inside Congress and outside, there is a widespread feeling that the exercise of this power in recent years has been a failure.

Members of the Senate (and House) feel frustrated in their attempts to get control of federal deficits, although they have had more effect in this regard than is generally accepted and were able to make commendable headway on deficit control as a result of the 1990 summit. Still, budget matters take up too much of the time and attention of individual members of Congress. Deadlines are commonly missed. The setting of national budget priorities has been difficult.

U.S. economic policy in recent times has produced highly unsatisfactory results. The maldistribution of wealth and income has worsened. Real wages have fallen. Investment in human and physical capital has lagged. The Federal Reserve has managed monetary policy in a way that has often been at cross purposes with fiscal policy (and should be brought more in line with it). Congress has not been sufficiently able, or consciously and fully informed, to make acceptable fiscal policy. Reforms of Senate (and House) budget processes and a fundamental change in Senate committees (and those of the House, too, if possible) are needed. After all, the basic function of Congress is policymaking. Congress should be able to do it better and more efficiently. That is true in regard to budget policy. It is also true in regard to national security and foreign policy.

8

Making National Security and Foreign Policy

It was one of those climactic, defining moments when the eminent role of the United States Senate in national affairs is most clear. The gravest possible national issue was to be decided. Would there be war or would there be peace? Senators rose one by one to announce their choices and their reasons, somber, yet with more feeling, more passion, in their words than was the custom.

On January 10, 1991, the Senate began debate on whether to authorize President George Bush to use force to drive Iraq out of Kuwait, or to continue on for a time with a United Nations-approved embargo and economic sanctions aimed at achieving the same purpose. It was not a debate, in the usual sense. Until toward the end, few members of the upper body were on the floor. Senators spoke less to each other than for the record and, through television, to the nation. They spoke to history, too.

"It may well become necessary to use force to expel Iraq from Kuwait, but because war is such a grave undertaking, with such serious consequences, we must make certain that war is employed only as a last resort," Democratic Majority Leader George Mitchell of Maine said.[1] But for Republican Senator Orrin Hatch of Utah, it seemed that the moment had come "for the Congress to join with the President and get behind him and our young men and women over there sitting in the sand and show that we're willing to back the use of force."[2]

Five months earlier, on August 8, 1990, President Bush had announced in a dramatic broadcast from the Oval Office that he had just dispatched U.S. air and ground forces to Saudi Arabia as a result of the invasion of Kuwait by neighboring Iraq five days earlier. Obliquely likening Iraq's ruler to Hitler, a comparison he was to resort to more explicitly in the weeks that followed, President Bush had said, "Appeasement does not work. As was the case in the 1930s, we see in Saddam Hussein an aggressive dictator threatening his neighbors."[3] The president called the "line in the sand" he had drawn "Operation Desert Shield"—and it was to be the most massive and rapid buildup of U.S. troops, tanks, planes, and ships since the Vietnam War.[4]

The United States was back in the Persian Gulf, but this time, in a sense, on a different side. Just three years earlier, President Reagan had sent a naval task force there to protect Kuwaiti shipping during the protracted Iran–Iraq War in which the United States had "tilted" toward Iraq.

Shortly before Hussein had ordered the invasion of Kuwait, Congress had been in a mood to condemn him and his country for warlike preparations and for menacing Kuwait. The Bush administration had opposed such congressional action.[5] In fact, after Iraq massed thirty thousand troops on the Kuwaiti border in July to underscore the seriousness of its dispute with that oil-rich country over control of two islands and an oil field, plus a complaint about Kuwait's running down the price of oil by overproduction, the Bush administration had signaled that the United States would not get involved. A U.S. State Department spokesperson in Washington announced: "We do not have any defense treaties with Kuwait, and there are no special defense or security commitments to Kuwait . . . [but] we remain strongly committed to supporting the individual and collective self-defense of our friends in the Gulf, with whom we have deep and long-standing ties."[6] In Baghdad, the U.S. ambassador, asked to elaborate, said that the Iraq–Kuwait dispute was an Arab, not an American, problem.[7]

All that changed after Iraq actually invaded Kuwait. In making his August troop-deployment announcement, President Bush said that he and his secretary of state had been in close touch with a great many other countries, including the Soviet Union, and that he had decided to deploy forces to the Persian Gulf only after "unparalleled international consultation." The president pointed to the fact that the United Nations Security Council had, earlier, unanimously condemned the Iraqi invasion and, following the U.S. lead on a full trade embargo, had voted mandatory sanctions against Iraq, the first such U.N. action in twenty-three years.

What President Bush did *not* say, of course, was that the American troop deployment had been formally approved in advance by the U.S. Senate and House. In fact, there had been no such action by Congress. And, though the president had consulted in general about the Gulf Crisis with selected congressional leaders, many, if not most, such leaders, including Majority Leader Mitchell and Sam Nunn (D., Ga.), chair of the Senate Armed Services Committee, were to say that they found out about the troop deployment "after the fact."[8] Republican Senator Mark Hatfield of Oregon made a statement implying that the August deployment had been purposely timed to find Congress away from Washington; he noted that Congress had been out of session four out of the five times that President Bush had ordered U.S. troops overseas.[9] (As a matter of fact, Secretary of Defense Richard Cheney was later to say that "it was an advantage that Congress was out of town. We could spend August doing what needed to be done rather than explaining it [to Congress]."[10]

Criticism of the president for lack of consultation and for failure to take Congress sufficiently into account was to grow sharper. The president's actions were to raise again old and serious constitutional and governmental questions concerning the shared and overlapping powers of the president and Congress in regard to national security and foreign policy.

President Bush said in his initial announcement of Desert Shield that its mission was "wholly defensive,"[11] that U.S. troops in Saudi Arabia were in a "defensive mode right now"; and he declared that "economic sanctions in this instance, if fully enforced, can be very, very effective."[12] A week or so later in two separate public speeches, one at the Pentagon and the other to a veterans' group, but both aimed at affecting public opinion, the president restated the reasons behind the deployment:

> Our action in the Gulf is about fighting aggression—and preserving the sovereignty of nations. And it is about our own national security interests and ensuring the peace and stability of the world. And we are also talking about maintaining access to energy resources that are key not just to the functioning of this country, but to the entire world.[13]

After the August recess, the Senate Foreign Relations Committee drafted a resolution stating that the president did not have the power to change the role of U.S. troops from defense to combat without approval of Congress. Senator Mitchell thought that the time was not right for formal Senate consideration of the resolution, and he persuaded the Committee to lay the measure aside.[14] Still, it was assumed by most senators that the resolution was correct in its assertion that the approval of Congress would be needed before any actual hostilities.

President Bush had talked in his initial announcement of a "new era." He spoke to Congress in September of a "new world order" as a result of the end of the Cold War and the new cooperation of the United States and the Soviet Union in the United Nations Security Council.[15] There were references to a "Bush Doctrine," the intention of the United States to act in concert with other nations in the future to protect economic interests against power seekers.[16] A good many members of the Senate became increasingly concerned about the long-term commitments that President Bush said he anticipated making to protect the security of the Persian Gulf, once Iraq had withdrawn from Kuwait.[17]

Public opinion polls showed, though, that at least the president's initial actions enjoyed the overwhelming approval of the American people—74 percent in a *New York Times* survey, 77 percent in one by *Newsweek*.[18] The situation was the same in Congress. Members were especially outspoken in their praise of Bush and his secretary of state, James A. Baker III, for their success in persuading other nations to join in the embargo, though some questions began to be voiced fairly early.[19] Republican Senator Richard Lugar of Indiana, a hawk who wanted Saddam Hussein removed from power, nevertheless thought that the United States should make a strong effort to get other nations to pay a part of the costs (and some did, later), and the Democratic chair of the Senate Foreign Relations Committee, Claiborne Pell of Rhode Island, a dove, signaled the limited nature of much of the Senate support for Bush's initial, defensive and nonviolent, actions in the Gulf by calling them "decisiveness with diplomacy" and "boldness with restraint."[20]

But what about the War Powers Resolution, which Congress had overridden President Richard Nixon's veto to pass in 1973?

No Invocation of the War Powers Resolution

The War Powers Resolution required presidents "in every possible instance" to consult with Congress before introducing U.S. armed forces "into hostilities or into situations where imminent involvement in hostilities is clearly indicated," to report to Congress within forty-eight hours of any such introduction of forces, and to withdraw them after sixty days unless Congress authorized their continued presence.[21] President Bush, like the other presidents before him, refused to invoke the War Powers Resolution and its sixty-day limitation, asserting in his written notification to Congress of the Persian Gulf troop deployment that "I do not believe involvement in hostilities is imminent; to the contrary, it is my belief that this deployment will facilitate a peaceful resolution of the crisis."[22] The president announced the call-up of the first forty thousand American reserves on August 22.

In his September address to a joint session of Congress, Bush said that the crisis in the Persian Gulf offered "a rare opportunity to move toward a historic period of cooperation." The cooperation the president was speaking of on that occasion was the unprecedented and impressive international cooperation he had been able to arrange in support of his Gulf policies. But there were growing complaints in Congress that similar cooperation by the president with the U.S. legislative branch was sadly deficient.

Senators complained of a lack of notice and consultation under the War Powers Resolution. Republican William Cohen of Maine said that the War Powers Resolution "happens to be the law of the land" and should be followed by the president, that the president should "formally submit the notification, and ask for approval or disapproval, to get Congress on record at this stage as to whether it supports the policy or not."[23] President Bush did not agree.

Majority Leader Mitchell and Republican leader Robert Dole of Kansas, both of whom for various reasons preferred not to have an early War Powers Resolution debate, worked out an alternative to a White House-drafted measure. The Senate resolution expressed support for Bush's actions of a purely deterrent and defensive nature and called on the president to seek a declaration of war under other circumstances, as well as to consult with a newly formalized "combined congressional leadership group"; the resolution zipped through the Senate on October 2 with only three votes against it.[24]

Not unnoticed in Washington, about this time, was a speech at home by Soviet Foreign Minister (later to resign) Eduard Shevardnadze, who, in admitting that his country had made a mistake in its earlier invasion of Afghanistan, went on to declare that "any use of Soviet troops outside the country demands a decision of the Soviet Parliament."[25] Was not that the way it was supposed to be in the United States also? Congressional leaders agreed to a procedure before ending the 1990 session (following the historically long budget struggle) that would permit House Speaker Thomas S. Foley of Washington and Senator Mitchell to reconvene Congress on their own, not waiting for a special-session call by President Bush, in case the president decided to go to war with Iraq.[26]

As the legislators prepared to adjourn, Secretary of Defense Richard Cheney announced that the administration was considering whether to increase the two hundred thousand troops then in the Persian Gulf, and Republican Senator Richard Lugar of Indiana said on a television talk show, "The fact is that we're headed toward conflict."[27]

A Change to Offense

On November 8, 1990, two days after the congressional elections, President Bush suddenly opted for offense. (The *Washington Post* was later to report that the decision had actually been made on October 31 but kept secret until after the elections.[28]) At least 150,000 more troops would be sent to the Persian Gulf, the president said, a figure that was eventually increased until total deployment came to over 400,000. This, incidentally, was a greater number than the total native population of Kuwait. The president also announced that the normal rotation of troops presently in the Gulf region would be indefinitely deferred.

Senate Majority Leader Mitchell was later to tell the Senate that this startling new buildup and change from a defensive to an offensive posture had been decided upon without consultation with Congress. "The President did not consult with Congress about that decision; he did not try to build support for it among the American people," Mitchell said. "He just did it."[29]

The shift to a war footing caused a "minifirestorm" of congressional and other criticism.[30] An antiwar movement began to develop in the country.[31] There were calls from both hawks and doves for a special session of Congress. Republican Senator Richard Lugar declared approvingly that the president had "set the United States on a collision course in which Iraq will either withdraw from Kuwait or be forced to do so by military means," and he said a special session was required to give Bush the explicit backing he needed.[32] Democratic Senator Edward Kennedy of Massachusetts, on the other hand, called for a special session in order to *stop* the president's "headlong course toward war."[33] He voiced the growing belief of many when he said that the troop buildup would make war "inevitable."[34]

Other important Democrats were also critical. It was an especially damaging blow to the president's policies when Senator Sam Nunn of Georgia, the heaviest of Senate heavyweights on military matters (who had received notice only an hour, at most, before the buildup announcement[35]) took to the television talk shows to declare that Bush's rush to go to war was a "mistake" and that the president should explain "why liberating Kuwait is in our vital interest."[36]

George Mitchell said flatly that if President Bush decided to attack Iraq, "he must come to Congress and ask for a declaration. If he does not get it, then there is no legal authority for the United States to go to war."[37] Mitchell, Nunn, and Senate Foreign Relations chair Claiborne Pell held a joint press conference to announce Senate committee hearings on the Persian Gulf situation.[38]

Mitchell said again that President Bush "has no legal authority, none whatever," to go on the offensive in the Gulf. "The Constitution clearly invests that great responsibility in the Congress and the Congress alone."[39]

The Controversy Grows

The words got rougher. One unidentified Democratic senator was quoted as saying, "If he [Bush] does not come to us for an authorization [before going to war], it's an impeachable offense."[40] Republican Senator Malcolm Wallop of Wyoming countered by accusing Congress and the Democrats of criticizing without themselves taking responsibility, of "taking a very typical Congress approach. That is, a position that is defensible on all grounds."[41] North Carolina Republican Jesse Helms suggested that the critics were giving aid and comfort to the enemy.[42]

The president called congressional leaders to the White House on November 14 and convinced some of them, at least, that despite the buildup, he had not abandoned economic sanctions or decided on war.[43] Though there would be no special session of Congress, the voices critical of presidential policy were not stilled. Public opinion continued to be split.[44] In the Senate, the Armed Services Committee opened hearings on November 27. The hearings were damaging for the president (who at first decided that administration witnesses would not appear before the committee); the initial witnesses, including two former chairmen of the Joint Chiefs of Staff, testified that the country should not go to war until economic sanctions had been given a longer time to work.[45] "The question is not whether military action is justified," committee chair Nunn said, adding, "It is. The question is whether military action is wise at this time."[46] Nunn thought it was not.

The Bush administration issued reports that indicated Iraq might be closer than previously thought to developing a nuclear capability to go with its existing weapons of chemical and, perhaps, biological warfare. Democratic Senator Albert Gore of Tennessee called this assertion "clearly misleading, unsupported by the evidence."[47]

Other Democratic senators questioned a policy that would put American lives in jeopardy to support the autocratic, even feudal, regimes in Kuwait and Saudi Arabia. And why, they asked, were not other nations, such as Japan and Germany, more dependent on Persian Gulf oil by far than the United States, doing more to help? U.S. demonstrators flashed signs that said, "No blood for oil."

On the other side, the "aid and comfort to the enemy" argument, charging that open debate on the matter showed Saddam Hussein a U.S. lack of resolve, was heard again in the Senate.[48] And Republican Senator John Warner of Virginia, the ranking member on the Senate Armed Services Committee, urged Democratic colleagues to follow what he said was the example of the late Senator Arthur Vandenberg (a Republican who had supported Democratic President Franklin Roosevelt's efforts in World War II and another Democrat, President Harry S. Truman, in regard to postwar Europe), and to "check politics at the water's edge."[49]

Republican senators William S. Cohen of Maine and Nancy Landon Kassebaum of Kansas, on the other hand, defended the prerogatives of Congress and the right of members to speak out, Kassebaum declaring that it was "a mistake to engage in a major military action or war without Congress."[50] President Bush's response to reporters was that he did not want to "end up where you have 435 voices in one house and 100 in the other, saying what not to do . . . kind of a handwringing operation that would send bad signals."[51] A Wasserman political cartoon in the *Boston Globe* had a war-helmeted Bush saying: "We will consult Congress on our Gulf plans . . . but a full, public debate is not appropriate. If every representative is throwing in his two cents worth . . . how can we defend democracy?"[52]

Juxtaposed with the persistent controversy at home was the president's continued success in winning unprecedented international backing for his policies. He and Secretary of State Baker were able to engineer a hugely important victory in the U.N. Security Council. That body, on

November 29, in a 12–2 vote, authorized the use of "all means necessary"—force, in plain language—if by January 15 Iraq had not voluntarily withdrawn from Kuwait. The next day the president proposed meetings between the United States and Iraq in order to go the "last mile" before war.[53]

These meetings were never to materialize because the first date proposed by Iraq, January 12, was, according to the United States, too close to the U.N. deadline. But the initial Bush offer of meetings, thought by some to be aimed more toward maintaining the international coalition behind Bush's policies and quieting opponents at home than toward actually getting Iraq out of Kuwait, had little effect on the president's senatorial critics and those who said that he could not go to war without congressional approval. A stunned Secretary of State Baker, unaccustomed to such rough treatment in Washington, underwent unusually harsh questioning when he appeared before the Senate Foreign Relations Committee on December 5 and said that there was now no assurance that economic sanctions would get Saddam Hussein out of Kuwait and that if the United States had to use force, which he indicated could be done without prior congressional approval, it would strike "suddenly, massively and decisively."[54]

Secretary of Defense Cheney also argued for the administration's shift to an offensive strategy in an appearance before the Senate Armed Services Committee. But when Cheney stated that there was no guarantee that sanctions would work against Iraq, Chairman Nunn responded caustically, "If we have a war, we're never going to know whether they would work, are we?"[55]

As the January 15 U.N. deadline drew nearer, public opinion was divided. A strong majority of Americans, surveys showed, supported going to war "at some point" after the deadline, if Iraq did not leave Kuwait by then. But they were almost equally split on whether to make war immediately thereafter or to give the sanctions more time.[56] A *Washington Post* poll showed that Americans, by a three-to-one margin, thought the president should seek formal permission from Congress before launching a military attack.[57]

What Role for Congress?

The 102nd Congress convened in Washington on January 3, 1991. On the same day, President Bush made a surprise proposal for a meeting between Secretary of State Baker and the Iraqi foreign minister, Tariq Aziz, in Geneva. Congressional leaders announced that Congress would not recess as it usually did to await the president's State of the Union Address, but would instead stay in session because of the urgency of the Persian Gulf situation.

Both houses scheduled debate on alternative resolutions regarding the U.S. use of force after President Bush, on January 8, wrote a letter to Congress specifically asking for authorization for war.[58] The president's request for congressional approval was the first such solicitation since President Johnson proposed the Tonkin Gulf Resolution at the beginning of the Vietnam War.[59]

President Bush had not changed his mind about whether congressional action was legally required, though; in a White House press conference the next day, in fact, the president said again that he did not need approval from Congress: "I feel that I have the authority to fully implement the United Nations resolution."[60] Instead, his decision to ask for a war resolution had probably grown out of assurances from Republican leaders that Congress would now pass it, Senate Republican leader Dole having estimated (too optimistically, as it turned out) that the war-authorization resolution would win sixty votes in the Senate.[61]

When it began to look like the vote might be close, especially in the Senate, the White House redoubled its lobbying efforts. Israel had, all along, urged war with Saddam Hussein, and Bush administration officials now worked closely with the American Israel Public Affairs Committee (AIPAC) to pressure Jewish members of Congress and others with close ties to Israel to support the administration position.[62] Defense contractors also helped with the last-minute lobbying drive.[63]

Iraq agreed on January 9 to Bush's proposal for a last-ditch Aziz–Baker meeting, and the meeting took place the next day. But it was a steely standoff.[64] Gloom quickly spread throughout Washington—and the world—with the first utterance of the word "regrettably," by Secretary Baker as he reported in his press conference that the six-hour meeting had ended without any change in either party's position.[65] "Regrettably," Baker said somberly in Geneva, "I heard nothing today that suggested to me any Iraqi flexibility."[66] The war clouds gathered. The United States planned for massive air strikes.[67] The secretary of defense announced that he would ask for longer terms for reservists.[68]

On the same day as the Baker–Aziz meeting (Thursday, January 10), the Senate and House of Representatives began three days of formal debate on the use of force by the United States in the Persian Gulf. In both houses, voting was set for the following Saturday. Asked by a reporter whether he had the power to order military action in the Gulf, even if Congress voted against the use of force, President Bush said, "I still feel that I have the constitutional authority, many attorneys having so advised me."[69]

Ninety-three senators spoke during the debate. Toward the end, the Senate president pro tempore, Democrat Robert Byrd of West Virginia, told the Senate that in his thirty-nine years in Congress he had cast 12,822 votes, but that in his mind this one would be the most important of all. Then, in support of the Mitchell–Nunn resolution to hold off on authorizing force, Byrd said, "A superpower doesn't have to feel rushed. We can afford to be patient and let sanctions work."[70]

Republican leader Dole, speaking against Mitchell–Nunn, said that what he and others who supported the president's position were "attempting to do in the Congress of the United States is to strengthen his hand for peace, not to give him the license to see how fast we can become engaged in armed conflict," and he added that "we're going to demonstrate today that the President of the United States is the Commander in Chief."[71]

The Senate Votes for War

Finally it was time to vote. No moment in the United States Senate is more electric, more drama charged, than when all the members have flooded into the chamber and taken their seats and, as they wait in tense silence, the clerk begins the call of the roll on an important question. The galleries are packed. Reporters in the front row of their section, above the presiding officer's desk, lean over the railing to hear each vote as it is cast on the floor below. The vice president is in the chair, his presence a certification of the seriousness of the occasion. Senators and staff members, heads down, anxiously keep score on long roll-call strips.

That was precisely the suspenseful scene when, a little before noon on Saturday, January 12, in accordance with an earlier unanimous consent agreement, all debate ended in the Senate and the call of the roll began on the first Persian Gulf measure, Senate Joint Resolution 1. This Mitchell–Nunn resolution proposed to withhold authorization for war and urged, instead, the continuation of economic and other sanctions against Iraq. Only ninety-nine senators would vote. Democrat Alan Cranston of California, a supporter of the resolution, was being treated for cancer in his home state and had been excused from voting.

There would be no surprises. Ten Democrats, mostly from the South, would vote against the resolution. Only one Republican, Charles Grassley of Iowa, would vote for it. All this was known before the roll call began. All senators had spoken by then or had announced their positions. Still, the atmosphere was tense and hushed when, after the last senator had voted, Vice President Dan Quayle declared the official result: "On this vote, there are forty-six yeas, fifty-three nays; Senate Joint Resolution 1 fails of passage."*

Under the terms of an earlier agreement, the Senate then turned, somewhat anticlimactically, to the alternative Senate Joint Resolution 2, the Warner–Dole resolution, to authorize the president to use force, if necessary, to dislodge Iraq from Kuwait. Senator Edward Kennedy led off the debate on this second measure. Speaking in opposition to it, he said that, now, "the last best hope for peace may be being played out elsewhere" (alluding to last-minute U.N. and European negotiating

---

* Forty-three Republicans and ten Democrats voted against the measure; forty-five Democrats and one Republican voted for it.

efforts still under way), and then added with a note of despondence in his voice, "It is sad that it is the United States that is driving the engine of war."[72]

The final outcome of the Warner–Dole resolution was not in doubt, of course. When the roll call on it came, a little over two hours following the first one, fifty-two senators voted in the affirmative, forty-seven in the negative (Republican Senator Hatfield being the only change and the only senator to vote against both measures).[73] The resolution was agreed to.

A similar result (with a little larger margin) in the House of Representatives, later on the same day, meant that President Bush had secured the formal congressional authorization for war that he had at the last decided he wanted, although the results had been close. Forty-seven members of the Senate were against the authorization; a four-vote change would have produced a different Senate result. It was clear that divisions in the Senate on the issue were as deep as those in the country.[74] There was no overwhelming support for beginning the war. Nevertheless, leading off with massive U.S. and coalition bombing runs in Iraq and Kuwait, the war—now called Desert Storm—began on January 16, one day following the U.N. deadline.[75]

It was clear that President Bush's unilateral action in sending such a huge force to the Gulf and in laying down unconditional nonnegotiable terms for Iraq's complete withdrawal from Kuwait had effectively preempted independent action by Congress, inexorably forcing a majority of the members to stand at the last with their president and with their flag.

Still, the formalities of concerted government action, action of the most serious kind that the founders thought would ever confront the nation, had been followed. It had been a close case, but the Constitution had been complied with.

An Invitation to Struggle

Both on matters of war and peace—in the making of both national security and foreign policy—those who wrote the U.S. Constitution intended, as Richard Neustadt put it some years ago, not really a government of strictly "separated powers" so much as a government of "separated institutions with *shared* powers."[76]

"Making peace and war are generally determined by writers on the Laws of Nations to be legislative powers," James Wilson argued at the Constitutional Convention in 1787. He was supported by James Madison in a successful effort to prevent these powers from being vested exclusively in the executive.[77] The decisions on these questions, though, came only after long and contentious debate,[78] and the final result was an overlapping of war and peace powers and a kind of encouragement to adversarial conflict between the president and Congress on their use—what has been called "an invitation to struggle."[79]

The Constitution gave the president the "executive power," the power to carry out the laws, but it gave Congress the "legislative powers," the power to make the laws, as well as the power of the purse. The president was permitted to name ministers, ambassadors, and other officials, but only "by and with the advice and consent of the Senate." The president was made the commander in chief of the armed forces, but Congress was given the power to "declare" war. The president could "by and with the advice and consent of the Senate," again, make treaties with foreign nations, "provided two thirds of the senators present concur."

Despite constitutional restrictions on the war and peace powers of the president, though, the chief executive, as one expert has put it, has always had "the authority and the special knowledge required to dominate the decision-making process in the field of foreign and military affairs."[80] That was certainly true during World War II and the immediate postwar period, a time when the making of U.S. national security and foreign policy was also distinguished by bipartisanship. This was the period, a kind of "golden era," to which Senator Warner was alluding when he spoke of the late Senator Arthur Vandenberg's dictum that, in the United States, politics should stop at the water's edge. As the hot Second World War came to an end, the Cold War between the Soviet Union and United States began, and close cooperation, accommodation, between the Congress and the president remained the norm for a time.[81]

With the commencement of the 1950s, some increased interbranch conflict in foreign affairs developed. These were the red-baiting years of Senator Joseph McCarthy (R., Wis.), who charged that the State Department, particularly, was full of Communists and their sympathizers; it was also the time of the Korean War (1950–1953), during which President Truman quite controversially dismissed the U.S. commander there, war hero General Douglas MacArthur.[82]

The Senate's "Mr. Republican," Senator Robert Taft of Ohio, took to the Senate floor in 1951 for a major speech in which he severely reproached Truman for sending American forces into combat in Korea and for stationing U.S. troops in Europe, both without prior approval of Congress. Taft expressly rejected the notion that such criticism "is an attack on the unity of the nation, that it gives aid and comfort to the enemy, and that it sabotages any idea of a bipartisan foreign policy for the national benefit."[83]

The rest of the 1950s, despite the fact that the president, Dwight Eisenhower, was a Republican while the Congress, except for a couple of years (1953–1954), was controlled by the Democrats, was a time of bipartisanship, cooperation, and acquiescence in national security and foreign policy by Congress.[84] This more or less continued into the 1960s, until the Vietnam War split the country apart.

In most of the 1950s, then, beginning with President Truman's institution of the policy of "containment" of communism and especially after President Eisenhower's ending of the Korean War, there was a strong policy-elite and president–Congress consensus on U.S. national security and foreign policy.[85]

Senators (and House members) found it easy to go along with the president; they could and did regularly vote against the opinions of their constituents in matters involving international affairs.[86] There was little open discussion and debate in the making of foreign policy, little presidential-congressional conflict, because there was general agreement on policy substance.

This consensus, as one authority put it, was sustained by an old foreign policy establishment of "relatively homogeneous, pragmatic, and mostly bipartisan East Coast diplomatic and financial figures," as well as by government leaders who "shared an internationalist and interventionist view of the U.S. role in world affairs" and "a poorly informed and largely inert mass public that tolerated official policy as long as it appeared to be working."[87]

A Changing Country, Senate, and World

American society changed greatly after the 1950s. As a result, the public became much more aware of, and attentive to, government and government policy, including foreign policy, and how these affected them.

The Senate changed. Senators became more national in their outlook and more individualistic in their behavior. They (and House members, too) found it more difficult to "go along in order to get along" in matters involving national security and foreign policy, just as in regard to domestic affairs. Partisanship in the Senate and House grew. Divided government—one party in control of the White House, another in the majority in one or both houses of Congress—became the norm.

All of these changes heightened the likelihood of presidential–Senate conflict concerning international affairs. The Vietnam War caused even more change, hastening the general breakdown of American consensus on the substance of national security and foreign policy. As Barbara Sinclair noted, opposition to the war caused people to reappraise all aspects of American defense and foreign policy, to question and heatedly debate previously accepted assumptions:

> The size of the military budget, the need for a variety of expensive and deadly new weapons systems, aid to repressive regimes, and the United States role in the Third World were brought into question. These issues did not fade with the end of the Vietnam War; no new consensus emerged. . . .

Thus, by the early 1970s, the agenda of issues concerning international involvement had become not only highly conflictual but also much broader than it had been in the 1950s and the 1960s.[88]

In the late 1980s the Cold War ended. The Soviet empire imploded. Under Mikhail Gorbachev, Soviet control over Eastern Europe collapsed faster than it could be dismantled. Internal Soviet politics and the Soviet economy began to transform themselves more rapidly than they could be opened up and restructured by conscious new efforts toward *glasnost* and *perestroika*.[89] A failed coup attempt in August 1991 hastened the fragmentation of the former U.S.S.R. into, at best, a confederation, and completed the change of its image in Congress from one of a military menace to that of a needy supplicant.[90]

Back in 1954 a Gallup poll had shown that only 23 percent of Americans thought that Western countries could continue to live more or less peacefully with the Soviets; 64 percent thought there was bound to be a war with them sooner or later. By contrast, in 1990, even before the failed Soviet coup, the percentages were nearly reversed: 63 percent now thought that we could live in peace with the Soviet Union, and 31 percent thought that there would sooner or later be a war between us.[91] Similarly, five and a half times as many Americans in a 1990 survey thought that drug traffickers and terrorism were the greatest threat to U.S. security as thought that Soviet military power was (83 to 15 percent and 82 to 15 percent, respectively), and three and a half times as many (75 to 21 percent) thought Japanese economic power was a greater threat than Soviet military might.[92] No longer could Americans—or the president and Congress—find common cause in a common enemy. American foreign policy was deprived of its most unifying, post–World War II theme.

The world, and the place of the United States in it, changed, and so did the principal issues. Charles William Maynes, editor of *Foreign Policy* magazine, has pointed out that America's post–World War II hegemony was based on the "three pillars" of nuclear superiority, control of the world's energy industry, and the preeminent role of the dollar.[93] Those pillars crumbled. Trade and economic issues, environmental problems, Third World debt, nuclear proliferation, and the spread of other modern and highly dangerous arms—these are the kinds of issues, Maynes has written, that will most likely dominate America's national security and foreign policy agenda for the 1990s and beyond.[94]

In foreign affairs particularly, then, the founders made the relationship between the president and the Congress an invitation to struggle. Still, as long as the world was simpler and bipolar and the American public was uncontentious and unassertive, a more locally and institutionally focused and less-partisan Senate could get along fairly well with the president in the making of U.S. national security and foreign policy. With all of the changes that occurred after the 1950s, the invitation to struggle was increasingly accepted by both sides. Nowhere was this more evident than in regard to the power to make war.

War Powers

As the 1991 Persian Gulf debate demonstrated again, something over two hundred years after the Constitution was written, the question of how and when the president can and should commit U.S. forces is still a matter of passionate disagreement in America.

In making the president the commander in chief while leaving it within the power of Congress to declare war, control the purse, and "raise and support armies" and "provide and maintain a navy," the founders felt that they were establishing four very important principles: civilian control of the military; the military as a creature of Congress; a limited army, except in wartime; and the power of Congress over the commencement of war.[95]

James Madison and most of the other founders (Alexander Hamilton was an exception) were against the United States maintaining a standing army.[96] The president could commit whatever armed force there was to repel sudden invasion or attack; he had the power to do that, it was clear, without advance approval of Congress. But, lacking a standing army, the president plainly would be unable to make war—until Congress declared war, created the army, and supplied the funds.

In 1793, James Madison authored an explanation of the rationale for separating the power to declare war from the power to command:

> Those who are to *conduct a war* cannot in the nature of things, be proper or safe judges, whether *a war ought* to be *commenced, continued,* or *concluded.* They are barred from the latter functions by a great principle in free government, analogous to that which separates the sword from the purse, or the power of executing from the power of enacting laws.[97]

All the early presidents recognized that it was the sole prerogative of Congress to decide when the nation would go to war.[98] This was true for the War of 1812. It was also true for the U.S.–Mexican War (1846–1848). Though President James K. Polk deliberately provoked the incident that brought on the war with Mexico, he nevertheless asked for and received from Congress formal approval for the war before proceeding with it. Abraham Lincoln, who as a one-term U.S. representative from Illinois held the view that President Polk's provocations themselves had been unconstitutional, wrote concerning the war powers:

> The provision of the Constitution giving the war-making power to Congress was dictated, as I understand it, by the following reasons: Kings have always been involving and impoverishing their people in wars, pretending generally, if not always, that the good of the people was the object. This our convention understood to be the most oppressive of all kingly oppressions,

and they resolved to frame the Constitution that no one man should hold the power of bringing oppression upon us.[99]

The Spanish-American War (1898), World War I (1917–1918), and World War II (1941–1945) were all declared by Congress.[100] But after World War II, the balance of war-making power shifted to the president. One principal reason was the establishment of a large standing army, a new development for the United States.[101] Congress approved the standing army, approved the stationing of forces abroad, approved the development of highly sophisticated arms of mass destruction, approved foreign alliances and agreements for mutual security, and encouraged the role of the United States as a superpower counterweight to the Soviet Union.[102]

Congress endorsed the concept of wielding this great new might and influential status for the achievement of U.S. foreign policy goals. And to manage such maneuvering, Congress helped institutionalize the "national security presidency," a chief executive not only in command of the armed services and aided by the State Department, but also backed by a great new national security bureaucracy, including particularly the Central Intelligence Agency and the National Security Council.[103] Both Congress and the president began to view war, or at least the ability and threat to wage it, as an instrument of diplomacy.

Little wonder, then, that presidents began to act on their own initiative. We have already noted how President Truman engaged the United States in the Korean War and stationed forces in Europe without prior approval of Congress. Despite stringent objections by Senator Robert Taft and other Republicans, the Senate passed a resolution that approved Truman's European troop deployment. Congress acquiesced in, and financed, other Truman actions, too, including the war in Korea.

President Dwight Eisenhower was more prone to seek the approval of Congress before committing U.S. forces abroad. But while maintaining that congressional-presidential cooperation was preferable, he asserted that the president had the authority as commander in chief to deploy U.S. forces and "take whatever emergency action [that] might be forced upon us in order to protect the rights and security of the United States."[104]

In 1961, President John F. Kennedy ordered the U.S. Navy to block Soviet ships from carrying missiles to Cuba, thus initiating a very tense and dangerous confrontation without formal consultation with Congress. Kennedy stated that as commander in chief, "I have full authority to take such [military] action."[105]

President Kennedy also sent military advisers to Vietnam without prior congressional approval.[106] After Kennedy's assassination, President Lyndon Johnson, in 1964, making highly doubtful assertions concerning a purported North Vietnamese attack on a U.S. Navy ship, asked

for and received from Congress an open-ended, ambiguous endorsement for military action in Vietnam.

The day after it was introduced, this Tonkin Gulf Resolution, as it
was called, passed the Senate by a vote of 82 to 2,[107] and the House of
Representatives approved it without a dissenting vote. But the war went
badly, and the longer it lasted the more public support it lost at home.
President Johnson was forced to relinquish any idea of seeking reelection in 1968.

President Richard Nixon continued the Vietnam War even after
Congress, in 1971, repealed the Tonkin Gulf Resolution. Finally,
backed by public opinion, Congress decided to reassert its war authority,
with the passage of the War Powers Resolution of 1973.

The War Powers Resolution

A principal author of the War Powers Resolution, the late Jacob Javits,
then a Republican senator from New York, wrote at the time that the
"undeclared war in Vietnam," which had cost "fifty thousand American
lives, and countless numbers of Vietnamese lives, bled away in Indo-
china," had finally forced Congress to "limit the president's power to
impose his military writ as he so chooses."[108]

The War Powers Resolution became the law in 1973. As noted earlier, it required presidents to consult with Congress before introducing
U.S. forces into hostile or imminently hostile situations, report to Congress within forty-eight hours of any such introduction, and withdraw
such forces after sixty days unless Congress authorized their continued
presence. But the Resolution has had very little effect. Presidents have
sometimes given notice under the Resolution, but there has been little
prior consultation with Congress and no invocation of the sixty-day
limit under the measure. Presidents have interpreted "hostilities" quite
narrowly so as not to trip the Resolution. They have sometimes gotten
U.S. forces in and out of a hostile situation so quickly as to make the
sixty-day limit meaningless. And, despite the fact that no court has ever
so held (and neither the chief executive nor Congress has ever filed a test
case on it), presidents have steadfastly maintained that the War Powers
Resolution is unconstitutional[109]—either because it impermissibly
interferes with the president's authority as commander in chief and chief
executive, thereby violating the principle of separation of powers, or
because, they argue, it provides for the kind of one-house legislative veto
that the Supreme Court struck down in the *Chadha* case,[110] or both.
That is why, in the 1990 Persian Gulf situation, Secretary of State Baker
told a congressional committee:

> What the executive branch has done on the basis of my experience over the
> last 10 years is to notify the Congress when the President dispatches troops,

tell the Congress whether the executive branch thinks hostilities are or are not imminent, but, even if hostilities are imminent, take the position that the 60-day time clock that I suppose you could say starts ticking . . . is unconstitutional.[111]

President Gerald Ford did not consult with Congress before sending forces in 1975 to rescue a U.S. ship, the *Mayaguez*, that Cambodians had captured, although after the brief incident in which forty U.S. Marines were killed was over, Ford did formally notify Congress. President Jimmy Carter sent military advisers to El Salvador (later increased in number by President Reagan) and armed forces to Iran in an unsuccessful attempt to rescue American hostages there, both without invoking the War Powers Resolution (and, in the Iran-hostage case, at least, without any prior consultation, either).

Congress itself has the power to invoke the sixty-day limit of the War Powers Resolution, but it has never done so. In 1982, a member of Congress sued President Ronald Reagan for sending military advisers to El Salvador, where they went into local combat zones wearing sidearms, without congressional approval. The court (in what was to be the standard federal court response to such suits, including one filed as a result of the 1990 dispatch of troops to the Persian Gulf by President Bush) held that since Congress had not acted to stop such presidential action, the courts would not do so either.[112]

In 1983 President Reagan sent two thousand U.S. Marines to Lebanon as part of a U.N. peacekeeping force. The president said that he could do this under his "constitutional authority with respect to the conduct of foreign relations and as Commander in Chief."[113] He claimed that the War Powers Resolution did not apply to this action, even though the marines were soon under fire and were calling down air strikes on their attackers. After Senator Robert Byrd introduced a measure to trigger the War Powers Resolution, the president and Congress worked out a compromise under which the president conceded that the War Powers measure was applicable and Congress extended to eighteen months the time during which the marines could remain in Lebanon. (Afterward, some two hundred of them were killed in a truck-bomb attack there.)

President Reagan did not invoke the War Powers Resolution when he ordered troops into war exercises with Honduran forces to intimidate the Sandinista government of Nicaragua, when he launched a series of bombings of Libya, or when he ordered an invasion of Grenada to rescue American students and derail a Marxist government there. In regard to the Grenada invasion, then Senator Charles McC. Mathias, Jr., Republican of Maryland, complained that "congressional leaders were simply called to the Oval Office and told that the troops were under way. That is not consultation. The Prime Minister of Great Britain was advised about the invasion before the President told the Speaker of the House of Representatives or the Majority Leader of the Senate."[114]

In 1987, during the Iran–Iraq War, to help Kuwait and to support Iraq, which depended on Kuwaiti oil, President Reagan allowed Kuwaiti oil tankers to be "reflagged" as U.S. ships and then dispatched American naval forces to the Persian Gulf to escort and protect them. When it was eventually too clear to be denied that the situation was one of "hostilities," President Reagan filed reports with Congress, but did not formally invoke the sixty-day limit under the War Powers Resolution. A lawsuit was instituted by a member of Congress to force the president to comply with the War Powers measure, but again, the federal court said that since Congress had not itself acted, the court would not do so.[115] A number of senators, including Republicans Lowell Weicker of Connecticut and Mark Hatfield of Oregon and Democrats Dale Bumpers of Arkansas and Brock Adams of Washington, urged the Senate to invoke the War Powers Resolution, but they were unsuccessful in this, primarily because a majority of the Senate agreed with the substance of Reagan policy.[116]

Presidential dominance in war-making and in the ability to commit U.S. armed forces abroad was enhanced by the conversion of the United States to an all-volunteer standing army in 1973. That change facilitated President George Bush's order for the invasion of Panama in late 1989,[117] for example. Similarly, his unprecedentedly rapid and massive deployment of forces to the Persian Gulf in 1990 and the war against Iraq that almost inevitably followed were politically, as well as physically, possible because President Bush had at his command a large all-volunteer army, national guard, and military reserve—and did not have to resort to the draft.

A provision added to the fiscal 1991 defense authorization bill that directed the military to restructure itself, so as to be able to rush forces to trouble spots in time of crisis while still being able to provide for national security generally, also moved the nation in the direction of increased presidential power.[118]

Something needs to be done. The intentions of the founders, based on their justified fears about what an untrammeled chief executive might do with the power to make war, show that. The president should not have such a free hand.

Foreign Policy magazine editor Charles William Maynes once wrote that the two most serious things a government can do to its citizens are to tax them and to send them into wars. It is an insupportable idea in a democratic system, he said, that it is only in regard to the first of these that there should be consultation with and debate and decision by Congress, the people's branch.[119] Congress should stand up to the president and struggle to win back its constitutional place in regard to war, the late Senator Javits declared, "because the interests of the people who suffer and die and pay for war are safeguarded best by joining the Congress in the war power."[120]

Congressional debate may be messy, and may sometimes become partisan, but, as I. M. Destler and his colleagues have correctly asserted, "Democratic debate is more likely than doctrine to produce sensible foreign policy decisions."[121] Regarding partisan politics, two political scientists have written:

> Bipartisanship should not be sought at the expense of policy debate, nor used as a tool to stifle the expression of divergent viewpoints that is the heart of democratic governance in foreign as well as domestic affairs. Too often the appeal for bipartisanship is designed to erode the separation of powers that assigns important foreign policy functions to both legislative and executive branches.[122]

To protect the people's interest and to comply with the Constitution, the Senate (and House, of course) must again be allowed to share fully in the exercise of the war powers.

Congressional Partnership in War Powers

Some have said that Congress has all the authority it needs in regard to war-making, as a result of its power of the purse. If it develops that Congress does not like a particular war, why can it not simply cut off the money? One major trouble with that course of action is that it is procedurally and practically quite difficult to take such action over the objections of a determined president.

The fund cut-off approach was tried by Congress during the Vietnam War, after a majority of the members of that body, like a majority of Americans, had turned against the war.[123] Twice, in 1971 and 1972, the Senate was able to adopt amendments for that purpose. But both times the amendments died in the House. Twice in 1973 Senate and House Democratic caucuses separately resolved to try to cut off war funds, but the Nixon administration was able to stymie the Democrats with public claims that such attempts would undermine ceasefire negotiations then going on in Paris.

After a Vietnam ceasefire *was* agreed to in 1973, President Nixon began the bombing of Cambodia without any authorization by Congress, at one point carpet-bombing the countryside there for fifty straight days to try to force the Lon Nol government of Cambodia into agreeing to a ceasefire with opposition forces. In response to congressional fund cut-off attempts, Secretary of Defense Elliot Richardson asserted that the administration would keep the bombing going, even without congressional authorization, and secure the necessary funds to do so by reducing U.S. troop levels in Europe, if necessary.[124]

Congress eventually attached a rider forbidding the use of funds for the bombing in Cambodia to an appropriations bill. President Nixon vetoed the bill. More riders were attached to more bills. At last, there was a Senate-House-presidential compromise to stop the funds for the bombing as of August 15, 1973—but this was some four months after the massive air offensive had commenced.

The after-the-fact, purse-strings power of Congress is not a very realistic way to stop a war or a troop commitment for another reason: the patriotic fervor that Congress must usually confront in such circumstances. Are you for the appeasement of Saddam Hussein or for President Bush's standing up to aggression? That was the type of question posed to members of Congress once the president had committed U.S. forces to the Persian Gulf in 1990 and drawn the "line in the sand." Are you for the peace demonstrators or for our brave men and women in uniform? Once our flag is committed, are you not required to support it? Do you not have to give the troops all the guns and ammunition they need for victory? Those are the kinds of questions that senators and House members had to answer during America's (and the United Nations') 1991 war with Iraq.

Americans rally around the president, at least at first, in times of international crisis, even if it is the president who precipitated the crisis. After the shooting started in the Iraqi war, for example, a January 1991 *New York Times*/CBS News poll showed that President Bush's job approval rating had soared to 86 percent, 20 points higher than it was just before the beginning of hostilities. This was the highest rating for any president in the preceding thirty years (since, incidentally, President John F. Kennedy's attempt to liberate Cuba with the Bay of Pigs invasion).[125] Seventy-five percent of those polled said that they now agreed with military action against Iraq, compared with only 47 percent just before the action started.[126]

It is extremely difficult for senators and House members to stand up against that kind of public feeling. Studies show that in the first month after the use of force by a president, Congress consistently supports the president on important international votes.[127]

Consultation with Congress should be required well in advance of military action by the president, not only because this would be the right and constitutional thing to do, but also because two (or a good many more) heads are, indeed, better than one. Consultation can help prevent mistaken decision and action.

So the first need is to make the prior-consultation provisions of the War Powers Resolution effective. Former President Gerald Ford, in asserting his view that the War Powers Resolution was both unconstitutional and impractical, once wrote that "if we, as a Nation, are to be respected by our adversaries, and if we can look forward to . . . cooperation from our allies, our President as Commander in Chief has to have the authority to act. We cannot take the time to have 535 secretaries of state or secretaries of defense."[128]

The first response to such an argument is: it sounds the same as saying that, if we are to have sound education and health policies, we cannot have 535 secretaries of education or secretaries of health; the Ford argument fuzzes over the constitutional division between policymaking and policy implementation. The second response is: prior war-powers consultation does not have to be, should not be required to be, with 535 people.

The War Powers Resolution should be amended to provide for a joint, bipartisan "National Security and Foreign Policy Council," as a permanent consultative group. This council should be made up of the party leaders in both houses, as well as of the chairs and ranking members of the House and Senate committees on Armed Services, Foreign Relations (called Foreign Affairs in the House), and Intelligence.

This very serious proposal has been put forward in the Senate by Senate Majority Leader Mitchell, Senate President Pro Tempore Byrd, chair of the Armed Services Committee Nunn, and chair of the Intelligence Committee Boren, all Democrats; as well as by Republicans John Warner and William Cohen, the ranking members of the Senate Armed Services and Intelligence committees, respectively.[129] The idea of establishing such a permanent consultative group in Congress has the backing of former Secretary of State Cyrus Vance,[130] among others, as well as that of former Assistant Secretary of State J. Brian Atwood, who helped draft the original War Powers Resolution.[131]

In setting up this new consultative mechanism, Congress should also amend the War Powers Resolution, as Robert Katzman, president of the Governance Institute, has proposed, to make clear that the definition of "consult" is more than "inform" and means "engage in discussion with" and "seek the advice of."[132] The president should have to consult with Congress before committing troops or starting a war.

Another matter for change is the sixty-day, automatic withdrawal of forces required if Congress fails to approve presidential action. This is the feature of the War Powers Resolution to which presidents have most objected and concerning which the most serious constitutional question has been raised. Senator Byrd and his co-authors would replace this part of the War Powers measure with a provision stating that upon the introduction of a joint resolution by the chair of the permanent consultative group, a measure that would receive expedited attention, Congress could approve or disapprove of troop commitment and military action.

This amendment to the War Powers Resolution should also be adopted. Congress would still have to summon the will to accept its part of the war-making responsibility. And it should do so. The president could still veto a joint resolution of disapproval, but would most likely be restrained from doing that because of the fact that such action would have serious consequences for the president—in the Congress and in the country.

Armaments and Arms Control

There are three main types of national security policy—crisis policy, structural policy, and strategic policy—and the congressional role in policymaking is different for each one.[133] President Bush's 1990 commitment of troops to the Persian Gulf and the subsequent 1991 decision for the use of force there involved crisis policy, of course. We have seen that there is a good deal of constitutional tension between the president and Congress in this field, and that that tension has mostly been resolved, since World War II, in favor of the president.

The role of Congress is much stronger in regard to structural and strategic national security policy. Structural policy has to do with government procurement, weapons contracts, and military bases, often approached on a kind of "pork barrel" or "bring home the bacon" basis, while strategic policy deals with overall defense spending and force levels, the appropriate mix and mission of military forces, types of weapons, and arms control.[134]

President Dwight Eisenhower originated the now-famous phrase "military-industrial complex" when he sounded a warning in his farewell address about the combined power of the government's defense establishment and American weapons manufacturers. Today a basic problem with the involvement of Congress in making structural national security policy, and sometimes in making strategic-policy choices among weapons systems, too, is the development since World War II of what a retired admiral once called a "congressional-industrial complex."[135]

There are some American states that would probably sink below sea level if any more military bases or weapons contractors were located in them. These are the states that have traditionally been strongly represented on the Senate and House armed services committees and defense appropriations subcommittees.

Structural national security policy has often been dominated by a local, back-home congressional outlook. Elimination of unneeded weapons systems has been made more difficult by this view of structural policy. Politically savvy weapons-system planners and defense contractors know how to maximize political support by farming out subcontracts over a wide number of states. One congressionally popular fighter plane, for example, was said to have components manufactured in forty-nine states.[136] Similarly, the Navy's decision some years ago to significantly enlarge the number of states in which it would locate "home ports" surely had more to do with increasing the Navy's congressional influence than with improving its preparedness and efficiency.

In structural policymaking, the representation function of a senator (or House member)—that of serving as an advocate for constituents—often comes into conflict with the law-making function, that of making decisions for the whole nation. In 1988 Congress used the device of a congressionally created, executive-appointed, independent commission to resolve the representation–law-making conflict in regard to military base closings, in effect nationalizing that issue for a time. This Commission on Base Realignment and Closure was authorized to lump all its base closings together in an order that would automatically go into effect unless rejected in toto by Congress, in a measure that could be vetoed. In that way, ninety-one bases were ordered to be closed and fifty-four others were directed to be "realigned," meaning that missions were switched from some to others.[137]

The House and Senate resorted to such an independent commission to close additional military bases again in 1991.[138] Such a device should be used again to make additional cuts and perhaps other related and politically difficult structural-policy decisions.

Another important step could be taken, too. Campaign financing plays too much of a part in the making of structural national security policy, and this should be stopped. The Senate Armed Services Committee and the corresponding House committee tend to attract members who are, or eventually become, supporters of military spending and installations and contracts back home.[139] The members are reinforced in this by the lobbying of defense contractors (who, incidentally, became almost frantically active in 1990 before the outbreak of hostilities in the Persian Gulf, when it looked like substantial cuts in military spending, and in whole weapons systems, were inevitable).[140]

A 1987 report showed that $230 billion, over 80 percent of military contracts that year, went to manufacturers and suppliers located in just a third of the states—the home states of the members of the Senate Armed Services Committee and the Senate Subcommittee on Defense Appropriations.[141] A writer for the citizen lobby Common Cause has called the money in the form of campaign contributions that is paid to these and other senators (and House members) by defense companies and their PACs "the glue that cements what's become known as a classic example of the 'Iron Triangle,' uniting Congress, the Pentagon and the defense industry." This money, the writer has asserted, "makes it possible for defense contractors to get the business they want, the Pentagon to get the weapons it wants, and members of Congress to get the jobs and federal money they want back home."[142]

To cut down the power of this congressional-industrial complex, Congress should adopt the fundamental reform of the campaign-finance system discussed in Chapter 3. This would reduce the pork-barrel appeal of military projects and programs and facilitate cuts in defense spending generally.

There is nothing terribly wrong with pork-barrel politics, so long as the pork is going to the right places and for the right purposes. The old-time pork used to include things like highways, water and sewer systems, airports, and school programs. This changed with the rise of the congressional-industrial complex and the increase in military spending.

Economist Robert Reich has pointed out that U.S. investment, at all levels of government, in infrastructure fell from 4 percent of gross national product in the 1960s to only 2 percent in the 1980s.[143] Reich noted that there has been a similar decline in U.S. investment in public elementary and secondary education and in research and development, other than for the military. He has added further that "As Western Europe and Japan lay plans for 'smart' roads, high-speed trains, and national information networks, America lies dormant; the nation has not even built a new airport since 1974."[144] The truth is that the power of the congressional-industrial complex has made the wrong kind of pork too succulent.

Strategic and Arms Control Policy

The Congress—and especially the Senate, because of its treaty-ratification power—is far more involved in the making of strategic policy than was the case twenty years ago. As one authority, Barry Blechman, has put it:

> The congressional role has gone from that of a relatively minor actor, frequently outspoken but only sporadically consulted, rarely involved in actual decisionmaking and never in policy execution, to that of a player with star billing in the making of U.S. national security policy and sometimes the lead role in U.S. government decisions.[145]

The increased role of Congress in strategic policy,[146] already noted, is a result of the nationalization of American society and Senate campaigns, which, together with the advocacy explosion, brought a simultaneous and interrelated nationalization and individualization of the U.S. Senate (to a lesser degree, also of the House). All of this produced, as the Democratic chair of the Senate Foreign Relations Committee, Claiborne Pell of Rhode Island, has expressed it, "an increasing interest of the American public in the details of foreign policy."[147] Similarly, the ranking Republican member of the Senate Finance Committee, Robert Packwood of Oregon, could just as well have been talking about national security matters when he said, "Senators are now much more subject to immediate public passion. We can finish [Finance Committee] mark-up with tentative, private agreement, and the next morning the telegrams will come in."[148]

National security issues, the same as domestic issues, can no longer be decided by a small group of leaders in their closed committees. Now nearly everything is out in the open, and the public is interested. Senators especially have become much more involved in the making of strategic national security policy because they have been forced to do so by a more attentive public and they have been permitted to do so by a more open and democratic Senate.

There are other reasons for the increased involvement of Congress in the making of strategic policy. Public concern about the Soviet security threat to the United States, as noted earlier, began to recede with the start of the 1980s (and dried up with the collapse of the Soviet Union). With the beginning of the 1980s, too, public opinion turned against the arms race and against the continued development of nuclear weapons.

In March 1982, Democratic Senator Edward Kennedy joined Republican Mark Hatfield in the introduction of a nuclear-freeze amendment in the Senate; in June of that year, nearly three-quarters of a million people marched in a pro-freeze demonstration in New York City. As the world and American public opinion changed, the Senate also changed. Presidents often did not.

In the Senate, the dominant role in regard to arms control shifted from the Foreign Relations Committee to the Armed Services Committee and, to a lesser extent, to the Intelligence Committee. This shift came about because of the relatively stronger leadership of senators like Sam Nunn and William Cohen, and their increased assertiveness.

In 1970 President Richard Nixon simply refused to inform Congress of the U.S. negotiating position in strategic arms limitation talks then under way. By 1983, in contrast, Congress had not only forced President Reagan to move toward arms control, which he resisted, but was actually laying down the U.S. negotiating position as well as placing an individual on the negotiating team to make certain that the congressional position would be respected.

Whereas President Nixon had rejected the proposal of then Senator John Sherman Cooper (R., Ky.) that the U.S. delegation in the first strategic arms limitation talks (SALT I) include key senators as observers, by 1985 President Reagan had been compelled to allow a Senate arms control observer group to sit in on similar negotiations and even to meet separately with Soviet negotiators and express their own viewpoints.

An assertive Senate joined the House in 1986 in adopting an amendment to a defense appropriations bill to require the Reagan administration to comply with the terms of the SALT II agreement that President Carter had negotiated with the Soviets, but that, after the Soviet invasion of Afghanistan, had never been ratified.

The Senate and House were poised in 1990 to make serious changes and cuts in the military budget (before the war in the Persian Gulf). But well after the Soviet retreat from Eastern Europe and the disappearance of the Soviet military threat, President Bush continued to support an outdated strategic policy, and the budget proposals to carry it out. In 1992 he proposed military budget cuts only half those of the most modest Democratic proposals.[149]

Senate Armed Services chair Sam Nunn had begun in 1990 to make a series of Senate-floor speeches aimed at crystallizing U.S. strategic policy in the light of changed circumstances. "The sudden collapse of the communist empire over the past year has created both a new requirement and a new opportunity for a defense strategy that responds to the changed threat and realistically relates our means to our ends," Senator Nunn said. Cuts in the defense budget would come, he said, but where they were made was very important. "The question is not whether we reduce military spending. . . . The question is whether we reduce military spending pursuant to a sensible strategy that meets the threats of today and tommorow."[150] Independently of each other, the White House and Congress, each with differing perspectives and objectives, clashed repeatedly over military strategy and cuts during consideration of the 1993 budget.[151]

On strategic policy, on arms control, and in regard to covert operations, the assertiveness of Congress is not going to disappear. There is a danger, though, that Congress will involve itself so much in the details of policy—in micromanagement—that it will not give sufficient attention to overall policymaking and will interfere with the efficient implementation of policy.[152] Better presidential-congressional consultation in the first instance would help guard against this danger, while allowing Congress to fulfill its responsibilities in strategic policymaking. Thus, the National Security and Foreign Policy Council, recommended earlier, could serve as a permanent consultative group on strategic and arms control policy as well as on war powers.

The Senate and Foreign Policy

The Senate and House have become more and more assertive in the foreign policy field. Responding to societal and internal changes, the Senate expanded its own foreign policy staff and bureaucracy, developed the tactic of using its control over the authorization and appropriation of funds to affect foreign policy, and demanded and received access to intelligence and other information.[153] So did the House.

This increased congressional assertiveness in foreign policy had been building since World War II, but the Vietnam War accelerated the trend.[154] Increased assertiveness has been accompanied by rising congressional partisanship in foreign policymaking during the last four decades, and the growing ideological disputes of recent years have reinforced, rather than bridged, partisan differences in foreign policy matters.[155]

The Reagan years of the 1980s were especially a time of distrust between the president and the Congress in the field of foreign policy. There was a high degree of presidential-congressional confrontation and institutional competititon.[156] A principal reason for this was the substantive policy disagreements between President Reagan and the Democrats in Congress on so many foreign policy issues, often with the president adamantly adhering to a position that was contrary to American public opinion.

The country was opposed to President Reagan's support for the counterrevolutionary *Contra* forces, or "freedom fighters," as he called them, fighting against the Sandinista government of Nicaragua.[157] A majority of the members of Congress, though they wavered some, were also opposed to U.S. aid for the *Contras*, and from time to time they restricted or prohibited it. The president's obdurate determination to help the *Contras* regardless of congressional restrictions resulted in the felonious and other illegal efforts that eventually festered and grew into the Iran-*Contra* scandal.[158]

President Reagan was also on the wrong side of public opinion when he opposed congressional attempts to enact antiapartheid sanctions against South Africa. There were serious presidential-congressional disagreements over arms sales to Saudi Arabia and Jordan, too, as well as over military aid to El Salvador; over President Reagan's nearly last-ditch backing of the Marcos regime in the Philippines, and, until Reagan eventually changed his position on the eve of the 1984 elections, over normalizing relations with the Soviet Union.

Congress—or at least most of the Democrats there—did not trust President Reagan to do the right thing, or even always to follow the law, in regard to many foreign policy issues. President Reagan, on the other hand, saw much of congressional assertiveness in foreign policy as nit-picking micromanagement that interfered with his constitutional powers and authority. The president tried to act without congressional oversight or control, which made Congress even more anxious to assert its authority.

All this was occurring at a time when the Senate Foreign Relations Committee was already losing a good deal of its traditional influence in the Senate and over foreign policy. Power, in general, became more diffused in the Senate, as we have seen; when it came to foreign policy, the Foreign Relations Committee increasingly had to share power with the Armed Services Committee, the newly established Intelligence Committee, the Appropriations Committee, Senate party leaders, and with the House appropriators, as well as with individual senators, who were increasingly willing and able to bring up amendments of all kinds on the Senate floor.[159] The decline in the Foreign Relations Committee's sway was also a result of leadership, or the lack of it. The chair of the Committee after 1986 was Democrat Claiborne Pell of Rhode Island, described as "genteel, deferential and often absent-minded." He was not a forceful leader, and worse, the ranking Republican on the committee was "wily, ideologically driven" Jesse Helms of North Carolina, a doctrinaire right-wing obstructionist.[160]

The result of all these Senate developments was that "The Foreign Relations Committee doesn't have much clout anymore," according to Republican Senator Mitch McConnell of Kentucky, a member of the committee. "Nongermane amendments get adopted on the floor; the Senate runs amok in this field. It is a post-Vietnam syndrome that everybody tries to be a Secretary of State—both within and without the Committee."[161]

A good example of this was the feeding frenzy of amending activity that developed when the Senate took up the 1987 State Department authorization aid bill. On that occasion, some eighty-six amendments—mostly of a nonbinding, "sense of the Senate" nature—were attached to the bill on the floor, including measures that called for: voiding of U.S.–Soviet agreements on new embassies; closing of the U.S. offices of the Palestine Liberation Organization; imposition of strict travel restrictions on Cuban and Eastern European diplomats in this country; condemnation of the Chinese government's actions in Tibet; a demand for an apology from the Soviet Union for ballistic-missile tests near Hawaii; an objection to the Soviet Union participating in a Middle East peace conference until it recognized Israel; creation of a new position of undersecretary for security and a new ambassador-at-large for Afghanistan; and ordering daily Voice of America broadcasts in Slovenian.[162]

The great battle over amending the 1987 State Department authorization bill was not an isolated event in the Senate. As a matter of fact, that fight occurred immediately after the Senate had spent five months on the defense authorization bill, debating, and finally adopting, Democratic amendments to prohibit the Reagan administration from deviating from the terms of the 1979 SALT II arms treaty or from the Senate's more restrictive interpretation of the unratified 1972 Antiballistic Missile Treaty.

When President George Bush was inaugurated in January 1989, he announced his desire to work more closely, and in a bipartisan way, with Congress on foreign policy matters. "A new breeze is blowing—and the old bipartisanship must be made new again," he said.[163] Bush's secretary of state, James Baker, was able to work out a bipartisan agreement with Congress on Nicaragua, ending U.S. assistance to the *Contras*. But a good deal of presidential-congressional conflict on other foreign policy issues, much of it along party lines, continued. Congress and the president disagreed over military aid to El Salvador, for example. Congress also began to question covert aid to particular conflicts, like those in Afghanistan, Angola, and Cambodia, as well as the whole idea of such secret activities, generally.

When the 1989 State Department authorization bill came up in the Senate, senators again took the opportunity, as a *Washington Post* article put it, "to play secretary of state for a day or two." That report continued:

> In the process, the Senate called on Cuban President Fidel Castro to hold a plebiscite on whether to relinquish power, declared its support for Soviet Armenia, condemned Bulgaria's treatment of its Turkish minority, urged more antidrug efforts by Mexico and offered scores of other proposals on issues ranging from Japan's fishing practices to food shortages in Romania.[164]

Between Congress and President Bush, as between that body and so many of Bush's predecessors, the element in the foreign policy relationship that continued to be missing was the kind of consultation that would allow the formation of a general consensus, an overall vision. Take foreign aid, for example, an area marked in recent years by a detailed and inflexible congressional earmarking of funds, tying the president's hands. Senate Democratic leader Mitchell declared in 1990:

> The [Bush] Administration does not have a five-year plan or a three-year plan or a two-year plan for foreign aid; indeed, they offer no plan at all. They have not explained how any plan would be affordable in relation to their other spending plans; they have not outlined the relationship of any plan to our national security objectives.[165]

Much of the congressional earmarking of foreign aid was attributable to the general lack of consultation of Congress by President Reagan and the resultant diminished trust of him by the legislative branch. In regard to foreign aid, and foreign policy generally, the Senate should stop its attempts at detailed micromanagement and concentrate on broader policy issues. But this is not going to happen until there is a broader consensus on policy and honest consultation with Congress by the president. In 1992, serious White House–congressional disputes erupted again over aid to the former Soviet Union.[166]

Richard Bissell, assistant director of the U.S. Agency for International Development, was speaking of detailed foreign-aid earmarking, but he might just as well have been talking about presidential-congressional conflict on foreign policy and congressional attempts at micromanagement, when he said in 1990: "My own personal feeling is that there will never be a deal [between the president and Congress] struck as such on earmarking. What can occur is an emerging consensus on a growing number of foreign policy issues between the Administration and Congress such that earmarks become irrelevant."[167]

In ceding Congress its rightful role in foreign policymaking, political scientist Larry George has written, there is no need to diminish the proper roles of the president and executive-department agencies "in gathering and processing information, conducting day-to-day diplomacy with foreign nations, or ensuring military security." He concluded:

> Rather what is being suggested is simply that final determinations of foreign policies are *political* decisions, to be decided wherever possible by democratically empowered political bodies. While mistakes, even egregious ones, may be made by elected leaders, it is no more reasonable to suggest that this is a fatal criticism of democracy in foreign policymaking than it would be of democratic policymaking on domestic issues.[168]

The end of the Cold War, according to Charles William Maynes, has deprived the United States not only of its enemy, but also of the "sextant by which the ship of state has been guided since 1945," and the times now demand more vigorous, open debate—in Congress and in the country—on foreign policy issues, once dominated by foreign policy elites.[169] According to Maynes:

> These elites complain of congressional micromanagement; but only in the field of foreign policy does the Congress repeatedly permit one administration after another to ignore laws such as the War Powers Act, spend large sums in unvouched CIA funds, and deny the public information vital to a rational discussion of particularly sensitive issues.[170]

A standardized, institutionalized mechanism for foreign-policy consultation between the president and Congress is what is needed, according to the former chair of the Senate Foreign Relations Committee, Republican Senator Richard Lugar of Indiana. "When there is a bipartisan, two-house consensus, then policies have more staying power," he has said. "When Ronald Reagan did it well, it worked; when, as in Irangate, he didn't, it didn't."[171]

The same permanent National Security and Foreign Policy Council discussed earlier would help to institutionalize and encourage foreign policy cooperation, instead of confrontation, between the president and Congress.

National security and foreign policy decisions are among the most important that a government can make—with great impact on the lives of its citizens. Their representatives in the Senate and House are entitled to be consulted, and to be listened to, in regard to these decisions and to be permitted to exercise their constitutional authority in respect to them. The creation of a permanent National Security and Foreign Policy Council could help to change a relationship between president and Congress that has almost always been an invitation to struggle, into more of an encouragement to cooperation.

Epilogue:
Toward a More
Effective Senate

A peculiar institution is that citadel of democracy, the Senate of the United States. Neither under the constitutional provisions that created it nor in the way it operates internally is it directly responsible to majority opinion or governed by majority rule.

Intended to be both a deliberative and a law-making body, the Senate, through the years, grew prestigious and powerful, developing a strong set of behavioral norms that facilitated Senate action and decision-making. But the Senate changed after the 1950s, primarily as a result of immense transformations that took place outside it. The Senate became more responsive to the people and more of an incubator and popularizer of ideas. But it also became less efficient in law-making.

First, there was a nationalization of American society because of vastly more rapid communications and transportation, greatly increased mobility and urbanization of the people, a higher general standard of living and educational level, enormous growth in the national government, and a great increase in media outlet numbers and the fervor and magnitude of their coverage.

Senatorial campaigns and elections became nationalized, too—"little presidential campaigns"—with national consultants, strategies, techniques, partisan sponsorship, and greatly increased national attention.

As a result of these external changes, senators became more outward-looking and national, but at the same time they became more accessible, more politically exposed, and more electorally vulnerable and insecure. The factors that contributed to the nationalized environment of the Senate are not likely to change much. There are things that can and should be done, though, to ameliorate their ill effects.

Reducing the Power of the Interests

A national advocacy explosion—an enormous increase in the number and diversity of interest groups, as well as a tremendous growth in the range, volume, and intensity of their activities—resulted in a nationalization of issues and interest groups. Senators now have both more opportunity and more demand to become national spokespersons.

Senate campaign financing was nationalized, also, and costs skyrocketed. Fund-raising pressures began to interfere with the work of senators and to affect their behavior. The need to raise great sums of campaign money became an ever more worrisome source of potentially corrupting influence.

Senate campaign costs and contributions should be strictly and enforceably limited, with provisions for free television and other subsidies in Senate campaigns, as well as the provision of public funding, to curb rich-candidate self-financing of Senate races and to circumscribe and offset independent expenditures. In these ways the power of money could be reduced and senators made more free to pursue the public interest.

Improving Senate Efficiency

As a result of the nationalization of the Senate's external environment, as well as of turnover in Senate membership (particularly with the influx of large numbers of majority-party senators during the critical-mass times of the early 1960s and the early 1980s), the strong norms of the 1950s were weakened and changed.

Having made themselves staff-rich, senators are able to give more attention to constituents while also acting as national advocates. They have opened the Senate up—mandating both public committee markup sessions and televised floor proceedings. These developments, added to the others, have helped to nationalize the Senate, making it more public regarding as well as more individualistic in the way power is wielded internally.

The Senate's *courtesy* norm is still important, though it has suffered some from increased Senate partisanship and a deepening of differing ideological commitments. Liking the Senate still, though it is not expected to be a central devotion of their lives, senators continue to respect the norm of *institutional patriotism*. The specialization norm of the Senate has been changed into a norm of *expertise*. Senators who would be respected and have influence are expected to know what they are talking about.

The norm of legislative work—paying attention to legislative details and being a work horse, not a show horse—has metamorphosed into a norm of *national advocacy* and a norm of *diligence*. Work horses and show horses are largely the same senators, now; a national advocate who works at the job almost invariably attracts media attention, and that is encouraged.

The Senate's *reciprocity* norm has been greatly weakened. Filibusters and other forms of obstructionism are more common and today are used on far less substantial or important issues than was true in the 1950s and earlier.

To combat this obstructionism, which has sometimes reached "gridlock" proportions, the Senate should adopt reforms to limit the force of the filibuster and related dilatory tactics. Amendment-rules reforms should be adopted, especially to prohibit nongermane amendments and to require that a bill be taken up, and amendments offered, on a section-by-section basis. Senate leaders should work to engender more support for the old Senate norm of reciprocity, and Senate rules should be changed to limit debate on motions to proceed to consideration of a measure to one hour, as recently proposed.

Senate subcommittees have proliferated, and the number of committee seats, especially on the most influential panels, has been increased, further fragmenting Senate power. Senate adoption of the kind of restructuring and reduction of Senate committees proposed by senators Nancy Kassebaum and Daniel Inouye, together with their plan for strictly limiting the number of committee and subcommittee seats and leadership positions that a senator could hold, would reverse some of the recent individualization in the Senate and improve Senate efficiency.

Senate committees with overlapping jurisdictions should hold more combined hearings, and there should be more joint hearings between House and Senate committees with the same jurisdictions.

Ameliorating Partisan Conflict

The political-party context of the Senate has been transformed. A party realignment has taken place in the South. There, and throughout the nation, Senate races have become more two-party competitive. Importantly, there has been a certain ideological realignment of the parties in the electorate. Republican identifiers now are more likely to be conservatives, and Democrats are more likely to be moderate to liberal. Each party in the electorate has become internally more homogeneous and, simultaneously, less like the other party. This has especially been true among party activists who have greatest influence on nominations, including U.S. Senate nominations.

National party organizations, including the senatorial campaign committees—the NRSC and the DSCC—are stronger today than ever, with big budgets and staff and great involvement in Senate races. They have helped to produce nationally partisan senatorial campaigns.

The Senate has become a more partisan place. As the numbers of Senate party countertypes—conservative Democrats and liberal Republicans—have diminished, each party in that body has become more internally homogeneous and less like the other party. Each Senate party has become more internally cohesive on the issues. Party-line voting, especially on the most important votes, has increased. Party leaders and the party conferences have become stronger, and the conferences have increasingly offered a needed mechanism for arching over Senate fragmentation.

Party-line filibusters and other obstructionist tactics, particularly by Senate Republicans, have become more common. Reforms to curb the filibuster and obstructionism generally would reduce these potentials for partisan deadlock.

A trend toward divided U.S. government has paralleled the increase in intraparty cohesion and unity in the Senate (and House), thus enlarging, rather than reducing, the possibilities of presidential-congressional conflict and government stalemate. These could be diminished by the institutionalization and greater use of the summit device to encourage cross-branch and cross-party cooperation on a range of issues.

Facilitating Budget Policymaking

The nationalization of the Senate has had important consequences in policymaking. Inside and outside Congress there is a widespread feeling that the exercise of congressional budget power in recent years has been a failure. Members of the Senate (and House) feel frustrated in their attempts to get control of federal deficits (although they have had more effect in this regard than is generally accepted and were able to make commendable headway on deficit control as a result of the 1990 summit). Budget matters take up too much of the time and attention of individual members of Congress. Deadlines are commonly missed. The setting of national budget priorities has been difficult.

U.S. economic policy in recent times has produced highly unsatisfactory results. The maldistribution of wealth and income has worsened. Investment in human and physical capital has lagged. The Federal Reserve has managed monetary policy in a way that has often been at cross purposes with fiscal policy (and should be brought more in line with it). Congress has not been sufficiently able, or consciously and fully informed enough, to make acceptable fiscal policy.

In line with the Kassebaum–Inouye proposal, the dichotomy between authorization and appropriations committees should be ended, and the Senate Budget Committee should be replaced by something like a Committee on National Priorities, made up of the leaders of the policy committees. Multiyear budgeting, already started with the 1990 budget deal, should be regularized, perhaps a nonbudget session alternating with a budget session. This would permit Congress to give greater attention to other important issues, as well as to oversight of the executive department.

Institutionalization of the budget summit, using the Kassebaum–Inouye Committee on National Priorities for the Senate half of it, would help to arch over U.S. presidential, congressional, two-house, and party divisions in budget-making.

Aiding National Security and Foreign Policymaking

From the end of World War II to the Vietnam War there was little open debate and little conflict between the president and Congress on national security and foreign policy matters because there was general agreement on policy substance. The president's dominance grew with the emergence of the United States as a world power in a nuclear age and with a standing, and later volunteer, army. Transformation of the Senate's external environment, as well as of the Senate itself, heightened presidential-Senate conflict. So did the increased public questioning of national security and foreign policy that developed during and after the Vietnam War. The end of the Cold War opened a new episode of national security debate.

Great fissures in the presidential-congressional relationship concerning war powers in particular developed as a result of the undeclared Vietnam War. Congress passed the 1973 War Powers Resolution over a veto by President Richard Nixon. The Resolution purported to require the president to consult with Congress prior to the introduction of U.S. armed forces into hostilities or a situation likely to lead to hostilities, to notify Congress within forty-eight hours after such a commitment of forces, and to withdraw the forces unless Congress gave its approval within sixty days.

The War Powers Resolution has been largely ignored by presidents, who have maintained that it is both unconstitutional and impractical. President Bush did not formally consult with Congress before sending troops to the Persian Gulf in 1990 nor before changing their mission from a defensive to an offensive one.

The Constitution and the interests of the people require that the war powers be truly shared between the executive and legislative branches of our government.

In regard to structural national security policy—that dealing with government procurement, weapons contracts, and military bases—it is clear that the elimination of unneeded weapons systems and military bases has sometimes been stalled by constituency considerations and lobbying by defense contractors. The Senate (and House) should use again, as it has before, the device of the independent, congressionally created, presidentially appointed commission to shut down surplus bases. The influence of the "congressional-industrial complex" should be reduced through the adoption of reforms in the way Senate campaigns are financed, permitting an indispensable reorientation of American priorities toward more investment in infrastructure and human capital.

There has been too much micromanagement by Congress concerning strategic national security and arms control policy, as well as foreign policy generally. But in this, Congress has largely been acting out of a growth in distrust of the president that sprang, particularly during the Reagan administration, from a studied presidential failure to consult with the legislative branch.

National security and foreign policy decisions are among the most important that a government can make. They often have great impact on the lives of citizens. Representatives of the people in the Senate (and House) should be permitted to exercise their constitutional authority in respect to such decisions.

A permanent National Security and Foreign Policy Council—in the Senate, made up of the leaders of the Foreign Relations, Armed Services, and Intelligence committees—should be created. It would encourage the president and Congress to cooperate in regard to national security and foreign policy; improve the consultation between the branches; reduce the causes of conflict; and cut down on the tendency of Congress toward excessive micromanagement.

The Senate in the American System

To match the growth in Senate responsiveness, then, an improvement in that body's efficiency and public accountability is needed. The reforms recommended here would make a significant difference.

Senatorial frustration has been on the increase in recent years. Public pressure for action has also mounted. Internal concern about the kind of image the Senate is projecting to the country through televised floor action has heightened. All this suggests that these reforms have become increasingly practical.

But despite all the changes that have occurred in the Senate, it still remains very much a unique institution, the citadel of Aaron Burr, a place where Webster, Clay, and Calhoun would still feel at home. Indeed, according to political scientist Samuel Patterson, if Henry Clay came back to the modern Senate (and House), even though he would encounter a Senate whose membership is almost twice what it was in his time and one that is a "more formidable" political body, "semi-sovereign," and facing a "far greater array of policy decisions," he "would certainly recognize where he was," because "the fundamental character of the institution has not changed so very much."[1] Historian Richard Allan Baker has written in much the same tone:

The United States Senate is the one institution within the federal government that the framers of the Constitution, after two centuries, would immediately recognize. They would understand its passion for deliberation, its untidiness, its aloofness from the House of Representatives, and its suspicion of the presidency. They would probably not comprehend the role of legislative political parties. They would wonder why its proceedings have been opened to the public, both in person and through the medium of radio and television. The framers would certainly be aghast at the three ornate office buildings and the 7,000 staff members who fill them, yet they would understand the Senate's capacity for meeting the changing circumstances inherent in the nation's twentieth-century rise to world-power status. They would sympathize with continuing calls for reform of Senate procedures just as they would acknowledge the force of precedent and tradition that make those changes so difficult. Above all, they would be delighted that the Senate, and the Constitution that created it, had endured for two centuries.[2]

It is still the same Senate of the United States, then, though it is both better and worse than it once was. It is more of a popular, plebiscitary body than either the founders intended or the times, through the 1950s, produced. The Senate speaks, today, more nearly in the voice of the people. It can be encouraged to act more as they would act, and do it more decisively as well.

Notes

Prologue

1. *Congressional Record* (October 15, 1991), pp. 14704–5.

2. See R. W. Apple, Jr., "Senate Confirms Thomas, 52–48, Ending Week of Bitter Battle; 'Time for Healing,' Judge Says," *New York Times* (October 16, 1991), p. A1.

3. See Joan Biskupic, "A Process Under Scrutiny," *Congressional Quarterly Weekly Report* (October 19, 1991), p. 3031.

4. See Richard E. Cohen, "Advice, Consent and Political Games," *National Journal* (October 5, 1991), p. 2436.

5. See Maureen Dowd, "Getting Nasty Early Helps G.O.P. Gain Edge on Thomas," *New York Times* (October 15, 1991), p. A1.

6. See *New York Times*/CBS News poll, reported in "Confirmation of Clarence Thomas: The Public's View," *New York Times* (October 15, 1991), p. A12.

7. See Apple, Jr., "Senate Confirms Thomas," p. A13.

8. Tom Wicker, "An Alienated Public," *New York Times* (October 13, 1991), p. 15.

9. James Reston, "More Than Just Up or Down," *New York Times* (October 15, 1991), p. A17.

10. *Congressional Record* (October 15, 1991), p. 14634.

11. Ibid., p. 14703.

12. See generally Richard E. Cohen, "Letter from Albany: Coming to Terms," *National Journal* (October 19, 1991), p. 2555; Ronald D. Elving, "National Drive to Limit Terms Casts Shadow over Congress," *Congressional Quarterly Weekly Report* (October 26, 1991), pp. 3101–5; and Ronald D. El-

ving, "Foley Helps Put the Brakes on Drive for Term Limits," *Congressional Quarterly Weekly Report* (November 9, 1991), pp. 3261–63.

13. See Adam Mitzner, "The Evolving Role of the Senate in Judicial Nominations," *Journal of Law and Politics* 5 (Winter 1989):387–428.

14. Donald R. Matthews, *U.S. Senators and Their World* (Chapel Hill: University of North Carolina Press, 1960), p. 5.

15. William S. White, *The Citadel: The Story of the U.S. Senate* (New York: Harper, 1956), pp. ix, x.

16. Matthews, *U.S. Senators and Their World*, p. 6.

Chapter 1

1. In regard to the early history of the Senate, see Robert C. Byrd, *The Senate, 1789–1989* (Washington, D.C.: U.S. Government Printing Office, 1988), pp. 1–20; George J. Schulz, *Creation of the Senate: From the Proceedings of the Federal Convention, Philadelphia, May-September, 1787* (Washington, D.C.: U.S. Government Printing Office, 1937), reprinted as Senate Document 100-7, 100th Congress, 1st Session, 1987; Richard Allan Baker, *The Senate of the United States: A Bicentennial History* (Malabar, Fla.: Robert E. Krieger Publishing Co., 1988); and Roy Swanstrom, *The United States Senate 1787–1801*, U.S. Senate Document No. 64, September 26, 1961 (Washington, D.C.: U.S. Government Printing Office, 1962).

2. See letter from Representative Henry Wynkoop to Dr. Reading Beattie, April 2, 1789, "The Letters of Judge Henry Wynkoop, etc.," *Pennsylvania Magazine of History and Biography* 38 (January 1914):48.

3. Madison's notes at the Constitutional Convention, quoted in Schulz, *Creation of the Senate*, pp. 4, 15.

4. Ibid., p. 32.

5. Ibid., p. 5.

6. Anecdote quoted in Richard F. Fenno, Jr., *The United States Senate: A Bicameral Perspective* (Washington, D.C.: American Enterprise Institute, 1982), p. 5:

"Why," asked Washington, "did you pour that coffee into your saucer?"
"To cool it," said Jefferson.
"Even so," said Washington, "we pour legislation into the senatorial saucer to cool it."

7. Madison's notes at the Constitutional Convention, quoted in Schulz, *Creation of the Senate*, p. 5.

8. Ibid., p. 28.

9. Quoted in Baker, *The Senate of the United States,* p. 21.

10. Quoted in James Q. Wilson, *American Government: Institutions and Policy* (Lexington, Mass.: D. C. Heath, 4th ed., 1989), p. 324.

11. Swanstrom, *The United States Senate 1787–1801*, p. 37.

12. "Condorcet" in the *National Gazette* (December 15, 1792), quoted in Swanstrom, *The United States Senate 1787–1801*, p. 68.

13. Everett Somerville Brown, *William Plumer's Memorandum of Proceedings in the United States Senate, 1803–1807* (New York, 1923), pp. 448–50, reprinted in Baker, *The Senate of the United States*, pp. 150–51.

14. *The Debates and Proceedings in the Congress of the United States*, 8th Congress, pp. 70–71, reprinted in Baker, *The Senate of the United States*, p. 148.

15. Baker, *The Senate of the United States*, p. 33.

16. Alexis de Tocqueville, *Democracy in America*, ed. J. P. Mayer (Garden City, N.Y.: Anchor Books, 1969), p. 201.

17. *The Committee System*, First Staff Report to the Temporary Select Committee to Study the Senate Committee System, U.S. Senate, 94th Congress, 2nd Session (Washington, D.C.: U.S. Government Printing Office, 1976), p. 11.

18. Quoted in Congressional Quarterly, *Origins and Development of Congress* (Washington, D.C.: Congressional Quarterly Press, 1876), p. 196.

19. Franklin L. Burdette, *Filibustering in the Senate* (Princeton, N.J.: Princeton University Press, 1940), p. 5.

20. Ibid., p. 13.

21. Hezekiah Niles, quoted in Burdette, *Filibustering in the Senate*, p. 18.

22. Quoted in David J. Rothman, *Politics and Power: The United States Senate 1869–1901* (Cambridge, Mass.: Harvard University Press), p. 146.

23. Quoted in Burdette, *Filibustering in the Senate*, pp. 121, 122.

24. William S. White, *The Citadel: The Story of the U.S. Senate* (New York: Harper, 1956).

25. Allen Drury, *Advise and Consent* (New York: Doubleday, 1959).

26. Except as otherwise noted, material in this section is taken from Donald R. Matthews, *U.S. Senators and Their World* (Chapel Hill: University of North Carolina Press, 1960), pp. 92–117. See also Donald R. Matthews, "The Folkways of the United States Senate: Conformity to Group Norms and Legislative Effectiveness," *American Political Science Review* 53 (December 1959):1965.

27. Material concerning the Senate Reception Room and the selection of the five outstanding senators is taken from Fred R. Harris, *Potomac Fever* (New York: W. W. Norton, 1977), pp. 74–87; "Proceedings at the Unveiling of the Portraits of Five Outstanding Senators," Senate Document No. 17, 86th Congress, 1st Session (Washington, D.C.: U.S. Government Printing Office, 1959); and two unpublished folders from the personal files of then U.S. Senator John F. Kennedy of Massachusetts, furnished to the author (through then U.S. Senator Robert F. Kennedy of New York) by the National Archives, Washington, D.C., and labeled "Reception Room Special Committee."

28. In June 1982, David L. Porter polled forty political historians and asked them to list the ten greatest senators of all time. On the resulting list, the first five, in order of their rank, were: Henry Clay; Robert M. La Follette; John C. Calhoun, tied for third with Daniel Webster; and George Norris. The next five, again in order, were: Robert Taft, Stephen Douglas, Hubert Humphrey, Lyndon Johnson, and Charles Sumner. David L. Porter, "America's Ten Greatest Senators," in William D. Pederson and Ann M. McLaurie, eds., *The Rating Game in Politics: An Introductory Approach* (New York: Irvington Publishers, 1987), pp. 110–30.

29. "Proceedings at the Unveiling of the Portraits of Five Outstanding Senators," p. 11.

30. Ibid., pp. 11, 12.

31. Randall B. Ripley, *Power in the Senate* (New York: St. Martin's Press, 1969), pp. 6–19.

32. An anonymous liberal Democratic senator, quoted in David W. Rohde, Norman J. Ornstein, and Robert L. Peabody, "Political Change and Legislative

Norms in the U.S. Senate, 1957–1974," in ed. Glen R. Parker, *Studies of Congress* (Washington, D.C.: Congressional Quarterly Press, 1985), p. 158.

33. Statement made at Hearing on Changes to Senate Rules, Rules Committee, U.S. Senate, December 2, 1987.

34. "In the Senate of 1980s, Teamwork Has Given Way to the Rule of Individuals," *Congressional Quarterly Weekly Report* (September 4, 1982), pp. 2175–82.

35. Statement made at Hearing on Changes to Senate Rules.

36. See, for example, Tim Hackler, "What's Gone Wrong with the U.S. Senate?" *American Politics* (January 1987), pp. 7–11; and Helen Dewar, "Frustration Without Achievement in the Senate," *Washington Post* (January 4, 1988), pp. 1, A4.

37. Senators Daniel J. Evans (R., Wash.) and Paul S. Trible, Jr. (R., Va.), respectively, quoted in Dewar, "Frustration Without Achievement in the Senate," p. 1.

38. Nelson W. Polsby, *Congress and the Presidency* (Englewood Cliffs, N.J.: Prentice-Hall, 4th ed., 1985), p. 89.

39. Leroy N. Rieselbach, *Congressional Reform* (Washington, D.C.: Congressional Quarterly Press, 1986), p. 121.

40. See Lawrence C. Dodd, "A Theory of Congressional Cycles: Solving the Puzzle of Change," in eds. Gerald C. Wright, Jr., Leroy N. Rieselbach, and Lawrence C. Dodd, *Congress and Policy Change* (New York: Agathon Press, 1986), pp. 3–44.

Chapter 2

1. On the life and Senate service of J. William Fulbright (D., Ark., 1945–1974), see Haynes Johnson and Bernard M. Gwertzman, *Fulbright: The Dissenter* (New York: Doubleday, 1968); and Eugene Brown, *J. William Fulbright: Advice and Dissent* (Iowa City: University of Iowa Press, 1985).

2. Personal interview with former Senator Fulbright, Washington, D.C., 1988.

3. Except as otherwise shown, information in this section is taken from the *U.S. Statistical Abstract* for the particular years involved.

4. William M. Lunch, *The Nationalization of American Politics* (Berkeley: University of California Press, 1987), p. 23.

5. One important measure, for example, of standard of living is the percentage of homes with complete indoor plumbing. In 1950, about 40 percent of homes in the nation and over 70 percent of those in Arkansas were *without* complete indoor plumbing. By 1960, ten years later, those figures had fallen to a little over 18 percent for the nation and about 40 percent in Arkansas, by 1970 to 6 percent and about 17 percent, and by 1980 down to only slightly over 4 percent in Arkansas.

6. U.S. Department of Education, U.S. Census Bureau, and Educational Testing Service, quoted in *Public Opinion* (November/December 1987), p. 33.

7. Television interview by Roger Mudd, MacNeil-Lehrer Report, public television, with Senator Stennis (D., Miss., 1947–1988), October 24, 1988.

8. Television interview by Roger Mudd, MacNeil-Lehrer Report, public television, with Senator Proxmire (D., Wis., 1957–1988), October 24, 1988.

9. Quoted by Senator J. James Exon (D., Neb.) in a personal interview with the author, Washington, D.C., September 22, 1988.

10. Norman J. Ornstein, Thomas E. Mann, and Michael J. Malbin, *Vital Statistics on Congress 1987-1988* (Washington, D.C.: Congressional Quarterly Press, 1987), p. 142.

11. Personal interview with Senator Thurmond (R., S.C.), Washington, D.C., August 31, 1988.

12. In Fall 1988, the author's survey of fifty-six Senate administrative assistants by questionnaire showed that the average senator had 38.78 personal staff members; the median was 39.

13. Senator Fulbright began making national news right away with his vocal and well-reported support for the creation of the United Nations. Journalist Allen Drury noted in a January 1945 diary entry that Fulbright, whom he called "a trim little fellow with an amiable face, little eyes and a little mouth that crinkles up when he smiles" had talked with reporters "for a long time this morning about foreign policy and politics and ticklish technique of a new man's baptism in the Senate. As a veteran of the House, however, he showed the same disregard as [Washington State's Warren] Magnuson for the tradition that new senators should be seen and not heard; he was willing to be quoted." Allen Drury, *A Senate Journal: 1943-1945* (New York: McGraw-Hill, 1963), p. 332ff.

14. Personal interview with Riddick, Washington, D.C., August 17, 1988.

15. Personal recollections of the author as a member of the Senate from Oklahoma from 1964 to 1973. For other accounts of unreported senatorial drunkenness, see Larry J. Sabato, *Feeding Frenzy: How Attack Journalism Has Transformed American Politics* (New York: Free Press, 1991), pp. 31–33.

16. Sabato, *Feeding Frenzy,* pp. 25, 26.

17. *Brown* v. *Board of Education of Topeka*, 347 U.S. 483 (1954). In regard to the Little Rock confrontation, see Fred R. Harris, Randy Roberts, and Margaret S. Elliston, *Understanding American Government* (Glenview, Ill.: Scott, Foresman/Little, Brown, 1988), p. 122.

18. On Senator Fulbright's opposition to civil rights legislation, and his justification of this position, see Brown, *J. William Fulbright,* p. 10.

19. The author's Fall 1988 survey of Senate administrative assistants by questionnaire in regard to senatorial office activity showed the following: the average number of mail items received daily by each of fifty-seven senators' offices was 690, and the median was 550; the average number of daily personal telephone conversations each of fifty senators had was 21.4, the median 17; and the average number of office appointments with persons or groups that each of fifty-three senators kept daily was 6.35, the median being 6.

20. The author's Fall 1988 survey by questionnaire of fifty-three Senate administrative assistants showed an average number of senatorial visits home each year of 29.33, the median being 27.

21. Roger H. Davidson, "Congressional Committees as Moving Targets," *Legislative Studies Quarterly* 7 (February 1986):21.

22. Personal conversation with Senator Russell, Washington, D.C., circa 1967.

23. Richard F. Fenno, Jr., *Home Style: House Members in Their Districts* (Boston: Little, Brown, 1978).

24. Personal conversation with Senator Russell, Washington, D.C., circa 1967.

25. Brown, *J. William Fulbright,* p. 122.

26. See Robert C. Byrd, "The Senate Class of 1958," *Congressional Record* (July 17, 1976), pp. 9221–25.

27. Except as otherwise shown, information in regard to campaign consultants comes from an interview by the author with polling expert Paul Harstad, Garin-Hart, Washington, D.C., February 10, 1989.

28. See, for example, *Campaigns & Elections* (August 1988), pp. 6–9.

29. Quoted in Kim Mattingly, "Sen. McConnell (R-Ky), Up in 1990, Is Said to Be 'Running Scared' in a Positive Sense," *Roll Call* (February 27–March 5, 1989), p. 4.

30. Jerry Hagstrom and Robert Guskind, "Calling the Races," *National Journal* (July 30, 1988), p. 1973.

31. "When Age Is the Issue," *National Journal* (November 5, 1988), p. 2785.

32. See Hagstrom and Guskind, "Calling the Races," pp. 1972–76.

33. Ibid., p. 1972.

34. Ibid., p. 1974.

35. Alan Ehrenhalt, "Technology, Strategy Bring New Campaign Era," *Congressional Quarterly Weekly Report* (December 7, 1985), p. 2560.

36. Ibid.

37. See Robert Guskind, "Digging Up Dirt," *National Journal* (October 27, 1990), p. 2593.

38. Ibid., pp. 2595, 2596.

39. See "The Green-Thumb Gang," *National Journal* (June 16, 1990), pp. 1464, 1465.

40. *Campaigns & Elections* (August 1988), p. 26.

41. Ibid., p. 30.

42. Ibid.

43. Quoted in Hagstrom and Guskind, "Calling the Races," p. 1973.

44. Quoted in Ehrenhalt, "Technology, Strategy Bring New Campaign Era," p. 2561.

45. Quoted in Michael Oreskes, "What Poison Politics Has Done to America," *New York Times* (October 29, 1989), p. E1.

46. Ibid.

47. Ibid.

48. Cliff Zukin, Eagleton Institute of Politics, Rutgers University, quoted in Lloyd Grove, "Mudslinging in New Jersey," *Washington Post National Weekly Edition* (February 13–19, 1989), p. 8, from which the other material in this section is also taken.

49. For details on the Lautenberg–Dawkins race, see "Polluting the Air in New Jersey," *National Journal* (November 5, 1988), p. 2786.

50. Quoted in David S. Broder, "Who Should Play Cop for Campaign Ads?," *Washington Post National Weekly Edition* (February 13–19, 1989), p. 8.

51. Jim Innocenzi and Charlie Black, quoted in Alan Ehrenhalt, "Technology, Strategy Bring New Campaign Era," p. 2563.

52. Broder, "Who Should Play Cop for Campaign Ads?," p. 8.

53. Quoted in C. C. Case, "No More Mr. Nasty," *Campaign Magazine* (December/January 1990–91), p. 7.

54. Grove, "Mudslinging in New Jersey," p. 8.

55. Quoted in Broder, "Who Should Play Cop for Campaign Ads?," p. 9; and in Oreskes, "What Poison Politics Has Done to America," p. E1.

56. Quoted in Paul Taylor, "Consultants Rise via the Low Road," *Washington Post* (January 17, 1989), p. 1.

57. Quoted in "Candidates and Process Wounded in 'Total War,'" *New York Times* (March 19, 1990), p. A14.

58. Personal interview with Senator Sarbanes, Washington, D.C., October 5, 1988.

59. Personal interview with Senator Bradley, Washington, D.C., September 26, 1989.

60. Harrison Hickman, quoted in Ehrenhalt, "Technology, Strategy Bring New Campaign Era," p. 2565.

61. Larry McCarthy, quoted in Dan Balz, "Campaign Dirt Is Muddying Congress," *Washington Post National Weekly Edition* (May 8–14, 1989), p. 8.

62. See Case, "No More Mr. Nasty," pp. 1, 7.

63. "The Press Plays Referee on Campaign Ads," *National Journal* (October 27, 1990), p. 2595.

64. E. J. Dionne, Jr., "Bill Would Curb Political Attack Ads," *New York Times* (May 17, 1989), p. A10.

65. See Neil Brown, "Senators Divide on Proposals to Tame Negative TV Ads," *National Journal* (July 22, 1989), p. 1890.

66. See Helen Dewar, "Senate Passes Campaign Bill After Vote to Ban Honoraria," *Washington Post* (August 2, 1990), p. A1; "Common Cause Wins Dramatic Campaign Finance and Honoraria Victories in Senate," *Common Cause Magazine* (July/August 1990), pp. i–iv.

67. For the pros and cons of this proposed legislation, see "Q: Should Legislation Be Enacted to Influence the Content of Political Advertising?," *Campaign Magazine* (June 1991), pp. 24, 25.

68. American Political Network, Inc., *Presidential Campaign Hotline* (July 26, 1988), item 95.

69. Government Research Corporation, *The GRC Cook Political Report* (February 13, 1989).

70. Personal interview with polling expert Paul Harstad, Garin-Hart, Washington, D.C., February 10, 1989.

71. See Harold W. Stanley, "Southern Partisan Changes: Dealignment, Realignment or Both?," *Journal of Politics* 50 (1988):64–88; John Petrocek, "Realignment: New Party Coalitions and the Nationalization of the South," *Journal of Politics* 49 (1987):348–75; and Earl Black and Merle Black, *Politics and Society in the South* (Cambridge, Mass.: Harvard University Press, 1987).

72. Thomas Byrne Edsall, "The Return of Inequality," *Atlantic Monthly* (June 1988), p. 93.

73. See *Public Opinion* (January/February 1987); and Vernon Bogdanor, ed., *Parties and Democracy in Britain and America* (New York: Praeger Publishers, 1984).

74. Dorothy Davidson Nesbit, "Changing Partisanship among Southern Party Activists," *Journal of Politics* (1987), pp. 322–34; *Public Opinion* (May/June 1988), p. 24, and (July/August 1988), p. 23; *Washington Post* (August 14, 1988), p. A30, citing *Washington Post*/ABC Poll; and Nelson W. Polsby and Aaron Wildavsky, *Presidential Elections* (New York: Scribners, 1980).

75. Keith T. Poole and Howard Rosenthal, "The Polarization of American Politics," *Journal of Politics* 46 (1984):1269–73; and James L. Sundquist, *The Decline and Resurgence of Congress* (Washington, D.C.: American Enterprise Institute, 1981), pp. 473, 478.

76. Barbara Sinclair, "The Distribution of Committee Positions in the U.S. Senate: Explaining Institutional Change," *American Journal of Political Science* 32 (May 1988):284.

77. Stephen Gettinger, "Partisanship Hits a New High in 99th Congress," *Congressional Quarterly Weekly Report* (November 15, 1986), pp. 2901–3; and *Congressional Quarterly Weekly Report* (June 6, 1987), pp. 873–85.

78. See, for example, Alan Ehrenhalt, "Changing South Perils Conservative Coalition," *Congressional Quarterly Weekly Report* (August 1, 1987), pp. 1699–1705.

79. Poole and Rosenthal, "The Polarization of American Politics," p. 1271.

80. Alan I. Abramowitz, "Explaining Senate Election Outcomes," *American Political Science Review* 82 (no. 2) (June 1988):398.

81. See Mark C. Westlye, *Senate Elections and Campaign Intensity* (Baltimore: Johns Hopkins University Press, 1991), p. 9.

82. Paul S. Herrnson, "National Organizations and the Postreform Congress," in ed. Roger H. Davidson, *The Postreform Congress* (New York: St. Martin's Press, 1992), p. 53. See also Paul S. Herrnson, *Party Campaigning in the 1980s* (Cambridge, Mass.: Harvard University Press, 1988), p. 39; and Paul S. Herrnson, "Reemergent National Party Organizations," in ed. L. Sandy Maisel, *The Parties Respond: Changes in the American Party System* (Boulder, Colo.: Westview Press, 1990), p. 51.

83. Michael Barone and Grant Ujifusa, *The Almanac of American Politics 1992* (Washington, D.C.: National Journal, 1992), p. 1338.

84. Personal interviews in Washington, D.C., with Jann Olsten, Executive Director, National Republican Senatorial Committee, August 25, 1988, and with Bob Chlopak, Executive Director, Democratic Senatorial Campaign Committee, August 24, 1988. See also Richard E. Cohen, "Party Help," *National Journal* (August 16, 1986), pp. 1998–2004; and Ronald Brownstein, "The Long Green Line," *National Journal* (May 3, 1986), pp. 1038–42.

85. "Memorandum to Democratic Vendors/Consultants," Democratic Senatorial Campaign Committee, May 5, 1989, Washington, D.C.

86. Gary C. Jacobson, "Parties and PACs in Congressional Elections," in eds. Lawrence C. Dodd and Bruce I. Oppenheimer, *Congress Reconsidered* (Washington, D.C.: Congressional Quarterly Press, 4th ed., 1989), p. 139.

87. See John R. Hibbing and Sara L. Brandes, "State Population and Electoral Success of U.S. Senators," *American Journal of Political Science* 27 (August–November 1983):808–19; and Peverill Squire, "Challengers in U.S. Senate Elections," in ed. John Hibbing, *The Changing World of the U.S. Senate* (Berkeley, Calif.: IGS Press, 1990), pp. 265–81.

88. James L. Sundquist and Richard M. Scammon, "The 1980 Election: Profile and Historical Perspective," in eds. Ellis Sandoz and Cecil V. Crabb, *A Tide of Discontent* (Washington, D.C.: Congressional Quarterly Press, 1981), p. 23.

89. See Westlye, *Senate Elections and Campaign Intensity*, p. 5; John R. Hibbing and John R. Alford, "Economic Conditions and the Forgotten Side of Con-

gress: A Foray into U.S. Senate Elections," *British Journal of Political Science* 12 (1982):506; and Charles O. Jones, "The New, New Senate," in eds. Sandoz and Crabb, *A Tide of Discontent*, p. 100.

90. Edie N. Goldberg and Michael W. Traugott, "Mass Media in U.S. Congressional Elections," *Legislative Studies Quarterly* 12 (August 1987):335.

91. Quoted in Richard F. Fenno, Jr., *The United States Senate: A Bicameral Perspective* (Washington, D.C.: American Enterprise Institute, 1982), p. 29.

92. Robert Peabody, "Senate Party Leadership from the 1950s to the 1980s," unpublished manuscript, quoted in Fenno, Jr., *The United States Senate*, p. 31.

93. Hibbing and Brandes, "State Population and Electoral Success of U.S. Senators," p. 808.

94. Westlye, *Senate Elections and Campaign Intensity*, p. 5.

95. Ibid.

96. Donald R. Matthews, *U.S. Senators and Their World* (Chapel Hill: University of North Carolina Press, 1960), p. 110.

97. Ibid., pp. 110, 111.

Chapter 3

1. See *Campaigns & Elections* (January 1989), pp. 21, 22. For a profile on Robert Kerrey, see *Congressional Quarterly Weekly Report* (November 12, 1988), p. 3261.

2. In regard to the Kerrey campaign, see Steven J. Jarding, "The Color of Money: The High Stakes of Financing a U. S. Senate Campaign," *Extensions* (Winter 1989), pp. 4–5, 20; and *Congressional Quarterly Weekly Report* (October 15, 1988), p. 2919.

3. Within months after his election to the Senate in 1988, both the *New York Times* and a capital magazine, *The Washingtonian*, picked him out as a national star, and the *Washington Post* ran a feature on him. See Robin Toner, "'Star' Girds at Home for Senate War," *New York Times* (September 3, 1989), p. 11; Fred Barnes, "Best, Mosts, Misfits, and Others," *The Washingtonian* (March 1989), p. 102; and Jim Naughton, "Have the Democrats Found Their Moses?" *Washington Post National Weekly Edition* (December 11–17, 1989), p. 10. See also Paul Taylor, "A New Star for the Democrats' Firmament," *Washington Post National Weekly Edition* (April 23–29, 1990), p. 14.

4. *Congressional Quarterly Weekly Report* (November 12, 1988), pp. 3252, 3304.

5. Information on the Kerrey campaign's budget, fund-raising efforts, and expenditures are taken from Jarding, "The Color of Money," pp. 4–5, 20.

6. See Richard Berke, "Most of the Senators Go Out of State for Contributors," *New York Times* (April 16, 1990), pp. 1, A10.

7. Jarding, "The Color of Money," p. 5.

8. Ibid.

9. Quoted in *Congressional Quarterly Weekly Report* (November 12, 1988), p. 3261.

10. *Congressional Quarterly Weekly Report* (October 15, 1988), p. 2919.

11. Jeffrey M. Berry, *The Interest Group Society* (Glenview, Ill.: Scott, Foresman/Little, Brown, 2nd ed., 1989), p. 17.

12. Kay Lehman Schlozman and John T. Tierney, "More of the Same:

Washington Pressure Group Activity in a Decade of Change," *Journal of Politics* 45 (1983):355.

13. Ibid., p. 356.

14. Hedrick Smith, *The Power Game: How Washington Works* (New York: Random House, 1988), p. 31. See also Berry, *The Interest Group Society*, p. 20.

15. Smith, *The Power Game,* p. 31; and Berry, *The Interest Group Society,* p. 20.

16. Berry, *The Interest Group Society,* p. 21.

17. Ibid.

18. See U. S. Senate, *Congress and Pressure Groups: Lobbying in a Modern Democracy,* report prepared for the Subcommittee on Intergovernmental Relations of the U.S. Senate Committee on Governmental Affairs, 99th Congress, 2nd Session, S. Prt. 99–161 (Washington, D.C.: U.S. Government Printing Office, 1986), pp. 43, 44.

19. Information furnished by Records Division, Office of the Secretary of the U.S. Senate, July 14, 1989.

20. Information furnished by Curtis McCormack with Columbia Books, Inc., Washington, D.C., publisher of *Washington Representatives* (July 14, 1989).

21. See Kirk Victor, "New Kids on the Block," *National Journal* (October 31, 1987), p. 2727.

22. Information for 1989 furnished by the Membership Office of the District of Columbia Bar Association, July 14, 1989. For the 1961 figure, see Smith, *The Power Game,* p. 31.

23. Paul Taylor, "Gladiators for Hire," *Washington Post* (July 31, 1983).

24. Kay Lehman Schlozman and John T. Tierney, *Organized Interests and American Democracy* (New York: Harper & Row, 1986), p. 56.

25. Quoted in U.S. Senate, *Congress and Pressure Groups,* p. 28.

26. Ibid., p. 54.

27. Justice Department report, cited in Fred R. Harris, Randy Roberts, and Margaret S. Elliston, *Understanding American Government* (Glenview, Ill.: Scott, Foresman/Little, Brown, 1988), p. 207.

28. Thomas B. Edsall, "In Position to Play: How a Lobbying Firm Makes Connections Pay Off," *Washington Post National Weekly Edition* (August 21–27, 1989), p. 6.

29. Rochelle L. Stanfield, "So That's How It Works," *National Journal* (June 24, 1989), pp. 1620–24.

30. R. Kent Weaver, "The Changing World of Think Tanks," *PS: Political Science and Politics* 22 (no. 3) (September 1989):563–78. See also James Allen Smith, *The Idea Brokers* (New York: Free Press, 1991).

31. See Carol Matlack, "Marketing Ideas," *National Journal* (June 22, 1991), pp. 1552–55.

32. Weaver, "The Changing World of Think Tanks," pp. 571, 572.

33. Matlack, "Marketing Ideas," pp. 1552–55.

34. U.S. Senate, *Congress and Pressure Groups,* p. 31.

35. Ibid., p. 30.

36. Ibid.; and Allan J. Cigler and Burdett A. Loomis, "Introduction: The Changing Nature of Interest Group Politics," in eds. Allan J. Cigler and Burdett A. Loomis, *Interest Group Politics* (Washington, D.C.: Congressional Quarterly Press, 1983), p. 23.

37. See Jack L. Walker, "The Origins and Maintenance of Interest Groups in America," *American Political Science Review* 77 (1983):390–406.

38. Berry, *The Interest Group Society,* p. 43. See also Paul Edward Johnson, "Organized Labor in an Era of Blue-Collar Decline," in Cigler and Loomis, *Interest Group Politics,* pp. 33–61; and Kirk Victor, "Labor Pains," *National Journal* (June 8, 1991), pp. 1336–39.

39. Barbara Sinclair, *The Transformation of the U.S. Senate* (Baltimore: Johns Hopkins University Press, 1989), p. 59.

40. U. S. Senate, *Congress and Pressure Groups,* p. 29.

41. Ibid., pp. 29, 30.

42. Quoted in Schlozman and Tierney, "More of the Same," p. 367.

43. U.S. Senate, *Congress and Pressure Groups,* p. 30.

44. Walker, "The Origins and Maintenance of Interest Groups in America," p. 398.

45. Ibid., p. 397.

46. Ann Cooper, "Middleman Mail," *National Journal* (September 14, 1985), p. 2036.

47. Ibid., p. 2038.

48. Advertisement of Bonner and Associates in *National Journal,* reprinted in Harris, Roberts, and Elliston, *Understanding American Government,* p. 210.

49. See Ann Cooper, "Lobbying in the '80s: High Tech Takes Hold," *National Journal* (September 14, 1985), p. 2032.

50. Sinclair, *The Transformation of the U. S. Senate,* p. 70.

51. Ibid.

52. Daniel A. Dutko, quoted in Victor, "New Kids on the Block," p. 2727.

53. Testimony of John T. Tierney, quoted in U. S. Senate, *Congress and Pressure Groups,* p. 32.

54. Federal Election Commission, "1987–88 Election Campaign Activity," Preliminary Report (February 24, 1989), p. 13.

55. Herbert E. Alexander, *Financing Politics* (Washington, D.C.: Congressional Quarterly Press, 3rd ed., 1984), fig. 1.2.

56. Joseph E. Cantor, "Campaign Financing in Federal Elections: A Guide to the Law and Its Operation," Congressional Research Service Report, Library of Congress, August 8, 1986, updated August 19, 1988, p. 59.

57. The rate of inflation between 1974 and 1986, measured by the Consumer Price Index, was 220 percent. Norman J. Ornstein, Thomas E. Mann, and Michael J. Malbin, *Vital Statistics on Congress, 1987–1988* (Washington, D.C.: American Enterprise Institute, 1987), p. 67.

58. Federal Election Commission, "$458 Million Spent by 1988 Congressional Campaigns," Press Release (February 24, 1988), p. 1.

59. Michael Oreskes, "Congress: Modern Politicking Is Forcing Lawmakers to Devote More Time to Money," *New York Times* (July 11, 1988), p. A12.

60. See Chuck Alston, "One Chamber's View of Reform Is Anathema in the Other," *Congressional Quarterly Weekly Report* (June 29, 1991), p. 1727.

61. See Richard L. Berke, "Enter Senate, Begin Chasing Money," *New York Times* (June 3, 1991), p. A8.

62. Quoted in Smith, *The Power Game,* p. 157.

63. From interviews with U.S. senators by the Center for Responsive Pol-

itics, Washington, D.C., compiled April 30, 1987, and made available to the author with the understanding that the identities of the senators would not be disclosed.

64. Comparisons between House and Senate campaigns in this section are based on Frank J. Sorauf, "Varieties of Experiences: Campaign Finance in the House and Senate," in ed. Kay Lehman Schlozman, *Elections in America* (Boston: Allen and Unwin, 1987), pp. 197–218.

65. Ibid., p. 205. Candidates for the House "represent districts that may only be a small part of a television market: in a metropolitan area of 3 million people a House candidate would have to buy television time reaching an audience of which only about a fifth live in the congressional district." Ibid., p. 202.

66. Ibid., pp. 207, 208.

67. Ibid., pp. 211–13.

68. See "Don't Look Homeward," *National Journal* (June 16, 1990), pp. 1458–60; "Senatorial Courtesy," *National Journal* (June 16, 1990), pp. 1462, 1463; and Berke, "Most of the Senators Go Out of State for Contributions," pp. 1, A10.

69. Berke, "Most of the Senators Go Out of State for Contributions," p. 1.

70. Quoted in Berke, "Most of the Senators Go Out of State for Contributions," p. A10. On the same point, see Sorauf, "Varieties of Experiences," p. 217; and Steven V. Roberts, "Politicking Goes High-Tech," *New York Times Magazine* (November 2, 1986), p. 42.

71. See Alston, "One Chamber's View of Reform Is Anathema in the Other," p. 1729.

72. Direct-mail letter of March 1983, quoted in Larry J. Sabato, *PAC Power: Inside the World of Political Action Committees* (New York: W. W. Norton, 1985), p. xi.

73. See generally Fred R. Harris, "Should We Limit the PACs?" in ed. Fred R. Harris, *Readings on the Body Politic* (Glenview, Ill.: Scott, Foresman, 1987), pp. 189, 190; Larry J. Sabato, *Paying for Elections* (New York: Priority Press Publications, 1989), pp. 9–18; Larry J. Sabato, *PAC Power: Inside the World of Political Action Committees* (New York: W. W. Norton, 1985), pp. 1–7; Berry, *The Interest Group Society*, pp. 118–20; and Elizabeth Drew, *Politics and Money: The New Road to Corruption* (New York: Macmillan, 1983), pp. 6–11.

74. Quoted in Drew, *Politics and Money,* p. 10.

75. Sabato, *PAC Power,* p. 9; and Stephen A. Salmore and Barbara G. Salmore, "Candidate-Centered Parties: Politics Without Intermediaries," in eds. Richard A. Harris and Sidney M. Milkis, *Remaking American Politics* (Boulder, Colo.: Westview Press), p. 228.

76. See "PACs Lead the Way," *National Journal* (June 16, 1990), p. 1471; and Ornstein, Mann, and Malbin, *Vital Statistics on Congress 1991–1992* (Washington, D.C.: American Enterprise Institute, 1990), p. 97.

77. Information in this paragraph on the number of PACs, as well as their contributions, comes from Federal Election Commission reports for the various years, digested in Joseph E. Cantor, "Campaign Financing in Federal Elections: A Guide to the Law and Its Operation," Congressional Research Service Report, Library of Congress, August 8, 1986, updated August 19, 1988; and *Washington Post* (April 10, 1989), p. A7.

78. The Federal Election Commission, using somewhat different final fig-

ures for those reported earlier elsewhere, reported that PAC spending for federal candidates in both 1988 and 1990 was $159 million. Federal Election Commission, reported in Richard L. Berke, "Donations to PACs Dropped for First Time in '90," *New York Times* (April 1, 1991), p. A6.

79. Study made by Common Cause and cited in Charles R. Babcock, "78 Senators Reach PACs' Million Mark," *Washington Post* (August 1, 1990), p. A17.

80. Information on 1988 election expenditures by PACs comes from Federal Election Commission reports, digested in *Washington Post National Weekly Edition* (May 22–28, 1989), pp. 14–15; and *Washington Post* (April 10, 1989), p. A7.

81. Reports of the Federal Election Commission, cited in "FEC Watch: They Still Get the Money," *Campaign Magazine* (December 1990), p. 10.

82. *Common Cause Magazine* (March/April 1987), p. 46; *Common Cause News* (January 1987), p. 1; and "PAC Flip-Flops Mean Cash to Senate Victors," *Campaigns & Elections* (July/August 1987), p. 10.

83. *Washington Post* (April 10, 1989), p. A7.

84. See "No Risk Investment," Common Cause press release, May 9, 1989, pp. 1–9.

85. Mark Green, *The New Republic* (December 13, 1982), p. 24.

86. Berry, *The Interest Group Society,* pp. 120, 125.

87. "PACs Lead the Way," p. 1471. See also Cantor, "Campaign Financing in Federal Elections," pp. 37–41; and the 1988 congressional campaign spending figures for particular PACs from Federal Election Commission reports digested in *Washington Post* (April 10, 1989), p. A7.

88. Norman J. Ornstein, Thomas E. Mann, and Michael J. Malbin, *Vital Statistics on Congress 1991–1992* (Washington, D.C.: Congressional Quarterly Press, 1992), pp. 87–92.

89. Alan I. Abramowitz, "Campaign Spending in U. S. Senate Elections," in ed. John Hibbing, *The Changing World of the U.S. Senate* (Berkeley, Calif.: IGS Press, 1990), p. 367.

90. Ornstein, Mann, and Malbin, *Vital Statistics on Congress 1991–1992,* pp. 101–3.

91. In regard to PAC solicitation rules and methods, see Larry J. Sabato, "Fundraising by the PACs," in ed. Larry J. Sabato, *Campaigns and Elections: A Reader in Modern American Politics* (Glenview, Ill.: Scott, Foresman/Little, Brown, 1989), pp. 145–51.

92. Richard L. Berke, "In New Ethics Climate, a Concern on How the Fund-Raisers Operate," *New York Times* (June 12, 1989), pp. 1, A12.

93. Glen Craney, "900 Calls May Transform Political Fund Raising," *Congressional Quarterly Weekly Report* (October 21, 1989), pp. 2820–21.

94. Berke, "In New Ethics Climate, a Concern on How the Fund-Raisers Operate," p. A12.

95. Ibid.

96. Federal Election Commission Final Report, press release, May 5, 1988; and "PACs Bearing Gifts," *National Journal* (June 16, 1990), p. 1470.

97. Investigation by *Campaign Practices Reports* (May 5, 1986), pp. 1–5, cited in Larry J. Sabato, *Paying for Elections: The Campaign Finance Thicket* (New York: Priority Press, 1989), p. 21.

98. Philip M. Stern, *The Best Congress Money Can Buy* (New York: Pantheon Books, 1988), p. 39.

99. Public Law 93–443, 88 Stat. 1263 (1974). See also Cantor, "Campaign Financing in Federal Elections."

100. Berry, *The Interest Group Society,* pp. 136, 137.

101. See Stern, *The Best Congress Money Can Buy,* pp. 170, 171.

102. See Ross K. Baker, *The New Fat Cats: Members of Congress as Benefactors* (New York: Priority Press Publications, 1989).

103. See Chuck Alston, "Senate GOP Has a Big Problem: How to Spend All That Cash," *Congressional Quarterly Weekly Report* (September 29, 1990), p. 3087.

104. Ornstein, Mann, and Malbin, *Vital Statistics on Congress 1991–1992,* pp. 94, 95.

105. Federal Election Commission figures reported in *National Journal* (February 10, 1990), p. 355.

106. *Federal Election Commission* v. *Democratic Senatorial Campaign Committee,* 343 U.S. 27 (1981).

107. See Alston, "Senate GOP Has a Big Problem," p. 3092; and "FEC Watch," p. 10.

108. Alston, "Senate GOP Has a Big Problem," p. 3092.

109. Center for Responsive Politics, *Campaign Practices Reports* (June 13, 1988); and various *Washington Post* reports, cited in Sabato, *Paying for Elections,* pp. 64–66.

110. Study released December 5, 1989, reported in Glen Craney, "'Soft Money' in Big States Exceeded $28 Million," *National Journal* (December 9, 1989), pp. 3389, 3390.

111. See Cantor, "Campaign Financing in Federal Elections," pp. 75–78; and Title 2, U.S. Code, Section 431.

112. *Buckley* v. *Valeo,* 424 U.S. 1 (1976).

113. Ornstein, Mann, and Malbin, *Vital Statistics on Congress 1991–1992,* p. 106.

114. Congressional Research Service Report, "Contributions and Loans by Major Party General Election Candidates," Library of Congress, 1984, updated by Kevin Coleman, July 25, 1986.

115. See Frank J. Sorauf, *Money in American Politics* (Glenview, Ill.: Scott, Foresman, 1988).

116. Gary J. Perkinson, quoted in Maxwell Glen, "Raising Money," *National Journal* (September 14, 1985), p. 2066.

117. Personal interview with J. D. Williams, Washington, D.C., August 17, 1988.

118. Gary J. Perkinson, quoted in Glen, "Raising Money," pp. 2068, 2076.

119. See, for example, Sabato, *Paying for Elections,* pp. 13–15.

120. Remarks made in the U.S. Senate, excerpted in Harris, *Readings on the Body Politic,* pp. 196–98.

121. From interviews with U.S. senators by the Center for Responsive Politics, Washington, D.C., compiled April 30, 1987, and made available to the author with the understanding that the identities of the senators would not be disclosed.

122. The author's own recollection.

123. Quoted in Stern, *The Best Congress Money Can Buy,* p. 105.

124. See study by the Center for Responsive Politics, reported in Richard L. Berke, "Study Confirms Interest Groups' Pattern of Giving," *New York Times* (September 16, 1990), p. 18; and "Quid Without Quo?" *National Journal* (June 16, 1990), pp. 1473–74, 1479.

125. Gary C. Jacobson, "Parties and PACs in Congressional Elections," in Lawrence C. Dodd and Bruce I. Oppenheimer, eds., *Congress Reconsidered* (Washington, D.C.: Congressional Quarterly Press, 4th ed., 1989), p. 142.

126. From interviews with U.S. senators by the Center for Responsive Politics, Washington, D.C., compiled April 30, 1987, and made available to the author with the understanding that the identities of the senators would not be disclosed.

127. Quoted in Jack W. Germond and Jules Witcover, "Inside Politics: Looking for a Smoking Gun on Campaign Funds?" *National Journal* (December 12, 1989), p. 2956.

128. Quoted in Donald R. Matthews, *U.S. Senators and Their World* (Chapel Hill: University of North Carolina Press, 1960), p. 193.

129. Quoted in Stern, *The Best Congress Money Can Buy*, pp. 103, 104.

130. Berry, *The Interest Group Society*, pp. 133, 134.

131. See Jacobson, "Parties and PACs in Congressional Elections," p. 141.

132. Former House Democratic Whip Tony Coehlo (D., Calif.), quoted in *Common Cause News* (January 1987), p. 2.

133. Quoted in Sabato, *Paying for Elections*, p. 16.

134. J. Skelley Wright, "Money and the Pollution of Politics: Is the First Amendment an Obstacle to Political Equality?" *Columbia Law Review* 822 (no. 4) (May 1982):645.

135. Jacobson, "Parties and PACs in Congressional Elections," p. 135.

136. Ibid., pp. 138, 143.

137. After hearings, the Senate Ethics Committee in February 1991 resolved the cases against senators Dennis DeConcini (D., Ariz.), John Glenn (D., Ohio), John McCain (R., Ariz.), and Donald W. Riegle (D., Mich.) with varying degrees of criticism for "poor judgment" and "inappropriate" conduct and with a "reprimand" for Senator Alan Cranston (D., Calif.). See John R. Cranford, with Janet Hook and Phil Kuntz, "Decision in Keating Five Case Settles Little for Senate," *Congressional Quarterly Weekly Report* (March 2, 1991), pp. 517–23; Richard E. Cohen, "Congressional Chronicle," *National Journal* (August 17, 1991), p. 2043; and Phil Kuntz, "Cranston Case Ends on Floor with a Murky Plea Bargain," *Congressional Quarterly Weekly Report* (November 23, 1991), pp. 3432–38.

138. For a comparison of the three plans, and others, see *Congressional Quarterly Weekly Report* (July 29, 1989), pp. 1919–25.

139. Chuck Alston and Janet Hook, "An Election Lesson: Money Can Be Dangerous," *Congressional Quarterly Weekly Report* (November 19, 1988), p. 3367. See also Chuck Alston and Glen Craney, "Bush Campaign-Reform Plan Takes Aim at Incumbents," *Congressional Quarterly Weekly Report* (July 1, 1989), p. 1649.

140. Terry Cooper, quoted in Chuck Alston, "Campaign Finance Gridlock Likely to Persist," *Congressional Quarterly Weekly Report* (December 17, 1988), p. 3529.

141. See Alston and Craney, "Bush Campaign-Reform Plan Takes Aim at Incumbents," pp. 1648–49.

142. In regard to the 1992 campaign finance reform bill passed by Congress and vetoed by President Bush, see Beth Donovan and Susan Kellam, "Campaign Finance: Overhaul Bill Heads for Veto Despite Senate Approval," *Congressional Quarterly Weekly Report* (May 2, 1992), p. 1133; Beth Donovan, "Campaign Finance: Overhaul of Election Funding Unlikely to Become Law," *Congressional Quarterly Weekly Report* (April 11, 1992), pp. 931, 932; and Beth Donovan, "Campaign Finance: Overhaul Plan Readied as Tool to Blunt Scandals' Effects," *Congressional Quarterly Weekly Report* (April 4, 1992), pp. 861–63.

143. Quoted in Chuck Alston, "The Election-Reform Debate Keeps Its Partisan Shape," *Congressional Quarterly Weekly Report* (March 23, 1991), p. 724.

144. U.S. Representative William H. Gray III, quoted in Janet Hook, "Members Are Looking Hard at Family Finances," *Congressional Quarterly Weekly Report* (February 4, 1989), p. 205.

145. Author's own experience.

146. Roger Craver, quoted in "Need for Money Tempers Ethics Drive," *New York Times* (June 5, 1989), p. 1.

147. Mike Mills, "Raising Members' Pay: A 200-Year Dilemma," *Congressional Quarterly Weekly Report* (February 4, 1989), pp. 209–12.

148. Ibid., p. 209.

149. See Janet Hook, "Proposal for 51 Percent Pay Hike Sets Up Fracas," *Congressional Quarterly Weekly Report* (December 17, 1988), pp. 3522–24; and Janet Hook, "Congress Wavering on 51 Percent Salary Hike," *Congressional Quarterly Weekly Report* (February 4, 1989), pp. 203–8.

150. See *Congressional Quarterly Weekly Report* (January 7, 1989), p. 17.

151. See Janet Hook, "Pay Raise Killed, but the Headaches Persist," *Congressional Quarterly Weekly Report* (February 11, 1989), pp. 261–63; Janet Hook, "How the Pay-Raise Strategy Came Unraveled," *Congressional Quarterly Weekly Report* (February 11, 1989), pp. 264–67; and Michael Oreskes, "Defeat of Congressional Pay Increase Demolishes No-Vote System," *New York Times* (February 9, 1989), p. 8.

152. Senator Lowell P. Weicker, Jr. (R., Conn., 1970–1988), quoted in Janet Hook, "Members Consider Honoraria Ban, Higher Pay," *Congressional Quarterly Weekly Report* (July 2, 1988), p. 1809.

Chapter 4

1. Quoted in Clifford D. May, "At Bradley Roast, Barbs Are Good-Natured (Kind of)," *New York Times* (June 25, 1988), pp. B1, B5.

2. Ibid., p. B1.

3. Ibid., p. B5.

4. From personal individual interviews with fifty sitting members of the Senate, Washington, D.C., September and October, 1988.

5. Quoted in Richard L. Berke, "Bradley Goes One-on-One with Gates," *New York Times* (September 18, 1991), p. A14. In regard to Senator Bradley's speaking out on Bush's civil rights record, see Jack W. Germond and Jules Whitcover, "Bradley Is Outfront on Civil Rights," *National Journal* (July 20, 1991), p. 1822; and Janet Hook, "Members Are Homeward Bound After Close Calls at Polls," *Congressional Quarterly Weekly Report* (February 23, 1991), pp. 445–49.

6. John R. Hibbing and Sue Thomas, "The Modern United States Sen-

ate: What is Accorded Respect," *Journal of Politics* 52 (no. 1) (February 1990):126–45.

7. Survey of 317 congressional staff members (50 from the Senate) by Washington public relations firm Fleischman-Hillard, Inc., reported in *Roll Call* (February 7, 1988), p. 1.

8. *Washington Magazine* (July 1988).

9. Michael Barone and Grant Ujifusa, *The Almanac of American Politics 1990* (Washington, D.C.: National Journal, 1989), pp. 744, 745.

10. Donald R. Matthews, *U.S. Senators and Their World* (Chapel Hill: University of North Carolina Press, 1960).

11. Ibid., p. 93.

12. Ibid., p. 114.

13. Ibid., pp. 115, 116.

14. Ibid., pp. 92–102.

15. Ibid., pp. 116, 117; and Barbara Sinclair, *The Transformation of the U.S. Senate* (Baltimore: Johns Hopkins University Press, 1989), p. 15.

16. See John E. Firling, *The First of Men: A Life of George Washington* (Knoxville: University of Tennessee Press, 1988), pp. 87, 88.

17. J. A. Carroll and M. W. Ashworth, *George Washington* (New York: C. Scribner's Sons, 1957), VII, p. 591, cited and quoted in Matthews, *U.S. Senators and Their World*, fn. 24, p. 117.

18. Matthews, *U.S. Senators and Their World*, p. 116.

19. See, generally, Sinclair, *The Transformation of the U.S. Senate*, pp. 14–22.

20. Nelson W. Polsby, "Tracking Changes in the U.S. Senate," *PS: Political Science and Politics* (December 1989), pp. 789–93.

21. Personal interview with Charles Ferris, Washington attorney, former staff member for the Senate Democratic majority, 1963–1977, Washington, D.C., August 11, 1988.

22. Randall B. Ripley, *Congress: Process and Policy* (New York: W. W. Norton, 3rd ed., 1983), p. 126.

23. See Sinclair, *The Transformation of the U.S. Senate*, pp. 9, 10.

24. Matthews, *U.S. Senators and Their World*, pp. 163, 164.

25. Norman J. Ornstein, Thomas E. Mann, and Michael J. Malbin, *Vital Statistics on Congress 1989–1990* (Washington, D.C.: American Enterprise Institute, 1990), p. 123.

26. Sinclair, *The Transformation of the U.S. Senate*, pp. 25, 26; and Matthews, *U.S. Senators and Their World*, pp. 163–65.

27. Michael Foley, *The New Senate: Liberal Influence on a Conservative Institution 1959–1972* (New Haven, Conn.: Yale University Press, 1980), p. 251.

28. Ralph K. Huitt, "The Internal Distribution of Influence: The Senate," in ed. David Truman, *The Congress and America's Future* (Englewood Cliffs, N.J.: Prentice-Hall, 1965), p. 89.

29. Nearly 98 percent of committee-recommended bills were approved by voice vote in the Senate during the mid-decade Eighty-fourth Congress (1955–1956), and only thirty-three floor amendments were adopted by roll-call vote during that same two-year period. See Sinclair, *The Transformation of the U.S. Senate*, p. 23.

30. See Randall B. Ripley, *Power in the Senate* (New York: St. Martin's Press, 1969), pp. 3–19.

31. Sinclair, *The Transformation of the U.S. Senate*, p. 15.

32. Ornstein, Mann, and Malbin, *Vital Statistics on Congress 1989–1990*, p. 19.

33. At the beginning of each Congress, in January, *Congressional Quarterly Weekly Report* lists the birthdates of senators. Ages of senators for 1955–1963 were compiled by Mildred Amer, Congressional Research Service, Government Division, Library of Congress. Ages of senators for the 89th through 101st Congress (as of November 12, 1988, for the latter) are found in Julie Rovner, "Record Number of Women, Blacks in Congress," *Congressional Quarterly Weekly Report* (November 12, 1988), p. 3293.

34. Information for the 1950s is taken from Matthews, *U.S. Senators and Their World*, p. 55; and for the 1980s from the *Congressional Directory* for the appropriate years.

35. Information for the 1950s is taken from Matthews, *U.S. Senators and Their World*, p. 55; for 1982 from Larry Sabato, *Goodbye to Goodtime Charlie* (Washington, D.C.: Congressional Quarterly Press, 2nd ed., 1983); and for later years from the *Congressional Directory* in the years indicated. As to the reason why fewer governors are running for the Senate, see Frank Codispoti, "The Governorship-Senate Connection: A Step in the Structure of Opportunities Grows Weaker," *Publius* (Spring 1987), pp. 41–52.

36. Codispoti, "The Governorship-Senate Connection," p. 52.

37. Generally, in regard to the 1958 Senate election, the events surrounding it, and the results of it, see U.S. Senator Robert C. Byrd (D., W. Va.), "The Senate Class of 1958," *Congressional Record* (July 17, 1986), pp. 9221–25.

38. Foley, *The New Senate*, p. 94.

39. Sinclair, *The Transformation of the U.S. Senate*, pp. 30–33.

40. Byrd, "The Senate Class of 1958," p. S 9222.

41. Foley, *The New Senate*, pp. 92–117.

42. Ibid., p. 244.

43. Ibid., pp. 150, 130, and 153 (in order of quotations cited).

44. See Christopher J. Bailey, *The Republican Party in the U.S. Senate 1974–1984: Party Change and Institutional Development* (Manchester, England: Manchester University Press, 1988); and Christopher J. Bailey, "The United States Senate: The New Individualism and the New Right," *Parliamentary Affairs* 39 (July 1986):354–67.

45. Ornstein, Mann, and Malbin, *Vital Statistics on Congress 1989–1990*, p. 19.

46. Ibid.

47. Bailey, *The Republican Party in the U.S. Senate*, p. 10.

48. Ibid., pp. 6, 7.

49. Personal interview with Senator Kassebaum, Washington, D.C., September 14, 1988.

50. Quoted in *New York Times* (November 26, 1984), p. A1.

51. Bailey, *The Republican Party in the U.S. Senate*, p. 4.

52. Bailey, "The United States Senate," pp. 357, 358.

53. Quoted in *New York Times* (November 26, 1984), p. A1.

54. Quoted in *New York Times* (March 21, 1983), p. B6.

55. Personal individual interviews with fifty sitting senators, Washington, D.C., September and October, 1988. Among other things, most of the senators were asked how important they felt the Senate norms of legislative work, cour-

tesy, specialization, and reciprocity (with an explanation of each, as established by Donald Matthews's work) currently were. A majority were also asked (without seeking names from them, unless volunteered) whether there were senators, or a senator, whom other senators voted against almost automatically, and why. A majority were asked, too, to say who were the most influential or most respected among their colleagues, and why.

56. Hibbing and Thomas, "The Modern United States Senate." See also the survey of 317 congressional staff members (50 from the Senate) by Washington public relations firm Fleishman-Hillard, Inc., reported in *Roll Call* (February 7, 1988), p. 1.

57. A brief biography of Senator Jack Miller is found in *Biographical Directory of the United States Congress 1774–1989* (Washington, D.C.: U.S. Government Printing Office, 1989), p. 1502. This section is derived from the author's own experience, serving in the Senate with Mr. Miller.

58. From personal individual interviews with fifty sitting members of the Senate, Washington, D.C., September and October, 1988. A majority were asked the question: "Without asking for names, are there senators, or a senator, who is almost a sure no vote? Why?"

59. Matthews, *U.S. Senators and Their World*, p. 97.

60. Sinclair, *The Transformation of the U.S. Senate*, pp. 20, 21.

61. Ibid., p. 99, citing William S. Cohen, *Roll Call: One Year in the United States Senate* (New York: Simon and Schuster, 1981), p. 238.

62. Quoted in R. W. Apple, "Senate Confirms Thomas, 52–48, Ending Week of Bitter Battle; 'Time for Healing,' Judge Says," *New York Times* (October 16, 1991), p. A13.

63. Franklin L. Burdette, *Filibustering in the Senate* (Princeton, N.J.: Princeton University Press, 1940), p. 22.

64. Ibid., p. 27.

65. Ibid., p. 32.

66. Ibid., pp. 37, 38.

67. Ibid., pp. 3, 4.

68. Ibid., p. 181.

69. Ibid., pp. 181, 182; and Robert C. Byrd, *The Senate 1789–1989* (Washington, D.C.: U.S. Government Printing Office, 1988), Vol. I, p. 508.

70. Byrd, *The Senate 1789–1989*, Vol. I, p. 507.

71. Statement of Senator William Proxmire (D., Wis., 1957–1989) in a personal interview, September 15, 1988, Washington, D.C., and confirmed by Washington, D.C., attorney and lobbyist Robert C. McCandless, then an Oklahoman on the payroll of the Senate as a part of Senator Kerr's patronage, in an interview by telephone, January 1989.

72. Statement of Senator Quentin Burdick, (D., N.D.) in a personal interview, Washington, D.C., September 29, 1988.

73. Statement of Hubert H. Humphrey, then vice president, to the author, Fall 1968, reported in Fred R. Harris, *Potomac Fever* (New York: W. W. Norton, 1977), p. 153.

74. Ibid.

75. Ross K. Baker, *House and Senate* (New York: W. W. Norton, 1989), pp. 181, 182.

76. Personal interview with Senator Bradley, Washington, D.C., September 26, 1988.

77. Baker, *House and Senate*, p. 180.

78. Ibid., p. 181.

79. From personal individual interviews with fifty sitting members of the Senate, Washington, D.C., September and October, 1988.

80. Personal interview with Senator Boren, Washington, D.C., October 5, 1988.

81. Ibid.

82. Personal interview with Senator Sasser, Washington, D.C., September 14, 1988.

83. Personal interview with Senator Thurmond, Washington, D.C., August 31, 1988.

84. From personal individual interviews with fifty sitting members of the Senate, Washington, D.C., September and October, 1988.

85. Mentioned by Senator Edwin Jacob (Jake) Garn (R., Utah) in a personal interview, Washington, D.C., September 21, 1988.

86. Personal interview with Senator Exon, Washington, D.C., September 27, 1988.

87. See Paul Starobin, "At the Races: In Pierre, Pressler's No Joke," *National Journal* (October 20, 1990), p. 2544.

88. Reported in *Campaigns & Elections* (December 1988), p. 20.

89. Personal interview with Senator Cohen, Washington, D.C., September 8, 1988.

90. Personal interview with Senator Boren, Washington, D.C., October 5, 1988.

91. Baker, *House and Senate*, p. 180.

92. Personal interview with Senator Packwood, Washington, D.C., September 26, 1988.

93. Personal interview with Senator Fowler, Washington, D.C., October 4, 1988.

94. Personal interview with Senator Mikulski, Washington, D.C., September 15, 1988.

95. Quoted in Baker, *House and Senate*, p. 183.

96. Personal interview with Senator Nickles, Washington, D.C., September 30, 1988.

97. Personal interview with Senator Mikulski, Washington, D.C., September 15, 1988.

98. Matthews, *U.S. Senators and Their World*, pp. 96, 97.

99. Ibid. p. 95.

100. Foley, *The New Senate*, p. 150.

101. Ornstein, Mann, and Malbin, *Vital Statistics on Congress 1989–1990*, p. 120; and Sinclair, *The Transformation of the U.S. Senate*, p. 73.

102. Personal interview with Senator Lugar, Washington, D.C., September 16, 1988.

103. Personal interview with Senator Boren, Washington, D.C., October 5, 1988.

104. Sinclair, *The Transformation of the U.S. Senate*, pp. 86–88.

105. Ibid., p. 92.

106. David W. Rohde, Norman J. Ornstein, and Robert L. Peabody, "Political Change and Legislative Norms in the U.S. Senate, 1957–1974," in ed.

Glenn R. Parker, *Studies of Congress* (Washington, D.C.: Congressional Quarterly Press, 1985), p. 177.

107. From personal individual interviews with fifty sitting members of the Senate, Washington, D.C., September and October, 1988.

108. Julie Rovner, "Rockefeller Begins to Edge into Senate Limelight," *Congressional Quarterly Weekly Report* (January 27, 1990), p. 237.

109. Ibid.

110. Steven S. Smith, "Informal Leadership in the Senate: Opportunities, Resources, and Motivations," in ed. John J. Kornacki, *Leading Congress: New Styles, New Strategies* (Washington, D.C.: Congressional Quarterly Press, 1990), p. 74.

111. Quoted in Matthews, *U.S. Senators and Their World*, p. 94.

112. Ibid., pp. 94, 95.

113. Ibid., p. 95.

114. Laura I. Langbein and Lee Sigelman, in a paper, "Show Horses, Work Horses, and Dead Horses," presented at the annual meeting of the American Political Science Association, Washington, D.C., August 1988, showed that this division cannot be made for House members either.

115. Personal interview in Washington, D.C., with, respectively, Senator Rudman, September 15, and Senator Hatch, September 8, 1988.

116. James A. Miller, *Running in Place: Inside the Senate* (New York: Simon and Schuster, 1986), p. 121.

117. Stephen Hess, *The Ultimate Insiders: U.S. Senators in the National Media* (Washington, D.C.: Brookings Institution, 1986), p. 6.

118. Smith, "Informal Leadership in the Senate," pp. 75, 76.

119. Steven V. Roberts, "The Congress, the Press, and the Public," paper delivered at the "Understanding Congress: A Bicentennial Conference," sponsored by the U.S. Senate Commission on the Bicentennial, the Commission of the Bicentenary of the U.S. House of Representatives, and the Congressional Research Service of the Library of Congress, Caucus Room, Russell Senate Office Building, Washington, D.C., February 9, 1988, p. 7.

120. Hess, *The Ultimate Insiders*, p. 11.

121. Sinclair, *The Transformation of the U.S. Senate*, pp. 189, 190. Sinclair excluded Republican leader Robert Dole of Kansas from her calculation of averages. She found that he was mentioned 125 times on television network evening news programs and 119 times in *New York Times* stories in 1985 and appeared 13 times on Sunday television interview programs in 1985 and 1986. Sinclair also excluded references to New York's own two senators in arriving at her averages of 1985 mentions of senators in stories in the *New York Times*.

122. Ibid., p. 190.

123. Miller, *Running in Place*, 159.

124. Quoted in Tomme Jeanne Fent, "The Eye of the Beholder: Beauty or the Beast? The United States Senate in the 100th Congress," paper done for the Carl Albert Congressional Research and Studies Center, University of Oklahoma, July 1988, p. 15.

125. See Richard F. Fenno, Jr., *Home Style: House Members in Their Districts* (Boston: Little, Brown, 1978). See also Glenn R. Parker, *Homeward Bound: Explaining Congressional Behavior* (Pittsburgh: University of Pittsburgh Press, 1986).

126. Miller, *Running in Place*, pp. 178, 179.

127. In Fall 1988, the author's survey by questionnaire of fifty-three Senate administrative assistants showed an average number of senatorial visits home each year of 29.33, the median being 27.

128. Parker, *Homeward Bound*, pp. 72–76; and Glenn R. Parker, "Stylistic Change in the U.S. Senate: 1959–1980," *Journal of Politics* 47 (1985):1190–97.

129. See Hibbing and Thomas, "The Modern United States Senate," p. 139.

130. Nelson W. Polsby, *Congress and the Presidency* (Englewood Cliffs, N.J.: Prentice-Hall, 4th ed., 1986), p. 89.

131. Foley, *The New Senate*, p. 153.

132. Polsby, "Tracking Changes in the U.S. Senate," p. 789.

133. Baker, *House and Senate*, p. 197.

134. Ibid., p. 132.

135. Ibid., p. 195.

136. Smith, "Informal Leadership in the Senate," p. 72.

137. Hibbing and Thomas, "The Modern United States Senate," pp. 126, 139–42.

138. Unidentified senator quoted in Norman J. Ornstein, Robert L. Peabody, and David W. Rohde, "Change in the Senate: Toward the 1990s," in eds. Lawrence C. Dodd and Bruce I. Oppenheimer, *Congress Reconsidered* (Washington, D.C.: Congressional Quarterly Press, 4th ed., 1989), p. 18.

139. Personal interview with Senator Cohen, Washington, D.C., September 8, 1988.

140. Sinclair, *The Transformation of the U.S. Senate*, pp. 201, 202.

141. Ibid., p. 193.

142. Statement of Senator Richard Lugar (R., Ind.) at a seminar sponsored by the Project on Congressional Leadership of the Everett McKinley Dirksen Congressional Center and the Congressional Research Service, September 30, 1987, Washington, D.C.

143. Quoted in Clifford D. May, "Learning to Govern the Bill Bradley Way," *New York Times* (March 1, 1988), pp. B1, B6.

144. Jacqueline Calmes and Rob Gurwitt, "Profiles in Power: Leaders Without Portfolio," *Congressional Quarterly Weekly Report* (January 3, 1987), pp. 11–13; and *Washington Post National Weekly Edition* (June 2, 1986), p. 24. See also, generally, Jeffrey H. Birnbaum and Alan S. Murray, *Showdown at Gucci Gulch* (New York: Random House, 1987).

145. Matthews, *U.S. Senators and Their World*, pp. 98–101.

146. Ibid., pp. 100, 101.

147. See Sinclair, *The Transformation of the U.S. Senate*, pp. 97, 98.

148. Ibid., pp. 80, 81.

149. Ibid., p. 82. Steven S. Smith has noted that it is "a little complicated" to discern "the effects of declining committee deference on patterns of floor participation"—because, for example, of the need to take changing committee cohesiveness into account, as well as the frequent lack of distinction between first- and second-degree amendments and friendly amendments offered to head off hostile ones. But Smith notes that, with the numbers of places on committees having increased, even a constant number of noncommittee amendments would, in effect, represent an increase in noncommittee activity on the floor. See Smith, *Call to Order*, p. 142.

150. For a history and explanation of the filibuster in the U.S. Senate, see Franklin L. Burdette, *Filibustering in the Senate* (Princeton, N.J.: Princeton University Press, 1940).

151. Sinclair, *The Transformation of the U.S. Senate*, p. 94.

152. Quoted in Julie Johnson, "A Tale of Two Legislators and How They View Their Institution," *New York Times* (December 21, 1987), p. 12.

153. See John E. Yang, "Senator No: Metzenbaum Amasses Power in the Senate by Blocking Action," *Wall Street Journal* (June 29, 1988), pp. 1, 17.

154. Ibid., p. 1.

155. See Helen Dewar, "Exasperated Senators Rebuff Helms," *Washington Post* (September 20, 1988), pp. A1, A6.

156. Statement to the author by a northern liberal senator who asked not to be identified, Washington, D.C., July 28, 1988.

157. See Janet Hook, "Mitchell Learns Inside Game; Is Cautious as Party Voice," *Congressional Quarterly Weekly Report* (September 9, 1989), p. 2294.

158. Personal individual interviews with fifty incumbent members of the Senate, Washington, D.C., Fall 1988.

159. Matthews, *U.S. Senators and Their World*, p. 93.

160. Ibid., p. 94.

161. See David W. Rohde, Norman J. Ornstein, and Robert L. Peabody, "Political Change and Legislative Norms in the U.S. Senate, 1957–1974," in ed. Glenn R. Parker, *Studies of Congress* (Washington, D.C.: Congressional Quarterly Press, 1985), pp. 158–67.

162. Author's own recollections from having served as a Democratic member of the Senate from 1964 to 1973.

163. Rohde, Ornstein, and Peabody, "Political Change and Legislative Norms in the U.S. Senate, 1957–1974," p. 175.

164. Sinclair, *The Transformation of the U.S. Senate*, p. 83; and Smith, *Call to Order*, p. 136.

165. Smith, *Call to Order*, p. 136.

166. Sinclair, *The Transformation of the U.S. Senate*, p. 77.

167. *Congressional Quarterly Weekly Report* (December 23, 1989), pp. 3374–88.

168. Ibid.

169. Author's own recollections as a Democratic member of the Senate from 1964 to 1973. See also Foley, *The New Senate*, p. 187.

170. Ripley, *Congress: Process and Policy*, p. 131.

171. Michael Barone and Grant Ujifusa, *The Almanac of American Politics* (Washington, D.C.: National Journal, 1988), pp. 875, 876.

172. Recollection of the author from a closed session of the Senate to consider the creation of a Senate Select Committee on Intelligence Oversight and methods for choosing its membership, circa 1967.

173. Hibbing and Thomas, "The Modern United States Senate," pp. 139, 142.

174. Ibid., p. 141.

175. Ibid., p. 140.

176. Matthews, *U.S. Senators and Their World*, pp. 101, 102.

177. Fenno, Jr., *Home Style*, pp. 164–68.

178. Personal interview with Senator Chiles, Washington, D.C., September 27, 1988.

179. Quoted in Johnson, "A Tale of Two Legislators and How They View Their Institution," p. 12.

180. Personal interview with Senator Evans, Washington, D.C., September 27, 1988.

181. Daniel J. Evans, "Why I'm Quitting the Senate," *New York Times Magazine* (April 17, 1988), p. 91.

182. See Baker, *House and Senate*, pp. 185, 186.

Chapter 5

1. Mark Hatfield, "The Senate of 1889," *Congressional Record* (April 6, 1989), pp. 3407, 3408.

2. Information in this section on the growth in Senate staff is taken from Norman J. Ornstein, Thomas E. Mann, and Michael J. Malbin, *Vital Statistics on Congress, 1987–1988* (Washington, D.C.: Congressional Quarterly Press, 1988), pp. 142, 146.

3. "The First Senate Office Building," *Senate History*, no. 11 (February 1986), Historical Office, Office of the Secretary, United States Senate, pp. 1, 9, 10.

4. Ibid., p. 9.

5. Information in this section concerning the construction, authorization, and cost of the three Senate office buildings is taken from *1987–1988 Official Congressional Directory* (Washington, D.C.: U.S. Government Printing Office, 1987), pp. 743, 744.

6. Quoted in 1972 *Congressional Quarterly Almanac* (Washington, D.C.: Congressional Quarterly Press, 1972), p. 691.

7. Andy Plattner, "New Quarters Await Members in Hart Senate Office Building," *Congressional Quarterly Weekly Report* (September 4, 1982), pp. 2203, 2204.

8. See Martin Tolchin, "19,000 Congressional Aides Discover Power but Little Glory on Capitol Hill," *New York Times* (November 12, 1991), p. A12; and *Special Report: The Cost of Congress*, Democratic Study Group, U.S. House of Representatives, no. 100–32 (May 17, 1988), p. 10.

9. Remarks of Senator Wyche Fowler (D., Ga.) at the "Understanding Congress: A Bicentennial Research Conference," sponsored by the U.S. Senate Commission on the Bicentennial, the Commission on the Bicentenary of the U.S. House of Representatives, and the Congressional Research Service of the Library of Congress, Caucus Room, Russell Senate Office Building, Washington, D.C., February 9, 1989.

10. Quoted in George B. Galloway, *Congress at the Crossroads* (New York: Thomas Y. Crowell, 1946), p. 157.

11. Peter Woll, *Congress* (Boston: Little, Brown, 1985), p. 191.

12. Congressional Management Project, Burdette A. Loomis, Director, *Setting Course: A Congressional Management Guide* (Washington, D.C.: American University, 1984), p. 151.

13. Barbara Sinclair, *The Transformation of the U.S. Senate* (Baltimore: Johns Hopkins University Press, 1989), p. 14.

14. Personal recollections of the author who, as a member of the Senate Finance Committee, made the motion mentioned.

15. Barbara Hinckley, *Stability and Change in Congress* (New York: Harper & Row, 4th ed., 1988), p. 88.

16. See *Congressional Record* (June 9, 1975), pp. 17836–63.

17. Norman J. Ornstein, Robert L. Peabody, and David W. Rohde, "Change in the Senate: Toward the 1990s," in eds. Lawrence C. Dodd and Bruce I. Oppenheimer, *Congress Reconsidered* (Washington, D.C.: Congressional Quarterly Press, 4th ed., 1989), p. 20.

18. Ibid.

19. See Steven S. Smith, "Taking It to the Floor," in eds. Dodd and Oppenheimer, *Congress Reconsidered*, p. 334.

20. See Michael J. Malbin, *Unelected Representatives* (New York: Basic Books, 1980).

21. Quoted in Harrison W. Fox, Jr., and Susan Webb Hammond, *Congressional Staffs: The Invisible Force in American Lawmaking* (New York: Free Press, 1977), p. 5.

22. Ibid., quoting Senator Ernest F. Hollings (D., S.C.).

23. From interviews with U.S. senators by the Center for Responsive Politics, Washington, D.C., compiled April 30, 1987, and made available to the author with the understanding that the identities of the senators would not be disclosed.

24. See David L. Boren, "Special Order: Congress in Trouble!" *Extensions* (Winter 1992), pp. 3–5.

25. Stephen Isaacs, *Washington Post* (February 16, 1975), quoted in Woll, *Congress*, p. 191.

26. Nelson W. Polsby, *Congress and the Presidency* (Englewood Cliffs, N.J.: Prentice-Hall, 4th ed., 1986), p. 89.

27. Norman J. Ornstein, Robert L. Peabody, and David W. Rohde, "Change in the Senate," p. 33.

28. Steven S. Smith, *Call to Order: Floor Politics in the House and Senate* (Washington, D.C.: Brookings Institution, 1989), p. 10.

29. Personal interview with Senator Mikulski, Washington, D.C., September 15, 1988.

30. Sinclair, *The Transformation of the U.S. Senate,* p. 146.

31. Ibid., pp. 79, 93, 94.

32. See Glenn R. Parker, "Members of Congress and Their Constituents: The Home-Style Connection," in eds. Dodd and Oppenheimer, *Congress Reconsidered,* p. 188.

33. See Barbara Sinclair, "Campaign Bonus or Kiss of Death: The Impact of National Prominence on Senate Elections," paper prepared for delivery at the annual meeting of the Western Political Science Association, March 22–24, 1990, Newport Beach, California.

34. *Congressional Quarterly Weekly Report* (1982), p. 2177.

35. For the figure of sixteen hundred staff members in home-state offices, see Walter Pincus, "New Computers Herald a New Kind of Senate," *Washington Post* (July 18, 1989), p. A21. The other figures are taken from Norman J. Ornstein, Thomas E. Mann, and Michael J. Malbin, *Vital Statistics on Congress 1989–1990* (Washington, D.C.: Congressional Quarterly Press, 1990), p. 135. As a result of a survey of the administrative assistants to fifty-six members of the U.S. Senate in 1988, the author found that the average senator stationed nearly 40 percent of his or her personal staff in the home-state offices.

36. See Glenn R. Parker, "Members of Congress and Their Constituents," p. 189.

37. See Janet Hook, "Mitchell Learns Inside Game; Is Cautious as Party Voice," *Congressional Quarterly Weekly Report* (September 9, 1989), p. 2294.

38. Walter Pincus, "New Computers Herald a New Kind of Senate," *Washington Post* (July 18, 1989), p. A21.

39. Ibid., p. A21.

40. See Walter Pincus, "Hill's Mass-Mailing Ban Falters," *Washington Post* (September 26, 1989), p. A1; and Walter Pincus, "$82 Million for Missives to 'Occupant,'" *Washington Post* (September 19, 1989), p. A25.

41. Walter Pincus, "Senate Seeks High-Tech Ways to Get Out the Word," *Washington Post* (January 28, 1989), p. A11.

42. William Boot, "Hustling the Folks Back Home," *Columbia Journalism Review* (November–December 1987), pp. 22–24.

43. Walter Pincus and Dan Morgan, "Using Public and Campaign Funds, Senators Beam Selves Back Home," *Washington Post* (November 4, 1988), p. A28. See also Teresa Riordan, "Beam Me Up Scotty: I Wanna Be on the Six O'Clock News," *Common Cause Magazine* (July/August 1986), pp. 13–15.

44. Pincus Morgan, "Using Public and Campaign Funds."

45. Woodrow Wilson, *Congressional Government* (New York: Meridian, 1956), pp. 69, 146. This book was first published by Houghton Mifflin in 1885.

46. "Report Together with Proposed Resolutions," Temporary Select Committee to Study the Senate Committee System, 98th Congress, 2nd Session (Washington, D.C.: U.S. Government Printing Office, 1984), p. 4.

47. Kenneth A. Shepsle and Barry R. Weingast, "The Institutional Foundations of Committee Power," *American Political Science Review* 81 (no. 1) (March 1987):85–104.

48. In regard to types of congressional committees, see Fred R. Harris and Paul L. Hain, *America's Legislative Processes: Congress and the States* (Glenview, Ill.: Scott, Foresman, 1983), pp. 252–55.

49. Woll, *Congress,* p. 313.

50. Sinclair, *The Transformation of the U.S. Senate*, p. 73.

51. *Congressional Quarterly Weekly Report* (December 23, 1989), pp. 3476–88.

52. See Dan Quayle, "Reform of the Senate Committee System—and More," *Congressional Record* (September 12, 1984), p. 10958.

53. Foley, *The New Senate,* p. 151.

54. See Ornstein, Mann, and Malbin, *Vital Statistics on Congress 1989–1990,* p. 119.

55. See Ornstein, Peabody, and Rohde, "Change in the Senate," p. 32.

56. Ibid.; and U.S. Senate Committee on Rules and Administration, Report on Senate Operations 1988, 100th Congress, 2nd Session, September 20, 1988, pp. 18–21.

57. See U.S. Senate Committee on Rules and Administration, Report on Senate Operations 1988, pp. 26–27.

58. Ibid., pp. 29–33.

59. Ibid., p. 47.

60. *Congressional Quarterly Weekly Report* (December 23, 1989), pp. 3476–88.

61. Ornstein, Mann, and Malbin, *Vital Statistics on Congress 1989–1990,* p. 120; and Sinclair, *The Transformation of the U.S. Senate*

62. *Congressional Quarterly Weekly Report* (December 23, 1989), pp. 3476–88.

63. Ibid.

64. Michael J. Barone and Grant Ujifusa, *The Almanac of American Politics 1990* (Washington, D.C.: National Journal, 1989), p. 986.

65. *Congressional Quarterly Weekly Report* (December 23, 1989), p. 3475.

66. Barone, *The Almanac of American Politics 1990,* p. 749.

67. *Congressional Quarterly Weekly Report* (December 23, 1989), p. 3474.

68. The average in 1985 was 148. Roger H. Davidson and Thomas Kephart, "Indicators of Senate Activity and Workload," Congressional Research Service Report No. 85–133S, 1985.

69. Senator Paul Simon (D., Ill.), quoted in Andy Plattner, "The Lure of the Senate: Influence and Prestige," *Congressional Quarterly Weekly Report* (May 25, 1985), p. 994.

70. Center for Responsive Politics, *Congressional Operations: Congress Speaks—A Survey of the 100th Congress* (Washington, D.C.: Center for Responsive Politics, 1988), pp. 200–202.

71. From interviews with U.S. Senators by the Center for Responsive Politics, Washington, D.C., compiled April 30, 1987, and made available to the author with the understanding that the identities of the senators would not be disclosed.

72. Senator Mark Andrews (R., N.D., 1981–1986), quoted in Plattner, "The Lure of the Senate," p. 993.

73. Center for Responsive Politics, *Congressional Operations: Congress Speaks—A Survey of the 100th Congress,* pp. 200–202.

74. Quoted in Sinclair, *The Transformation of the U.S. Senate,* p. 153.

75. Center for Responsive Politics, *Congressional Operations,* pp. 168–75.

76. See S. Res. 260, U.S. Senate, 100th Congress, June 1987; Opening Statement of Senator Nancy Landon Kassebaum (R., Kan.) before the Senate Committee on Rules and Administration, February 25, 1988; and Senate Committee on Rules and Administration, Report on Senate Operations 1988, 100th Congress, 2nd Session, September 20, 1988, pp. 45–49.

77. Opening Statement of Senator Nancy Landon Kassebaum (R., Kan.) before the Senate Committee on Rules and Administration, February 25, 1988.

78. Nancy Landon Kassebaum, "The Senate Is Not in Order," *Washington Post* (January 27, 1988).

79. Senate Committee on Rules and Administration, Report on Senate Operations 1988, pp. 48, 49.

80. Ibid., p. 49.

81. Personal recollections of the author who, as a member of the Senate Finance Committee, made the motion mentioned.

82. See Leroy N. Rieselbach, *Congressional Reform in the Seventies* (Morristown, N.J.: General Learning Press, 1977), p. 48; and *Congressional Quarterly Weekly Report* (November 8, 1975), pp. 2413–14.

83. Sinclair, *The Transformation of the U.S. Senate,* pp. 105, 106.

84. Ronald Garay, *Congressional Television: A Legislative History* (Westport, Conn.: Greenwood Press, 1984), p. 36. Except as otherwise indicated,

material concerning the broadcasting of Senate proceedings is taken from this source.

85. See William S. White, *The Citadel: The Story of the U.S. Senate* (New York: Harper and Brothers, 1957), pp. 78, 113; and Donald R. Matthews, *U.S. Senators and Their World* (Chapel Hill: University of North Carolina Press, 1959), p. 109.

86. Virgil C. McClintock, "Congressional Inquisition by Television," *Oklahoma Law Review* 5 (May 1952):235.

87. See Garay, *Congressional Television,* pp. 46, 47.

88. Joint Committee on Congressional Operations, Hearings on "Congress and Mass Communications," 93rd Congress, 2nd Session, 1974, p. 2.

89. See Len Allen, "Television from the Senate Floor," in *Commission on the Operation of the Senate, Senate Communications with the Public,* Committee Print, 94th Congress, 2nd Session (Washington, D.C.: U.S. Government Printing Office, 1977), p. 91.

90. See Richard F. Fenno, Jr., "The Senate Through the Looking Glass: The Debate over Television," in ed. John Hibbing, *The Changing World of the U.S. Senate* (Berkeley, Calif.: IGS Press, 1990), pp. 183–215; and Garay, *Congressional Television,* pp. 118–29.

91. Hearings on S. Res. 20, Committee on Rules and Administration, U.S. Senate, Washington, D.C., April 8, 9, and May 5, 1981, pp. 4, 5.

92. Ibid., p. 9.

93. *Congressional Record* (February 2, 1982), p. 273.

94. Ibid., pp. 280, 282.

95. Hearings on S. Res. 20, Committee on Rules and Administration, p. 81.

96. *Congressional Record* (February 20, 1986), p. 1449.

97. *Congressional Record* (April 15, 1982), p. 3581.

98. Ibid.; and *Congressional Record* (February 2, 1982), p. 280.

99. *Congressional Record* (February 23, 1986), p. 843.

100. See *Congressional Record* (February 2, 1982), p. 281; and *Congressional Record* (February 3, 1982), p. 347.

101. *Congressional Record* (February 27, 1986), p. 1736.

102. *Congressional Record* (February 24, 1986), p. 1511; and *Congressional Record* (February 27, 1986), pp. 1731–33.

103. *Congressional Record* (February 27, 1986), p. 1756; and *Congressional Record* (July 29, 1986), p. 9775.

104. Paul S. Reundquist and Llona B. Nickles, "Senate Television: Its Impact on Senate Floor Proceedings," Congressional Research Service, Washington, D.C., July 21, 1986, reprinted in *Congressional Record* (July 22, 1986), pp. 9505–10.

105. *Congressional Record* (July 29, 1986), pp. 9766, 9767.

106. *Congressional Record* (July 29, 1986), p. 9766.

107. Brian Nutting, "After One Year of Television, Senate Is Basically Unchanged," *Congressional Quarterly Weekly Report* (May 30, 1987), p. 1140.

108. Quoted in Alan Ehrenhalt, "In the Senate of the '80s, Team Spirit Has Given Way to the Rule of Individuals," *Congressional Quarterly Weekly Report* (September 4, 1982), p. 2178.

109. Ibid.

110. *Congressional Record* (August 7, 1987), p. 11602.

111. U.S. Senate Committee on Rules and Administration, Report on Senate Operations 1988, p. 9.

112. Steven S. Smith, "Taking It to the Floor," pp. 334–36.

113. Smith, *Call to Order,* pp. 141, 142.

114. Sinclair, *The Transformation of the U.S. Senate,* p. 80.

115. Ibid., pp. 80, 81.

116. Steven S. Smith, "Informal Leadership in the Senate: Opportunities, Resources, and Motivations," in ed. John J. Kornacki, *Leading Congress: New Styles, New Strategies* (Washington, D.C.: Congressional Quarterly Press, 1990), p. 81.

117. Sinclair, *The Transformation of the U.S. Senate,* p. 82. Steven S. Smith has noted that it is "a little complicated" to discern "the effects of declining committee deference on patterns of floor participation"—because, for example, of the need to take changing committee cohesiveness into account, as well as the frequent lack of distinction between first- and second-degree amendments and friendly amendments offered to head off hostile ones. But Smith notes that, with the numbers of places on committees having increased, even a constant number of noncommittee amendments would, in effect, represent an increase in noncommittee activity on the floor. See Smith, *Call to Order,* p. 142.

118. Sinclair, *The Transformation of the U.S. Senate,* p. 82.

119. See Ilona B. Nickels, Issue Brief, "Senate Procedure, Rules, and Organization: Proposals for Change in the 100th Congress," Congressional Research Service, July 11, 1988, p. 6.

120. *Congressional Record* (December 16, 1982), p. 15285.

121. U.S. Senate Committee on Rules and Administration, *Report on Senate Operations 1988,* p. 50.

122. See U.S. Senate Committee on Rules and Administration, *Report on Senate Operations 1988,* pp. 52, 53.

123. Senator Daniel Evans (R., Wash.), *Congressional Record* (October 8, 1987), pp. 13869, 13870.

124. U.S. Senate Committee on Rules and Administration, *Report on Senate Operations 1988,* p. 54.

125. *Congressional Record* (August 7, 1987), p. 11602.

126. Ibid.

127. Statement made at "Understanding Congress: A Bicentennial Research Conference," February 9, 1989, Senate Caucus Room, Washington, D.C.

128. See Sinclair, *The Transformation of the U.S. Senate,* pp. 127, 128.

129. See Michael Foley, *The New Senate: Liberal Influence on a Conservative Institution 1959–1972* (New Haven, Conn.: Yale University Press, 1980), p. 177.

130. Randall B. Ripley, *Congress: Process and Policy* (New York: W. W. Norton, 1983), pp. 148, 149.

131. Jacqueline Calmes, "'Trivialized' Filibuster Is Still a Potent Tool," *Congressional Quarterly Weekly Report* (September 5, 1987), p. 2115.

132. See Sinclair, *The Transformation of the U.S. Senate,* pp. 129, 130.

133. See Calmes, "'Trivialized' Filibuster Is Still a Potent Tool," pp. 2115–19; and Sinclair, *The Transformation of the U.S. Senate,* pp. 126–29.

134. Information compiled by the office of Senator David Pryor (D., Ark.), January 27, 1988.

135. Ibid.

136. See Richard E. Cohen, "World's Greatest Non-Debating Society," *National Journal* (August 3, 1991), p. 1940.

137. See Calmes, "'Trivialized' Filibuster Is Still a Potent Tool," pp. 2115–19.

138. See Irvin Molotsky, "Blaming G.O.P., Democrats Drop Effort to Raise the Minimum Wage," *New York Times* (September 27, 1988), p. A22.

139. See Franklin L. Burdette, *Filibustering in the Senate* (Princeton: Princeton University Press, 1940), pp. 43, 44.

140. In regard to the postcloture filibuster generally, see Christopher J. Bailey, "The United States Senate: The New Individualism and the New Right," *Parliamentary Affairs* 39 (July 1986):356, 357.

141. James G. Abourezk, *Advise & Dissent: Memoirs of South Dakota and the U.S. Senate* (Chicago: Lawrence Hill Books, 1989), p. 134.

142. Calmes, "'Trivialized' Filibuster Is Still a Potent Tool," pp. 2115–19; and Ann Cooper, "Conservatives Learning How to Run a Filibuster on Labor Revision Bill," *Congressional Quarterly Weekly Report* (June 17, 1978), pp. 1519–22.

143. Abourezk, *Advise & Dissent,* pp. 133–45.

144. Calmes, "'Trivialized" Filibuster Is Still a Potent Tool," p. 2118.

145. Linda Greenhouse, "The New Improved (?) Filibuster in Action," *New York Times* (May 20, 1987), p. B10.

146. *Congressional Record* (August 7, 1987), p. 11602.

147. Nelson W. Polsby, "Postwar Institutional Development," in ed. John Hibbing, *The Changing World of the U.S. Senate* (Berkeley: Institute of Governmental Studies Press, 1990), p. 384.

148. Comments of Nelson Polsby as a panelist at the Hendricks Symposium on the U.S. Senate, University of Nebraska, Lincoln, October 6–8, 1988.

149. Sinclair, *The Transformation of the U.S. Senate,* p. 139.

Chapter 6

1. For a review of the Tower confirmation fight in the Senate, see Elizabeth Drew, "Letter from Washington," *New Yorker* (March 20, 1989), pp. 97–106; Pat Towell, "Senate Panel Deals Bush His First Defeat," *Congressional Quarterly Weekly Report* (February 25, 1989), pp. 396–99; Pat Towell, "Senate Wages Partisan Duel on Fitness, Prerogatives," *Congressional Quarterly Weekly Report* (March 4, 1989), pp. 461–67; and Pat Towell, "Senate Spurns Bush's Choice in a Partisan Tug of War," *Congressional Quarterly Weekly Report* (March 11, 1989), pp. 530–34.

2. See "Rejected Nominees," *Congressional Quarterly Weekly Report* (March 4, 1989), p. 465.

3. Drew, "Letter from Washington," p. 99.

4. Ibid., p. 98.

5. Statement of the late Senator Robert S. Kerr (D., Okla.), related to the author by Kerr's administrative assistant, Burl Hays, in 1964.

6. Helen Dewar, "Defeat of Tower Reflects Dramatic Changes in Senate," *Washington Post* (March 13, 1990), pp. A1, A6, A7.

7. Drew, "Letter from Washington," p. 98. See also Sam Nunn, "Sen.

Nunn Replies: We Are Not Being Unfair," *Washington Post* (March 3, 1989), p. A19.

8. Nunn, "Sen. Nunn Replies."

9. Pat Towell, "Vote on Tower Nomination Delayed by New Allegations," *Congressional Quarterly Weekly Report* (February 11, 1989), p. 259.

10. Towell, "Senate Panel Deals Bush His First Defeat," p. 396.

11. Bernard Weintraub, "Senate Opposition to Tower Nomination Marks a Personal Defeat for Bush," *New York Times* (February 24, 1989), p. A10.

12. Drew, "Letter from Washington," p. 98.

13. Ibid.

14. Robin Toner, "Nunn Bipartisan Image Fades in Tower Fight," *New York Times* (March 21, 1989), p. A22.

15. John R. Hibbing and Sue Thomas, "The Modern United States Senate: What Is Accorded Respect," *Journal of Politics* 52 (no. 1) (February 1990):127–45. See also a survey of 317 congressional staff members by the Washington public relations firm of Fleishman-Hillard, Inc., reported in *Roll Call* (February 7, 1987), p. 1.

16. "The Teflon Senator Slugs It Out," *Congressional Quarterly Weekly Report* (March 4, 1989), p. 462.

17. See Martin Schram, "Towering Inferno," *Washingtonian* (May 1989), pp. 109–16.

18. Statement made at presidential press conference, quoted in *New York Times* (March 8, 1989), p. A12.

19. Quoted in Towell, "Senate Panel Deals Bush His First Defeat," p. 396.

20. Quoted in Dan Balz, "How Tower Used His 'No' Vote," *Washington Post* (March 8, 1989), p. A21.

21. See Michael Oreskes, "Senate Rejects Tower 53–47; First Cabinet Veto Since '59; Bush Confers on New Choice," *New York Times* (March 10, 1989), p. 1.

22. The changeable senators were Republican Nancy Landon Kassebaum of Kansas and Democrat Lloyd Bentsen of Texas. See Drew, "Letter from Washington," p. 105; and Don Phillips, "Kassebaum Told Bush of Decision," *Washington Post* (March 10, 1989), p. A18.

23. Quoted in Towell, "Senate Spurns Bush's Choice in a Partisan Tug of War," p. 530.

24. Quoted in Oreskes, "Senate Rejects Tower 53–47," p. A10.

25. Ibid.

26. Quoted in Drew, "Letter from Washington," p. 106.

27. In regard to the Clarence Thomas confirmation generally, see Andrew Rosenthal, "Thomas's Edge Steady, Vote Due Today," *New York Times* (October 14, 1991), p. A1; R. W. Apple, Jr., "Senate Confirms Thomas, 52–48, Ending Week of Bitter Battle; 'Time for Healing,' Judge Says," *New York Times* (October 16, 1991), p. A1; and Joan Biskupic, "Thomas' Victory Puts Icing on Reagan-Bush Court," *Congressional Quarterly Weekly Report* (October 19, 1991), pp. 3026–33.

28. See Richard E. Cohen, "Advice, Consent and Political Games," *National Journal* (October 5, 1991), p. 2436.

29. See Adam Mitzner, "The Evolving Role of the Senate in Judicial Nominations," *Journal of Law and Politics* 5 (Winter 1989):387–428.

30. See Maureen Dowd, "Getting Nasty Early Helps G.O.P. Gain Edge on Thomas," *New York Times* (October 15, 1991), p. A1.

31. Quoted in Biskupic, "Thomas' Victory Puts Icing on Reagan-Bush Court," p. 3033.

32. A *New York Times* poll showed that 58 percent believed Thomas, 24 percent Hill, and that a plurality—45 percent to 20 percent—thought he should be confirmed. Elizabeth Kilbert, "Survey Finds Most of Public Believes Nominee's Account," *New York Times* (October 15, 1991), p. A1.

33. See Biskupic, "Thomas' Victory Puts Icing on Reagan-Bush Court," p. 3026.

34. For a discussion of political parties in America see Fred R. Harris and Gary Wasserman, *America's Government* (Glenview, Ill.: Scott Foresman/Little, Brown, 1990), pp. 209–43.

35. David S. Broder, *The Party's Over: The Failure of Politics in America* (New York: Harper & Row, 1971).

36. Larry J. Sabato, *The Party's Just Begun: Shaping Political Parties for America's Future* (Glenview, Ill.: Scott, Foresman, 1988).

37. John R. Petrocik, "Realignment: The South, New Party Coalitions and the Elections of 1984 and 1986," in eds. Warren E. Miller and John R. Petrocik, *Where's the Party? An Assessment of Changes in Party Loyalty and Party Coalitions in the 1980s* (Washington, D.C.: Center for National Policy, 1987), p. 52; and Sabato, *The Party's Just Begun*, p. 161.

38. William J. Keefe, *Parties, Politics, and Public Policy in America* (New York: Holt, Rinehart, and Winston, 4th ed., 1984), p. 114, and Gallup polls through the years cited there; Andrew Kohut and Larry Hugick, "Republican Party's Image at a High Point," *Gallup Report* (July 1989), p. 2; and *Gallup Poll Monthly* (May 1990), p. 36. Note that the Gallup poll on party affiliation excludes persons who say "don't know" or "other party."

39. Ibid.

40. Ibid.

41. *Gallup Poll Monthly* (May 1990), p. 36.

42. Ibid.

43. Data from the Michigan Survey Research Center of the Center for Political Studies, *National Election Studies* series, cited in Warren E. Miller, "The Electorate's View of Parties," in ed. L. Sandy Maisel, *The Parties Respond: Changes in the American Party System* (Boulder, Colo.: Westview Press, 1990), p. 108. These data show among voters a Democratic advantage over Republicans in 1956, for example, of 15 percent, compared with only 3 percent in 1988.

44. See Rhodes Cook, "Republican Fountain of Youth May Be Springing a Leak," *Congressional Quarterly Weekly Report* (November 25, 1989), pp. 3263–65.

45. Ibid., p. 155.

46. Warren E. Miller, "Party Identification Re-Examined: The Reagan Era," in eds. Miller and Petrocik, *Where's the Party?*, p. 22.

47. Warren E. Miller, "The Electorate's View of Parties," in ed. Maisel, *The Parties Respond*, p. 109.

48. Cook, "Republican Fountain of Youth May Be Springing a Leak," p. 3265.

49. Miller, "The Electorate's View of Parties," pp. 113, 114.

50. See, for example, Seymour Martin Lipset, "The Elections, the Economy, and Public Opinion," *PS, Political Science and Politics* (no. 1) (Winter 1985):28–38.

51. Martin P. Wattenberg, *The Decline of American Political Parties 1952–1988* (Cambridge, Mass.: Harvard University Press, 1990), p. 145.

52. Ibid., p. 155.

53. Sabato, *The Party's Just Begun,* pp. 112, 114.

54. Ibid., p. 133.

55. Wattenberg, *The Decline of American Political Parties,* p. 165.

56. See Stephen C. Craig, "The Decline of Partisanship in the United States: A Reexamination of the Neutrality Hypothesis," *Political Behavior* 7 (1985):57–78; and John E. Stanga and James F. Sheffield, "The Myth of Zero Partisanship: Attitudes Toward American Political Parties, 1964–1984," *American Journal of Politics* 31 (1987):829–55.

57. Miller, "The Electorate's View of Parties," pp. 104–7.

58. Sabato, *The Party's Just Begun,* p. 113.

59. Wattenberg, *The Decline of American Political Parties 1952–1988,* pp. 156, 163.

60. See Ray Wolfinger, "Concern About the Growth of Independent Voters Is Much Ado About Nothing," *Public Affairs Report,* Institute of Governmental Studies, University of California at Berkeley (March 1992), p. 5.

61. Ibid., p. 4; Norman J. Ornstein, Thomas E. Mann, and Michael J. Malbin, *Vital Statistics on Congress 1991–1992* (Washington, D.C.: Congressional Quarterly, 1992), p. 67; and Morris P. Fiorina, "The Electorate in the Voting Booth," in ed. Maisel, *The Parties Respond,* p. 121.

62. See discussion in James L. Sundquist, "Strengthening the National Parties," in ed. A. James Reichley, *Elections American Style* (Washington, D.C.: Brookings Institution, 1987), p. 205.

63. James E. Campbell and Joe A. Sumners, "Presidential Coattails in Senate Elections," *American Political Science Review* 84 (no. 2) (June 1990):521.

64. John Petrocik, "Realignment: New Party Coalitions and the Nationalization of the South," *Journal of Politics* 49 (1987):348–75.

65. Harold W. Stanley, "Southern Partisan Changes: Dealignment, Realignment or Both?" *Journal of Politics* 50 (1988):64–88.

66. Earl Black and Merle Black, *Politics and Society in the South* (Cambridge, Mass.: Harvard University Press, 1987), pp. 312–16.

67. William M. Lunch, *The Nationalization of American Politics* (Berkeley: University of California Press, 1987), p. 110.

68. John R. Petrocik, "Realignment: The South, New Party Coalitions and the Elections of 1984 and 1986," in eds. Miller and Petrocik, *Where's the Party?,* p. 48.

69. Sabato, *The Party's Just Begun,* p. 122.

70. See Thomas Byrne Edsall, "The Return of Inequality," *Atlantic Monthly* (June 1988), p. 93.

71. Sabato, *The Party's Just Begun,* p. 123.

72. See *Public Opinion* (January/February 1987), p. 34.

73. Sabato, *The Party's Just Begun,* p. 142.

74. James Allan Davis and Tom W. Smith, *General Survey, 1972–1988* (Chicago: National Opinion Research Center, 1988).

75. Sabato, *The Party's Just Begun,* p. 124.

76. Gerald M. Pomper, *Voters, Elections, and Parties: The Practice of Democratic Theory* (New Brunswick, N.J.: Transaction Books, 1988), p. 389.

77. Ibid.

78. *New York Times*/CBS News poll, reported in *New York Times* (August 14, 1988), p. 14.

79. Wattenberg, *The Decline of American Political Parties 1952–1988,* p. 144.

80. See Thomas B. Edsall, "Introduction," in eds. Miller and Petrocik, *Where's the Party?* pp. 7, 8.

81. Dorothy Davidson Nesbit, "Changing Partisanship among Southern Party Activists," *Journal of Politics* 50 (1987):322–34; *Public Opinion* (May/June 1988), p. 24; *Public Opinion* (June/July 1988), p. 23; *Washington Post*/ABC poll, reported in *Washington Post* (August 14, 1988), p. A30; and Nelson W. Polsby and Aaron Wildavsky, *Presidential Elections* (New York: Scribners, 1980).

82. Nesbit, "Changing Partisanship Among Southern Party Activists," pp. 322–34.

83. *Washington Post* poll, reported in *Washington Post* (August 14, 1988), p. A30.

84. Lunch, *The Nationalization of American Politics,* p. 37. See also Everett C. Ladd, *Where Have All the Voters Gone* (New York: W. W. Norton, 1982), p. 66.

85. Walter J. Stone, Ronald B. Rapoport, and Alan I. Rabinowitz, "The Reagan Revolution and Party Polarization in the 1980s," in ed. Maisel, *The Parties Respond,* pp. 67–93.

86. Keith T. Poole and Howard Rosenthal, "The Polarization of American Politics," *Journal of Politics* 46 (1984):1075.

87. Donald R. Matthews, *U.S. Senators and Their World* (Chapel Hill: University of North Carolina Press, 1960), p. 119.

88. Ibid.

89. Quoted in V. O. Key, Jr., *Politics, Parties, and Pressure Groups* (New York: Thomas Y. Crowell, 4th ed., 1958), p. 361.

90. See John H. Bibby, "Party Organization at the State Level," in ed. Maisel, *The Parties Respond,* pp. 21, 22; Paul S. Herrnson, *Party Campaigning in the 1980s* (Cambridge, Mass.: Harvard University Press, 1988), p. 120; Lunch, *The Nationalization of American Politics,* p. 241; and Ruth K. Scott and Ronald J. Krebner, *Parties in Crisis: Party Politics in America* (New York: John Wiley, 2nd ed., 1984), pp. 119–24.

91. Leon D. Epstein, *Political Parties in the American Mold* (Madison: University of Wisconsin Press, 1986), pp. 144–53.

92. See Bibby, "Political Organization at the State Level," p. 22.

93. Ibid., p. 27.

94. Ibid.

95. Joseph A. Schlesinger, "The New American Political Party," *American Political Science Review* 79 (December 1985):1152–69.

96. Ibid., p. 1168.

97. Paul S. Herrnson, "Reemergent National Party Organizations," in ed. Maisel, *The Parties Respond,* p. 41.

98. Herrnson, *Party Campaigning in the 1980s,* p. 30.

99. Ibid., p. 36.

100. Herrnson, "Reemergent National Party Organizations," pp. 46–54.

101. Federal Election Commission figures, reported in Herrnson, "Reemergent National Party Organizations," p. 49.

102. See Paul S. Herrnson, "National Party Organizations and the Postreform Congress," in ed. Roger H. Davidson, *The Postreform Congress* (New York: St. Martin's Press, 1992), p. 53.

103. Ibid.; Herrnson, *Party Campaigning in the 1980s,* p. 39; and Herrnson, "Reemergent National Party Organizations," p. 51.

104. Bibby, "Party Organization at the State Level," p. 37.

105. Herrnson, *Party Campaigning in the 1980s,* p. 39; Herrnson, "Reemergent National Party Organizations," p. 51; and Herrnson, "National Party Organizations and the Postreform Congress," p. 53.

106. Herrnson, *Party Campaigning in the 1980s,* p. 49; and interviews by the author with Jann Olsten, executive director, National Republican Senatorial Committee, Washington, D.C., August 25, 1988, and with Bob Chlopak, executive director, Democratic Senatorial Campaign Committee, Washington, D.C., August 24, 1988.

107. See Barnes, "Four for the Money," *National Journal* (March 16, 1991), p. 638; and Ornstein, Mann, and Malbin, *Vital Statistics on Congress 1991–1992,* p. 95.

108. Ornstein, Mann, and Malbin, *Vital Statistics on Congress 1991–1992,* pp. 48–63; and personal interviews by the author with Jann Olsten, executive director, National Republican Senatorial Committee, and with Bob Chlopak, executive director, Senate Democratic Campaign Committee. See also Richard E. Cohen, "Party Help," *National Journal* (August 16, 1986), pp. 1998–2004; and Ronald Brownstein, "The Long Green Line," *National Journal* (May 3, 1986), pp. 1038–42.

109. Herrnson, *Party Campaigning in the 1980s,* pp. 121, 122.

110. Ibid., pp. 126, 127; Gary Jacobson, "Parties and PACs in Congressional Elections," in eds. Lawrence C. Dodd and Bruce I. Oppenheimer, *Congress Reconsidered* (Washington, D.C.: Congressional Quarterly Press, 4th ed., 1989), pp. 138, 139; eds. Dodd and Oppenheimer, "The New Congress: Fluidity and Oscillation," in eds. Lawrence C. Dodd and Bruce I. Oppenheimer, *Congress Reconsidered,* pp. 444–49; Richard E. Cohen, "Assertive Freshman," *National Journal* (May 2, 1987), p. 1060; and Alan Ehrenhalt, "Changing South Perils Conservative Coalition," *Congressional Quarterly Weekly Report* (August 1, 1987), pp. 1699–1704.

111. See Herrnson, *Party Campaigning in the 1980s,* p. 127.

112. James MacGregor Burns, *The Deadlock of Democracy* (Englewood Cliffs, N.J.: Prentice-Hall, 1963).

113. See Gerald C. Wright, "Policy Voting in the U.S. Senate: Who Is Represented?" in ed. Hibbing, *The Changing World of the U.S. Senate,* pp. 221–42.

114. David W. Rohde, "Election Forces, Political Agendas, and Partisanship in Congress," in ed. Roger W. Davidson, *The Postreform Congress* (New York: St. Martin's Press, 1992), p. 29.

115. Joseph A. Davis, "Conservative Coalition No Longer a Force," *Congressional Quarterly Weekly Report* (January 16, 1988), pp. 110–15; and Alan Ehrenhalt, "Changing South Perils Conservative Coalition," *Congressional Quarterly Weekly Report* (August 1, 1987), pp. 1699–1704.

116. William R. Shaffer, "Ideological Trends Among Southern U.S. Democratic Senators: Race, Generation, and Political Climate," *American Politics Quarterly* 15 (no. 3) (July 1987):299–324.

117. Richard E. Cohen and William Schneider, "Shift to the Left," *National Journal* (April 2, 1988), pp. 873–85.

118. See Apple, Jr., "Senate Confirms Thomas, 52–48, Ending Week of Bitter Battle," pp. A1, A13.

119. William Schneider, "Politics of the 80s Widens the Gap Between the Two Parties in Congress," *National Journal* (June 1, 1985), pp. 1268–82.

120. Sundquist, "Strengthening the National Parties," p. 214.

121. See Stephen Gettinger, "Partisanship Hit New High in 99th Congress," *Congressional Quarterly Weekly Report* (November 15, 1986), pp. 2901–3.

122. See Sinclair, "The Congressional Party: Evolving Organizational, Agenda-Setting, and Policy Roles," in ed. Maisel, *The Parties Respond,* pp. 245, 247; Sundquist, "Strengthening the National Parties," pp. 214, 215; and Roger H. Davidson and Walter J. Oleszek, *Congress and Its Members* (Washington, D.C.: Congressional Quarterly Press, 3rd ed., 1990), pp. 345–49.

123. Personal interview with Senator Bradley, Washington, D.C., September 26, 1988.

124. Personal interview with Senator McConnell, Washington, D.C., September 27, 1988.

125. Personal interview with Senator Kassebaum, Washington, D.C., September 14, 1988.

126. Quoted in Hayes Gorey, "Kings of the Hill," *Time* (October 3, 1988), p. 26.

127. Samuel C. Patterson, "Party Leadership in the United States Senate," in ed. John Hibbing, *The Changing World of the U.S. Senate* (Berkeley, Calif.: IGS Press, 1990), p. 98.

128. See John R. Cranford, "Party Unity Scores Slip in 1988, but Overall Pattern is Upward," *Congressional Quarterly Weekly Report* (November 19, 1988), pp. 3334–39; and *Congressional Quarterly Weekly Report* (January 6, 1990), pp. 54, 55.

129. See Jill Zuckman, "Thirty-Year High in Partisanship Marked 1990 Senate Votes," *Congressional Quarterly Weekly Report* (December 22, 1990), pp. 4188–91.

130. See *Congressional Quarterly Weekly Report* (December 28, 1991), p. 3788.

131. See Patricia A. Hurley, "Parties and Coalitions in Congress," in ed. Christopher J. Deering, *Congressional Politics* (Chicago: Dorsey Press, 1989), p. 131 (in regard to "important" 1981–1985 Senate votes, as selected by *National Journal*); and Patterson, "Party Leadership in the United States Senate," p. 18 (in regard to "key" votes identified by *Congressional Quarterly Weekly Report* for 1983–1986).

132. See Richard E. Cohen and William Schneider, "Partisan Politics," *National Journal* (January 19, 1991), pp. 134–47; and Chuck Alston and CQ Staff, "Key Votes Inspired More Heat Than New Policies in 1990," *Congressional Quarterly Weekly Report* (November 24, 1990), pp. 3991–4019.

133. Cranford, "Party Unity Scores Slip in 1988, but Overall Pattern Is Upward," *Congressional Quarterly Weekly Report* (January 6, 1990), pp. 54, 55;

for 1990, Zuckman, "Thirty-Year High in Partisanship Marked 1990 Senate Votes," p. 4190; and for 1991, *Congressional Quarterly Weekly Report* (December 28, 1991), p. 3792.

134. Daniel S. Ward, "Alive and Well in the U.S. Senate: Parties in the Committee Setting," a paper prepared for presentation at the Southwest Political Science Association meeting, Austin, Texas, March 18–21, 1992.

135. Alan I. Abramowitz, "Explaining Senate Election Outcomes," *American Political Science Review* 82 (no. 2) (June 1988):398.

136. Poole and Rosenthal, "The Polarization of American Politics," p. 1061.

137. Ibid., pp. 1061–79.

138. Schlesinger, "The New American Political Party," p. 1168.

139. Norman J. Ornstein, Robert L. Peabody, and David W. Rohde, "Change in the Senate Toward the 1990s," in eds. Dodd and Oppenheimer, *Congress Reconsidered,* p. 35.

140. See Bruce I. Oppenheimer, "Split Party Control of Congress, 1981–86: Exploring Electoral and Apportionment Explanations," *American Journal of Political Science* 33 (no. 3) (August 1989):653–59.

141. Quoted in Jacqueline Calmes, "Byrd Struggles to Lead Deeply Divided Senate," *Congressional Quarterly Weekly Report* (July 4, 1987), p. 1420.

142. Personal interview with Senator Bradley, Washington, D.C., September 26, 1988.

143. Personal interview with Senator Domenici, Washington, D.C., September 16, 1988.

144. Quoted in Calmes, "Byrd Struggles to Lead Deeply Divided Senate," p. 1420.

145. Personal interview with Senator Symms, Washington, D.C., September 12, 1988.

146. Personal interview with Senator Nickles, Washington, D.C., September 30, 1988.

147. Personal interview with Senator Packwood, Washington, D.C., September 26, 1988.

148. Sidney M. Milkis, "The Presidency, Policy Reform, and the Rise of Administrative Politics," in eds. Richard A. Harris and Sidney M. Milkis, *Remaking American Politics* (Boulder, Colo.: Westview Press, 1989), p. 175. See also L. Sandy Maisel, "The Evolution of Political Parties; Toward the Twenty-First Century," in ed. Maisel, *The Parties Respond,* p. 318.

149. Quoted in Tim Hackler, "What's Gone Wrong with the U.S. Senate?" *American Politics* (January 1987), p. 9.

150. Personal interview with Senator Chiles, Washington, D.C., September 27, 1988.

151. Quoted in Calmes, "Byrd Struggles to Lead Deeply Divided Senate," p. 1422.

152. Personal interview with Senator Proxmire, Washington, D.C., September 15, 1988.

153. See Calmes, "Byrd Struggles to Lead Deeply Divided Senate," pp. 1419–23; Janet Hook, "Byrd Will Give Up Senate Majority Leadership," *Congressional Quarterly Weekly Report* (April 16, 1988), p. 978; Donald C. Baumer, "Senate Democratic Leadership in the 101st Congress," in eds. Allen D. Hertzke and Ronald M. Peters, Jr., *The Atomistic Congress: An Interpretation of*

Congressional Change (Armonk, N.Y.: M. E. Sharpe, 1992), pp. 293–332; and Donald C. Baumer, "An Update on the Senate Democratic Policy Committee," *PS: Political Science and Politics* (June 1991), pp. 174–79.

154. See, for example, Michael Oreskes, "Congress Leaders Scramble to Forge New Budget Deal in Move to Break Deadlock," *New York Times* (October 8, 1990), p. 1.

155. See Calmes, "Byrd Struggles to Lead Deeply Divided Senate," p. 1422.

156. See Jacqueline Calmes, "'Trivialized' Filibuster Is Still a Potent Tool," *Congressional Quarterly Weekly Report* (September 5, 1987), pp. 2115–18.

157. See *Congressional Quarterly Weekly Report* (July 21, 1990), p. 2314.

158. See Helen Dewar, "The 'Era of Good Feeling' May Be Drawing to a Close," *Washington Post National Weekly Edition* (July 30–August 5, 1990), p. 15; and Joan Biskupic, "Partisan Rancor Marks Vote on Civil Rights Measure," *Congressional Quarterly Weekly Report* (July 21, 1990), pp. 2312–16.

159. Quoted in *Congressional Quarterly Weekly Report* (July 21, 1990), p. 2315.

160. Ibid.

161. For a good history of these leadership positions see Floyd M. Riddick, "Majority and Minority Leaders of the Senate," Senate Document 100–29, 100th Congress, 2nd Session (U.S. Government Printing Office, Washington, D.C., 1988).

162. Roger H. Davidson, "The Senate: If Everyone Leads, Who Follows?" in Dodd and Oppenheimer, *Congress Reconsidered*, p. 280.

163. See Robert W. Merry, "The Prism of History: Johnson Was King Among 20th-Century Senate Leaders," *Congressional Quarterly Weekly Report* (April 16, 1988), p. 983.

164. Ibid.

165. Senator Nancy Landon Kassebaum, quoted in Andy Plattner, "Dole on the Job: Keeping the Senate Running," *Congressional Quarterly Weekly Report* (June 29, 1985), p. 1270.

166. See Ronald V. Elving, "Part Judge, Part Pol: Mitchell Blends Judicial Mien, Hard-Nosed Partisanship," *Congressional Quarterly Weekly Report* (April 16, 1988), p. 981.

167. See Helen Dewar, "Senate Democrats Unveil Agenda, but Few Details," *Washington Post* (March 18, 1989), p. A3.

168. See Richard E. Cohen, "Crumbling Committees," *National Journal* (August 4, 1990), p. 1881.

169. Personal interview with Senator Bob Graham (D., Fla.), Washington, D.C., September 20, 1988; and Sinclair, "The Congressional Party," pp. 244, 245.

170. See Baumer, "An Update on the Senate Democratic Policy Committee," pp. 174–79.

171. See Susan F. Rasky, "The Target of Choice for the GOP: Mitchell," *New York Times* (August 3, 1990), p. A9.

172. Personal interviews in Washington, D.C., with senators Bingaman, Exon, Sasser, and Boren, October 5, 1988.

173. Except as otherwise indicated, the material in this section is taken from Davidson, "The Senate: If Everyone Leads, Who Follows?" pp. 275–305.

174. Ibid., p. 281.

175. Former Senator Abraham Ribicoff (D., Conn.), quoted in Ross K. Baker, *House and Senate* (New York: W. W. Norton, 1989), p. 86.

176. See Helen Dewar, "Mitchell's Orderly, Consultative Style Gets Early Plaudits in Senate," *Washington Post* (May 5, 1989), p. A11; and Helen Dewar, "On Capitol Hill as Well: A Changing of the Guard," *Washington Post National Weekly Edition* (January 16–22, 1989), p. 13.

177. Steven S. Smith, "Informal Leadership in the Senate: Opportunities, Resources, and Motivations," in ed. John J. Kornacki, *Leading Congress: New Systems, New Strategies* (Congressional Quarterly Press, Washington, D.C.: 1990), p. 73.

178. "Committee on Political Parties, Toward a More Responsible Two-Party System," Special Supplement to *American Political Science Review* 44 (September 1950).

179. Ibid., pp. v, 1.

180. Sundquist, "Strengthening the National Parties," p. 204.

181. On this point and for a general discussion of party responsibility in the context of divided government, see Sundquist, "Strengthening the National Parties," pp. 195–221.

182. Davidson and Oleszek, *Congress and Its Members,* pp. 350–52.

183. Ibid., p. 352.

184. See *Congressional Quarterly Weekly Report* (December 28, 1991), p. 3784.

185. See George Hager, "Defiant House Rebukes Leaders; New Round of Fights Begins," *Congressional Quarterly Weekly Report* (October 6, 1990), pp. 3183–88.

Chapter 7

1. See R. W. Apple, Jr., "Government Itself Is the Big Casualty," *New York Times* (October 25, 1990), p. A15.

2. See Andrew Rosenthal, "Campaigning for the G.O.P., Bush Discovers He's an Issue," *New York Times* (October 24, 1990), pp. A1, A13.

3. See David E. Rosenbaum, "Once Near Accord, Budget Talks Lapse into Partisan Squabbling," *New York Times* (October 24, 1990), A1, A12.

4. The chronological report throughout this chapter on the fiscal 1991 budget struggle is based on Pamela Fessler: "Recapping the Budget Talks," *National Journal* (September 15, 1990), p. 2896; "Countdown to Crisis: Reaching a 1991 Budget Agreement," *New York Times* (October 9, 1990), p. A12; "The Long Road to Stalemate," *New York Times* (October 17, 1990), p. A14; and George Hager, "Recapping the Budget Struggle," *National Journal* (October 20, 1990), p. 3478.

5. See Lawrence J. Haas, "Budgeting for Savings and Cover," *National Journal* (April 28, 1990), pp. 1035, 1036.

6. See Alan Murray and Jackie Calmes, "The Great Debate: How the Democrats, with Rare Cunning, Won the Budget War," *Wall Street Journal* (November 5, 1990), p. 1.

7. Richard E. Cohen, "Congressional Chronicle: Spending Control is GOP's Summit Goal," *National Journal* (May 19, 1990), p. 1239.

8. Lawrence J. Haas, "Scoring the Summit," *National Journal* (September 29, 1990), p. 2329.

9. Richard E. Cohen, "Summiteer's Challenge Now Even More Taxing," *National Journal* (July 7, 1990), p. 1670.

10. Pamela Fessler, "Summit Talks Go in Circles as Partisan Tensions Rise," *Congressional Quarterly Weekly Report* (July 21, 1990), p. 2276.

11. See John R. Cranford and Pamela Fessler, "Recession Fears, Gulf Threats Raise Hurdles for Summit," *Congressional Quarterly Weekly Report* (August 11, 1990), pp. 2583–86.

12. Ibid., p. 2584.

13. Ibid., p. 2585.

14. George Hager, "Outline Begins to Take Shape as Deadlines Come and Go," *Congressional Quarterly Weekly Report* (September 15, 1990), p. 2897.

15. Ibid.

16. See George Hager, "Huge Automatic Cuts Loom as Summit Talks Stall," *Congressional Quarterly Weekly Report* (September 22, 1990), pp. 2995–99.

17. Ibid.

18. See Paul Taylor, "A Newer Way of Saying 'Soak the Rich,'" *Washington Post National Weekly Edition* (February 26–March 4, 1990), p. 12; Lawrence Mishel and David M. Frankel, *The State of Working America 1990–91* (Washington, D.C.: Economic Policy Institute, 1990); and Kevin Phillips, *The Politics of Rich and Poor: Wealth and the American Electorate in the Reagan Aftermath* (New York: Random House, 1990).

19. Ibid., pp. 2996, 2997; and see Phillips, *The Politics of Rich and Poor*.

20. See Elizabeth Drew, "Letter from Washington," *New Yorker* (October 15, 1990), p. 121.

21. "Countdown to Crisis."

22. See George Hager, "Defiant House Rebukes Leaders; New Round of Fights Begins," *Congressional Quarterly Weekly Report* (October 6, 1990), p. 3186.

23. "President's Address to Nation on Federal Deficit and the Budget," *New York Times* (October 3, 1990), p. C22.

24. See Hager, "Defiant House Rebukes Leaders," pp. 3183–88; Janet Hook, "Anatomy of a Budget Showdown: The Limits of Leaders' Clout," *Congressional Quarterly Weekly Report* (October 6, 1990), pp. 3189–91; Michael Oreskes, "Budget Boomerang," *New York Times* (October 6, 1990), p. 11; and Tom Kenworthy, "The House GOP Can't Make Up Its Many Minds," *Washington Post National Weekly Edition* (October 15–21, 1990), p. 7.

25. In regard to these chronological developments, particularly, see Hager, "Recapping the Budget Struggle," p. 3478.

26. *Congressional Record* (October 8, 1990), p. 14747.

27. Ibid., pp. 14748, 14749.

28. Except as otherwise indicated, reports and quotations concerning President Bush's "flip-flops" are taken from Hager, "Recapping the Budget Struggle," p. 3478; and George Hager, "Parties Angle for Advantage as White House Falters," *Congressional Quarterly Weekly Report* (October 13, 1990), pp. 3391–99.

29. Quoted in Ann Devroy and John E. Yang, "Bush Wavers on Taxing Rich as Senators Protest," *Washington Post* (October 10, 1990), pp. 1, A4.

30. See Hager, "Parties Angle for Advantage as White House Falters," pp. 3391–98.

31. Reported in Andrew Rosenthal, "Campaigning for the G.O.P.," p. A13, and in "The Art of the Deal," *Newsweek* (November 5, 1990), p. 21.

32. Quoted in Maureen Dowd, "Lost for Words: George Bush's Communication Breakdown on the Budget," *New York Times* (October 21, 1990), p. E1.

33. See David E. Rosenbaum, "Senators Vow to Form Plan Less Favorable to Wealthy," *New York Times* (October 16, 1990), p. A12.

34. Ibid.

35. George Hager and Pamela Fessler, "Negotiators Walk Fine Line to Satisfy Both Chambers," *Congressional Quarterly Weekly Report* (October 20, 1990), p. 3479.

36. Elizabeth Drew, "Letter from Washington," *New Yorker* (November 12, 1990), p. 113.

37. See David E. Rosenbaum, "President's Aides Quit Budget Talks in Tax Stalemate," *New York Times* (October 22, 1990), p. 1.

38. See David E. Rosenbaum, "Bush Termed Open to Tax Rise on Upper Income," *New York Times* (October 21, 1990), p. 1.

39. In regard to the ultimate agreement, see George Hager, "Deficit Deal Ever So Fragile as Hours Dwindle Away," *Congressional Quarterly Weekly Report* (October 27, 1990), pp. 3574–82; and David E. Rosenbaum, "Leaders Reach a Tax Deal and Predict Its Approval; Bush Awaits Final Details," *New York Times* (October 25, 1990), p. 1.

40. Quoted in David E. Rosenbaum, "Once Near Accord, Budget Talks Lapse into Partisan Squabbling," *New York Times* (October 24, 1990), p. A12.

41. See Rosenthal, "Campaigning for the G.O.P.," p. 1.

42. See Susan F. Rasky, "Aides Say Bush Faced Choice: A Deal on Taxes, or a Fiasco," *New York Times* (October 25, 1990), p. 1.

43. Joint Committee on Taxation, reported in "Taxes: Who Pays?" *Congressional Quarterly Weekly Report* (November 3, 1990), p. 3715.

44. "Fourth Stopgap Bill Enacted," *Congressional Quarterly Weekly Report* (October 27, 1990), p. 3579.

45. See Andrew Rosenthal, "Bush Mounts Effort to Quell G.O.P. Rebellion over Taxes," *New York Times* (October 26, 1990), pp. 1, A11.

46. See David E. Rosenbaum, "Relief Is Visible," *New York Times* (October 25, 1990), p. 1.

47. House Budget Committee, reported in "The Deficit-Reduction Pie," *Congressional Quarterly Weekly Report* (November 3, 1990), p. 3711.

48. See Richard L. Berke, "101st Congress Wraps Up Work Belatedly and a Little Battered," *New York Times* (October 29, 1990), pp. 1, A15.

49. House Budget Committee, reported in "The Deficit-Reduction Pie," p. 3711.

50. See John R. Cranford, "New Budget Process for Congress," *Congressional Quarterly Weekly Report* (November 3, 1990), p. 3712; John E. Yang, "Gramm–Rudman–Hollings Redux," *Washington Post* (October 30, 1990), p. A19; Susan F. Rasky, "Substantial Power on Spending Is Shifted from Congress to Bush," *New York Times* (October 30, 1990), pp. 1, A13; George Hager, "Huge Deficit Adds Pressure for Spending Rules Changes," *Congressional Quarterly Weekly Report* (December 21, 1991), pp. 3728–32; and "Appropriations in Perspective," *Congressional Quarterly Special Report* (December 7, 1991), p. 17.

51. See Robin Toner, "Republicans from Afar Glare at Washington," *New*

York Times (October 28, 1990), p. 5; and Maureen Dowd, "From President to Politician; Bush Attacks the Democrats," *New York Times* (October 30, 1990), pp. 1, A13.

52. See Michael Oreskes, "Advantage: Democrats," *New York Times* (October 29, 1990), pp. 1, A14.

53. See Apple, Jr., "Government Itself Is the Big Casualty," p. A15; and Thomas B. Edsall, "The Gridlock of Government," *Washington Post National Weekly Edition* (October 15–21, 1990), p. 6.

54. Nelson W. Polsby, "A Healthy Duel," *New York Times* (October 25, 1990), p. A19.

55. Harold Lasswell, *Politics: Who Gets What, When, How?* (New York: World, 1936).

56. Aaron Wildavsky, *The New Politics of the Budgetary Process* (Glenview, Ill.: Scott, Foresman, 1988), p. vii.

57. For a brief history of the development of the budget process see Joseph White and Aaron Wildavsky, *The Deficit and the Public Interest: The Search for Responsible Budgeting in the 1980s* (Berkeley: University of California Press, 1989), pp. 7–17.

58. On this early history see Howard E. Shuman, *Politics and the Budget: The Struggle Between the President and the Congress* (Englewood Cliffs, N.J.: Prentice-Hall, 1988), pp. 62–64.

59. See Roger H. Davidson and Walter J. Oleszek, *Congress and Its Members* (Washington, D.C.: Congressional Quarterly Press, 3rd ed., 1990), pp. 377, 378; Shuman, *Politics and the Budget*, p. 63; and White and Wildavsky, *The Deficit and the Public Interest*, pp. 8–11.

60. Shuman, *Politics and the Budget*, pp. 62–73.

61. Ibid., p. 69. The study was made by the Commission on the Operation of the Senate (the Hughes Commission).

62. Shuman, *Politics and the Budget*, p. 103.

63. See White and Wildavsky, *The Deficit and the Public Interest*, p. 5.

64. Shuman, *Politics and the Budget*, p. 103.

65. Quoted in Lance T. LeLoup, *Budgetary Politics* (Brunswick, Ohio: King's Court Communications, 4th ed., 1988), p. 153.

66. See Lance T. LeLoup, "Fiscal Policy and Congressional Politics," in ed. Christopher J. Deering, *Congressional Politics* (Chicago: Dorsey Press, 1989), p. 275.

67. John Maynard Keynes, *The General Theory of Employment, Interest, and Money* (New York: Harcourt Brace Jovanovich, 1964).

68. White and Wildavsky, *The Deficit and the Public Interest*, pp. 7–17.

69. Ibid.; and LeLoup, *Budgetary Politics*, p. 155.

70. Allen Schick, *Congress and Money: Budgeting, Spending, and Taxing* (Washington, D.C.: The Urban Institute, 1980), p. 17.

71. In regard to the Budget and Impoundment Control Act of 1974 and events leading up to it see LeLoup, "Fiscal Policy and Congressional Politics," pp. 276–78; LeLoup, *Budgetary Politics*, pp. 154–56; and White and Wildavsky, *The Deficit and the Public Interest*, pp. 11–15.

72. The figures here are fictitious, of course, but the overall point is accurate and based on the author's own recollections from his years in the Senate, 1964–1973.

73. See Robert Eisner, *How Real Is the Federal Deficit?* (New York: The

Free Press, 1986), p. 85; and James D. Savage, *Balanced Budgets & American Politics* (Ithaca, N.Y.: Cornell University Press, 1988), p. 162.

74. The following figures in regard to deficits are taken from LeLoup, *Budgetary Politics*, p. 36.

75. *Public Papers of the President of the United States, Richard Nixon, Containing the Public Messages, Speeches, and Statements of the President, 1972* (Washington, D.C.: U.S. Government Printing Office, 1974), pp. 965, 966.

76. See LeLoup, *Budgetary Politics*, p. 159.

77. For these votes and for the general provisions of the Budget and Impoundment Control Act of 1974, see Shuman, *Politics and the Budget,* pp. 217–44.

78. Public Law 93–344, 88 Stat. 297 (July 12, 1974), Section 301 (a), 3.

79. Dennis S. Ippolito, "Reform, Congress, and the President," in eds. W. Thomas Wander, F. Ted Herbert, and Gary W. Copeland, *Congressional Budgeting: Politics, Process, and Power* (Baltimore: Johns Hopkins University Press, 1984), pp. 147–48.

80. Ibid.

81. See LeLoup, "Fiscal Policy and Congressional Politics," p. 277.

82. See William Greider, "The Education of David Stockman," *Atlantic Monthly* (December 1981), pp. 27–54.

83. In regard to 1981 see Louis Fisher, "The Budget Act of 1974: A Further Loss of Spending Control," in eds. Wander, Herbert, and Copeland, *Congressional Budgeting,* p. 184. In regard to 1990 see Elizabeth Drew, "Letter from Washington," *New Yorker* (October 15, 1990), p. 118.

84. See Roger H. Davidson, "The Congressional Budget: How Much Change? How Much Reform?" in eds. Wander, Herbert, and Copeland, *Congressional Budgeting,* p. 166.

85. See White and Wildavsky, *The Deficit and the Public Interest,* pp. 12–17.

86. See LeLoup, "The Impact of the Budget Reform of 1974," pp. 79–85. In the House, by contrast, where with a weaker Budget Committee party leaders were more involved in budget-making from the first, the budget process was always highly partisan. House Republicans saw the budget as a prime focal point for emphasizing party differences on governmental policy and philosophy. House Democrats necessarily put together a more liberal budget in order to hold the support of most of their members on the floor, and budgets regularly passed the House with a majority of Republicans in that body in opposition. Fisher, "The Budget Act of 1974," pp. 178–80.

87. Except as otherwise indicated, information in this section is taken from Shuman, *Politics and the Budget,* pp. 245–73; and White and Wildavsky, *The Deficit and the Public Interest,* pp. 66–354.

88. See LeLoup, "The Impact of Budget Reform on the Senate," pp. 85, 86.

89. Ibid., p. 88; and Catherine E. Rudder, "Fiscal Responsibility, Fairness, and the Revenue Committees," in eds. Lawrence C. Dodd and Bruce I. Oppenheimer, *Congress Reconsidered* (Washington, D.C.: Congressional Quarterly Press, 4th ed.), p. 232.

90. LeLoup, "The Impact of Budget Reform on the Senate," p. 88.

91. See LeLoup, *Budgetary Politics*, p. 172.

92. See LeLoup, "Fiscal Policy and Congressional Politics," p. 277.

93. Davidson and Oleszek, *Congress and Its Members*, p. 385.

94. LeLoup, *Budgetary Politics*, p. 186.

95. Steven S. Smith, *Call to Order: Floor Politics in the House and Senate* (Washington, D.C.: Brookings Institution, 1989), p. 123.

96. *Congressional Record* (October 16, 1986), p. 16656.

97. For a discussion of a switch to majoritarian congressional decision-making on budget issues see John B. Gilmour, *Reconcilable Differences? Congress, the Budget Process, and the Deficit* (Berkeley: University of California Press, 1990), pp. 93–138.

98. Ibid., pp. 4, 5.

99. James A. Thurber, "The Impact of Budget Reform on Presidential and Congressional Governance," in ed. James A. Thurber, *Divided Democracy: Cooperation and Conflict Between the President and Congress* (Washington, D.C.: Congressional Quarterly Press, 1991), p. 166.

100. See LeLoup, *Budgetary Politics*, pp. 186, 187.

101. Fisher, "The Budget Act of 1974," pp. 175–85.

102. Smith, *Call to Order*, p. 55.

103. Schick, *Congress and Money*, p. 332.

104. See Ippolito, "Reform, Congress, and the President," pp. 144–47.

105. Charles Wolf, Jr., "Scoring the Economic Forecasters," *Public Interest* (Summer 1987), pp. 48–55.

106. Mark S. Kamlet, David C. Mowrey, and Tsai-Tsu Su, "Whom Do You Trust? An Analysis of Executive and Congressional Economic Forecasts," *Journal of Policy Analysis and Management* 60 (Spring 1987):365–84; and *National Journal* (March 7, 1987), p. 552.

107. For an excellent report by an outsider on the way the Federal Reserve System works see William Greider, *Secrets of the Temple: How the Federal Reserve Runs the Country* (New York: Simon and Schuster, 1987).

108. H. R. 2795, 101st Congress, 2nd Session, June 29, 1989; and statement by Representative Lee Hamilton (D., Ind.), *Congressional Record* (November 16, 1989).

109. See Ippolito, "Reform, Congress, and the President," p. 145.

110. Savage, *Balanced Budgets & American Politics*, p. 241.

111. The material in this section dealing with the enormous growth of federal deficits during the Reagan administration is taken from Fred R. Harris, "Deficits, Debt—and Gramm-Rudman," in ed. Fred R. Harris, *Readings on the Body Politic* (Glenview, Ill.: Scott, Foresman, 1987), pp. 492–99.

112. Shuman, *Politics and the Budget*, p. 243.

113. See White and Wildavsky, *The Deficit and the Public Interest*, p. 110.

114. Testimony of Secretary of the Treasury Donald Regan, reported in *Congressional Quarterly Weekly Report* (February 21, 1981), p. 335.

115. See White and Wildavsky, *The Deficit and the Public Interest*, pp. 77, 99, 111, 183–203.

116. *Congressional Record* (May 14, 1982), p. 5321.

117. See Savage, *Balanced Budgets & American Politics*, p. 263.

118. Secretary of the Treasury Donald Regan told reporters that "you cannot get inflation under control without having high interest rates." See "Reagan Says U.S. Will Adhere to Tough Monetary Policy," *Washington Post* (June 14, 1981).

119. White and Wildavsky, *The Deficit and the Public Interest*, pp. 183–86.

120. See Robert Heilbroner and Peter Bernstein, *The Debt and the Deficit: False Alarms/Real Possibilities* (New York: W. W. Norton, 1989), pp. 22–27.

121. See LeLoup, *Budgetary Politics*, p. 36.

122. *Economic Report of the President, 1989*, reported in Heilbroner and Bernstein, *The Debt and the Deficit*, p. 48; and White and Wildavsky, *The Deficit and the Public Interest*, p. 571.

123. "President's News Conference on Foreign and Domestic Matters," *New York Times* (December 18, 1981), p. B6.

124. Text of the President's Address Reporting on the State of the Nation's Economy, *New York Times* (February 6, 1981), p. A12.

125. Hobart Rowen, "Reagan Aides Abandon GOP Rhetoric on Deficits," *Washington Post* (December 9, 1981), p. A1; and "Reagan Aides Defend Deficit," *New York Times* (December 9, 1981).

126. Francis X. Clines, "Puzzle of the 'Structural Deficit' Penned by Reagan," *New York Times* (December 3, 1982), p. A26.

127. See White and Wildavsky, *The Deficit and the Public Interest*, pp. 346, 347.

128. See Eisner, *How Real Is the Federal Deficit?*, particularly pp. 176–80.

129. Quoted in White and Wildavsky, *The Deficit and the Public Interest*, p. 349; and "Reagan Doesn't Know Which Aide Is Right on Deficits and Interest Rates," *San Francisco Examiner* (September 1983).

130. See reference to a 1984 Treasury Department study that found "no systematic relationship between government budget deficits and interest rates" and a study by economist Paul Evans, Ohio State University, that found no consistent correlation between government deficits and inflation or interest rates, in Charles R. Morris, "Deficit Figuring Doesn't Add Up," *New York Times Magazine* (February 12, 1989), pp. 36–40.

131. See Morris, "Deficit Figuring Doesn't Add Up," p. 38.

132. See the studies and statements of economists Benjamin Friedman, V. Vance Roley, James Tobin, and Lawrence Klein referred to in Savage, *Balanced Budgets & American Politics*, pp. 36–39.

133. Eisner, *How Real Is the Federal Deficit?*, pp. 17, 20.

134. See Savage, *Balanced Budgets & American Politics*, p. 13.

135. Heilbroner and Bernstein, *The Debt and the Deficit*, pp. 35–37, citing *Statistical Abstract*, Table 722.

136. Eisner, *How Real Is the Federal Deficit?*, p. 179.

137. Quoted in Peter T. Kilborn, "Is the Deficit Really a Threat? Maybe Not, Some Are Saying," *New York Times* (January 23, 1989), pp. A1, A22.

138. See Harris, "Deficits, Debt—and Gramm-Rudman," p. 495; and Office of Management and Budget calculations, reported in Steven Mufson, "The Thing That Wouldn't Die," *Washington Post National Weekly Edition* (February 11–19, 1991), p. 7.

139. Milton Friedman, "Why the Twin Deficits Are a Blessing," *Wall Street Journal* (December 14, 1988).

140. *Economic Report of the President, 1989*, reported in Heilbroner and Bernstein, *The Debt and the Deficit*, p. 42.

141. For the percentages to 1984, see Savage, *Balanced Budgets & American Politics*, p. 12; and for 1988, see Robert Eisner, Letter to the Editor, "Let's Stop Worrying About the Budget Deficit," *New York Times* (February 19, 1989), p. E18.

142. See George Hager, "The Budget: Deficit Hits Record Level," *Congres-

sional Quarterly Weekly Report (November 2, 1991), p. 3188; and George Hager, "Deficit: Leadership Needed to Erase Red Ink," *Congressional Quarterly Weekly Report* (September 26, 1992), pp. 2890–92.

143. Savage, *Balanced Budgets & American Politics*, p. 39.

144. White and Wildavsky, *The Deficit and the Public Interest*, p. 570.

145. Ibid., p. 573.

146. Quoted in Savage, *Balanced Budgets & American Politics*, pp. 229, 232.

147. Eisner, *How Real Is the Federal Deficit?*, p. 81.

148. Hager, "Deficit: Leadership Needed to Erase Red Ink."

149. See Savage, *Balanced Budgets & American Politics*, p. 230.

150. *Rebuilding the Road to Opportunity: A Democratic Direction for the 1980s*, Democratic Caucus, U.S. House of Representatives, September 1982, p. 19.

151. Reported in Savage, *Balanced Budgets & American Politics*, p. 276.

152. Quoted in Timothy B. Clark, "Promises, Promises—the Presidential Candidates and Their Budget Plans," *National Journal* (March 10, 1984), pp. 452–57.

153. See "Excerpts from Platform the Democratic Convention Adopted," *New York Times* (July 19, 1984).

154. Quoted in Dan Balz and Milton Coleman, "Accepting Nomination, Mondale Offers Voters Era of 'New Realism,'" *Washington Post* (July 20, 1984), pp. A1, A15.

155. CBS/*New York Times* poll, reported in David E. Rosenbaum, "Poll Shows Many Choose Reagan Even if They Disagree with Him," *New York Times* (September 19, 1984), p. A1.

156. CBS/*New York Times* poll, reported in Howell Raines, "Reagan's Policies Lose Favor in Poll," *New York Times* (January 25, 1983), pp. A1, B4; and Gallup poll, 1985, cited in Harris, "Deficits, Debt—and Gramm-Rudman," p. 497.

157. Except as otherwise indicated, material in this section is taken from White and Wildavsky, *The Deficit and the Public Interest*, pp. 432–39; and Shuman, *Politics and the Budget*, pp. 274–80.

158. Quoted in *Washington Post* (June 22, 1985).

159. Quoted in *Washington Post* (July 12, 1985), p. A1.

160. Quoted in *Quarterly Almanac 1985*, Vol. 41 (Washington, D.C.: Congressional Quarterly, Inc., 1985), p. 459.

161. Except as otherwise indicated, information concerning the history, provisions, and operation of the Gramm–Rudman–Hollings Balanced Budget and Emergency Deficit Control Act of 1985 is taken from Shuman, *Politics and the Budget*, pp. 280–92; LeLoup, *Budgetary Politics*, pp. 174–91; and White and Wildavsky, *The Deficit and the Public Interest*, pp. 439–67, 506–29.

162. Quoted in Harris, "Deficits, Debt—and Gramm–Rudman," p. 498.

163. Quoted in *Washington Post* (October 5, 1985), p. A6.

164. *Bowser* v. *Synar*, 106 S. Ct. 3181 (1986).

165. Quoted in Committee on the Budget, U.S. Senate, *Transcript of Hearings*, February 19, 1987.

166. See LeLoup, *Budgetary Politics*, pp. 180–81, 186–87; and Steven Smith and Christopher J. Deering, *Committees in Congress* (Washington, D.C.: Congressional Quarterly Press, 2nd ed., 1990), p. 209.

167. Quoted in Steven Mufson, "Missing the Broad Side of the Barn," *Washington Post National Weekly Edition* (May 28–June 3, 1990), p. 13.

168. Quoted in Helen Dewar, "The Proud Fathers of a 'Monster' Brainchild," *Washington Post National Weekly Edition* (October 22–28, 1990), p. 12.

169. John B. Gilmour, *Reconcilable Differences?*, p. 190.

170. White and Wildavsky, *The Deficit and the Public Interest*, p. 525.

171. Ibid.

172. See Hager, "The Budget," p. 3188; and Hager, "Deficit Report Shows No Gain from Pain of Spending Rules," p. 1963.

173. White and Wildavsky, *The Deficit and the Public Interest*, p. 527.

174. Dewar, "The Proud Fathers of a 'Monster' Brainchild," p. 12.

175. Quoted in Jonathan Rauch, "Zero-Sum Budget Game," *National Journal* (May 10, 1986), p. 1097.

176. LeLoup, *Budgetary Politics*, p. 181; and White and Wildavsky, *The Deficit and the Public Interest*, pp. 468–505.

177. From interviews with U.S. senators by the Center for Responsive Politics, Washington, D.C., compiled April 30, 1987, and made available to the author with the understanding that the identities of the senators would not be disclosed.

178. Personal interview with Senator Chiles, Washington, D.C., September 27, 1988.

179. Alice M. Rivlin, "The Need for a Better Budget Process," *The Brookings Review* (Summer 1986), p. 10.

180. Ibid.

181. Personal interviews in Washington, D.C., with senators DeConcini, September 14, 1988; Nickles, September 30, 1988; Reid, September 20, 1988; and Domenici, September 16, 1988.

182. Except as otherwise indicated, discussion of a balanced-budget amendment is based on information in LeLoup, *Budgetary Politics*, pp. 288–90; Raymond J. Saulnier, "Do We Need a Balanced-Budget Amendment to the Constitution?" *Presidential Studies Quarterly* 18 (no. 1) (Winter 1988):157–60; and Rivlin, "The Need for a Better Budget Process," p. 10.

183. See Saulnier, "Do We Need a Balanced-Budget Amendment to the Constitution?" pp. 157–60.

184. See John R. Cranford, "Balanced-Budget Amendment Suddenly Comes to Life," *Congressional Quarterly Weekly Report* (May 9, 1992), pp. 1233–37.

185. Commentary of John McEvoy in ed. Rudolph G. Penner, *The Congressional Budget Process After Five Years* (Washington, D.C.: American Enterprise Institute, 1981), p. 81.

186. Except as otherwise indicated, discussion in this section of the line-item veto is drawn from LeLoup, *Budgetary Politics*, p. 292; Rivlin, "The Need for a Better Budget Process," p. 10; Ronald C. Moe, "Prospects for the Item Veto at the Federal Level: Lessons from the States," an occasional paper of the National Academy of Public Administration, 1988; and Lawrence J. Haas, "Line-Item Logic," *National Journal* (June 9, 1990), pp. 1391–94.

187. John C. Dill, aide to former U.S. Representative James R. Jones (D., Okla.), quoted in Haas, "Line-Item Logic," p. 1394.

188. See "Seven Ways to Cut the Deficit . . . Easier Said than Done," *Congressional Quarterly Weekly Report* (May 2, 1992), p. 1144.

189. Haas, "Line-Item Logic," p. 1391.

190. See White and Wildavsky, *The Deficit and the Public Interest,* pp. 518–22.

191. Personal interview with Senator Johnston, Washington, D.C., September 28, 1988.

192. Personal interview with Senator Domenici, Washington, D.C., September 16, 1988.

193. See Peter V. Domenici (R., N.M.) and J. Bennett Johnston (D., La.), "Congress Must Strengthen the Budget Process, Not Junk It," *Albuquerque Journal* (January 16, 1989), p. A9.

194. Rivlin, "The Need for a Better Budget Process," pp. 3–10; and personal interview with former Senator Chiles (D., Fla., 1970–1988), Washington, D.C., September 27, 1988.

195. Rivlin, "The Need for a Better Budget Process," p. 9.

196. Personal interviews in Washington, D.C., with senators Mitchell, September 20, 1988; Sasser, September 14, 1988; Fowler, October 4, 1988; Kassebaum, September 14, 1988; Leahy, September 27, 1988; Nickles, September 30, 1988; Johnston, September 28, 1988; and Boren, October 5, 1988.

197. See Susan F. Rasky, "Substantial Power on Spending Is Shifted from Congress to Bush," *New York Times* (October 30, 1990), p. 1, A18; John E. Yang, "Gramm-Rudman-Hollings Redux," *Washington Post* (October 30, 1990), p. A19; and George Hager, "One Outcome of Budget Package: Higher Deficits on the Way," *Congressional Quarterly Weekly Report* (November 3, 1990), pp. 3710–13.

198. Dennis DeConcini (D., Ariz.), for example, a member of the Senate Appropriations Committee, and Thomas Daschle (D., S.D.), a member of the Senate Finance Committee. From personal interviews, Washington, D.C., September 14, 1988, and September 15, 1988, respectively.

199. Bill Bradley (D., N.J.), for example, a member of the Senate Finance Committee. Personal interview, Washington, D.C., September 26, 1988.

200. Harry Reid (D., Nev.), for example, a member of the Senate Appropriations Committee, and Sam Nunn (D., Ga.), chair of the Senate Armed Services Committee. From personal interviews, Washington, D.C., September 20, 1988, and September 27, 1988, respectively.

201. Warren Rudman (R., N.H.), for example, a member of the Senate Appropriations Committee, and John W. Warner (R., Va.), for example, a member of the Senate Armed Services Committee. From personal interviews, Washington, D.C., September 15, 1988, and September 29, 1988, respectively.

202. Jacob ("Jake") Garn (R., Utah), for example, a member of the Senate Appropriations Committee. From personal interview, Washington, D.C., September 21, 1988.

203. Nancy Landon Kassebaum (R., Kan.), for example, a member of the Senate Budget Committee, as well as Robert Packwood (R., Ore.), a member of the Senate Finance Committee, John W. Warner (R., Va.), a member of the Senate Armed Services Committee, and David Boren (D., Okla.), a member of the Senate Finance Committee. From personal interviews, Washington, D.C., September 14, 1988, September 28, 1988, September 29, 1988, and October 5, 1988, respectively.

204. S. Res. 260, 100th Congress, 1st Session.

Chapter 8

1. *Congressional Record* (January 10, 1991), p. 101.

2. *Congressional Record* (January 10, 1991), p. 141.

3. For the text of President Bush's August 8, 1990, announcement that troops were being sent to Saudi Arabia, see "Presidential Address: Bush Announces Deployment of Forces to Saudi Arabia," *Congressional Quarterly Weekly Report* (August 11, 1990), pp. 2614, 2615.

4. "News Conference: Bush Emphasizes Sanctions, Defense of Saudi Border," *Congressional Quarterly Weekly Report* (August 11, 1990), pp. 2615, 2616.

5. See Christopher Madison, "No Blank Check," *National Journal* (October 6, 1990), p. 2398.

6. Quoted in Milton Viorst, "A Reporter at Large: The House of Hashem," *New Yorker* (January 7, 1991), p. 44.

7. Ibid.

8. Carroll J. Doherty, "Consultation on the Gulf Crisis Is Hit-or-Miss for Congress," *Congressional Quarterly Weekly Report* (October 13, 1990), pp. 3440, 3441.

9. Ibid., p. 3441.

10. Quoted in Doherty, "Consultation on the Gulf Crisis Is Hit-or-Miss for Congress," p. 3441.

11. Ibid.

12. "News Conference: Bush Emphasizes Sanctions, Defense of Saudi Border," pp. 2615, 2616.

13. Quoted in Rochelle L. Stanfield, "A Bush Doctrine?" *National Journal* (September 1, 1990), p. 2063.

14. Madison, "No Blank Check," p. 2395.

15. For a critique of the "new world order" after the Soviet Union used its armed forces against the independence movement in Lithuania, see A. M. Rosenthal, "The New World Order Dies," *New York Times* (January 15, 1991), p. A17.

16. See Stanfield, "A Bush Doctrine?" pp. 2062–66.

17. Madison, "No Blank Check," p. 2398.

18. Stanfield, "A Bush Doctrine?" p. 2063.

19. Carroll J. Doherty, "Hill Support Remains Firm, but Questions Surface," *Congressional Quarterly Weekly Report* (September 8, 1990), pp. 2838, 2839.

20. Ibid., p. 2839.

21. For a discussion of the War Powers Resolution by one of its principal authors, see Jacob K. Javits, "War Powers Reconsidered," *Foreign Affairs* (Fall 1988), pp. 130–40.

22. Quoted in Madison, "No Blank Check," p. 2396.

23. Ibid., p. 2397.

24. See Nathaniel C. Nashi, "Congress and the Crisis: To Intervene or Not," *New York Times* (September 13, 1990), p. A8; and Richard E. Cohen, "Another Show of Cold Feet and Hot Air," *National Journal* (November 24, 1990), p. 2878.

25. Quoted in Gary M. Stern, "Put War to a Vote," *New York Times* (October 30, 1990), p. A21.

26. Michael R. Gordon, "In Case of War, Congress Wants Right to Meet," *New York Times* (October 25, 1990), p. A8.

27. Ibid.

28. See Rick Atkinson and Bob Woodward, "The Doctrine of Invincible Force," *Washington Post National Weekly Edition* (December 10–16, 1990), pp. 6, 7.

29. *Congressional Record* (January 10, 1991), p. 101.

30. See Carroll J. Doherty, "Uncertain Congress Confronts President's Gulf Strategy," *Congressional Quarterly Weekly Report* (November 17, 1990), pp. 3879–82.

31. See *New York Times* (November 13, 1990), p. A11; and Michael deCourcy Hinds, "Drawing on Vietnam Legacy, Antiwar Effort Buds Quickly," *New York Times* (January 11, 1991), p. 1.

32. Quoted in Andrew Rosenthal, "Senators Asking President to Call a Session on the Gulf," *New York Times* (November 14, 1990), p. 1.

33. Ibid.

34. Quoted in Doherty, "Uncertain Congress Confronts President's Gulf Strategy," p. 3880.

35. See Christopher Madison, "Sideline Players," *National Journal* (December 15, 1990), p. 3025.

36. Quoted in Michael R. Gordon, "Nunn, Assailing a 'Rush' to War, Criticizes Troop Rotation Decision," *New York Times* (November 12, 1990), p. A9.

37. Quoted in Gordon, "Nunn, Assailing a 'Rush' to War," p. A9.

38. Ibid., pp. 1, A8.

39. Ibid., p. 1.

40. Quoted in Doherty, "Uncertain Congress Confronts President's Gulf Strategy," p. 3879.

41. Quoted in "What They're Saying," *Congressional Quarterly Weekly Report* (November 17, 1990), p. 3881.

42. Ibid.

43. Doherty, "Uncertain Congress Confronts President's Gulf Strategy," pp. 3880, 3881.

44. See polls reported in Richard Marin, "Fish or Cut Bait, Mr. President," *Washington Post National Weekly Edition* (November 26–December 2, 1990), p. 37.

45. See Pamela Fessler, "Bush Quiets His Critics on the Hill by Sending Baker to Iraq," *Congressional Quarterly Weekly Report* (December 1, 1990), pp. 4006–8.

46. Ibid., p. 4007.

47. Ibid., p. 4008.

48. Republican senators Trent Lott of Mississippi and Dan Coats of Indiana, quoted in Carroll J. Doherty, "Public Debate on Persian Gulf Poses Challenge for Members," *Congressional Quarterly Weekly Report* (December 1, 1990), p. 4004.

49. Ibid., p. 4005.

50. Ibid., p. 4005.

51. Quoted in Fessler, "Bush Quiets His Critics on Hill," p. 4007.

52. Reprinted in *Washington Post National Weekly Edition* (November 26–December 2, 1990), p. 37.

53. Pamela Fessler, "Bush Quiets His Critics on Hill," pp. 4006–8.

54. See Carroll J. Doherty, "Administration Makes Its Case but Fails to

Sway Skeptics," *Congressional Quarterly Weekly Report* (December 8, 1990), pp. 4082–85; R. W. Apple, Jr., "The Collapse of a Coalition," *New York Times* (December 6, 1990), p. 1; and Madison, "Sideline Players," pp. 3024–26.

55. Quoted in Doherty, "Administration Makes Its Case," p. 4084.

56. See a reference to a *Washington Post*/ABC News poll, showing 63 percent supporting war "at some point" after January 15, in Richard Marin, "Do People Want a War? Yes and No," *Washington Post National Weekly Edition* (December 10–16, 1990), p. 37; a reference to an Associated Press poll showing 44 percent favoring going to war after January 15, 50 percent for letting sanctions work for a longer period, in *Albuquerque Journal* (January 10, 1991), p. A7; and a reference to a *New York Times*/CBS News poll, showing almost as much support for immediate war as for giving sanctions more time, in Hinds, "Drawing on Vietnam Legacy," p. 1.

57. See Marin, "Do People Want a War?," p. 37.

58. See Carroll J. Doherty, "Bush Is Given Authorization to Use Force Against Iraq," *Congressional Quarterly Weekly Report* (January 12, 1991), pp. 67, 70.

59. Ibid., p. 67.

60. "War Drums," *Congressional Quarterly Weekly Report* (January 12, 1991), p. 71.

61. Doherty, "Bush Is Given Authorization to Use Force Against Iraq," p. 68.

62. See Elizabeth Drew, "Letter from Washington," *New Yorker* (February 4, 1991), p. 86.

63. Ibid.

64. See Thomas L. Friedman, "Baker-Aziz Talks on Gulf Fail: Fears of War Rise; Bush Is Firm; Diplomatic Effort to Continue," *New York Times* (January 10, 1991), pp. 1, A4.

65. See R. W. Apple, Jr., "Gloom in Washington," *New York Times* (January 10, 1991), pp. 1, A5.

66. See "The Language of Impasse," *New York Times* (January 10, 1991), p. 1; and "Remarks by Baker at News Conference in Geneva on Standoff in the Gulf," *New York Times* (January 10, 1991), pp. A4, A5.

67. See Eric Schmitt, "U. S. Battle Plan: Massive Air Strikes," *New York Times* (January 10, 1991), p. A7.

68. See Michael R. Gordon, "Cheney Says He Will Seek Longer Terms for Reservists," *New York Times* (January 10, 1991), p. A7.

69. Quoted in Anthony Lewis, "Presidential Power," *New York Times* (January 14, 1991), p. A15.

70. *Congressional Record* (January 12, 1991), pp. 357, 359.

71. *Congressional Record* (January 12, 1991), pp. 366, 367.

72. *Congressional Record* (January 12, 1991), p. 371.

73. *Congressional Record* (January 12, 1991), p. 403.

74. A *New York Times*/CBS News poll, conducted by telephone from January 11 to 13, showed that 47 percent of respondents felt that if Iraq did not withdraw from Kuwait by January 15, the United States should start military action, whereas 46 percent said that the United States should wait longer to see if the trade embargo and economic sanctions would work. "The Split on Iraq: Sanctions vs. War," *New York Times* (January 15, 1991), p. A7.

75. See Andrew Rosenthal, "U. S. and Allies Open Air War on Iraq, Bomb

Baghdad and Kuwaiti Targets; 'No Choice' but Force, Bush Declares," *New York Times* (January 17, 1991), p. 1.

76. Richard E. Neustadt, *Presidential Power: The Politics of Leadership* (New York: John Wiley & Sons, 1960), p. 33. Emphasis added.

77. See David Gray Adler, "The Constitution and Presidential Warmaking: The Enduring Debate," *Political Science Quarterly* 103 (no. 1) (Spring 1988):2.

78. See Norman A. Graebner, "Foreign Policy Under the Constitution: Should the President's Power Be Curbed?" *News Letter,* University of Virginia Institute of Government (March 1986), p. 46.

79. Edwin S. Corwin, *The President: Office and Powers, 1787–1957* (New York: New York University Press, 4th rev. ed., 1957), p. 171.

80. Ibid., p. 45.

81. Roger H. Davidson and Walter J. Oleszek, *Congress and Its Members* (Washington, D.C.: Congressional Quarterly Press, 3rd ed., 1990), p. 410.

82. Ibid., pp. 410, 411.

83. Quoted in Louis Fisher, "War Powers: The Need for Collective Judgment," in ed. James A. Thurber, *Divided Democracy: Cooperation and Conflict Between the President and Congress* (Washington, D.C.: Congressional Quarterly Press, 1991), p. 204.

84. Davidson and Oleszek, *Congress and Its Members,* p. 411; and James M. McCormick and Eugene R. Wittkopf, "Bush and Bipartisanship: The Past as Prologue?" *Washington Quarterly* (Winter 1990), pp. 5–16.

85. See Barbara Sinclair, *The Transformation of the U.S. Senate* (Baltimore: Johns Hopkins University Press, 1989), p. 52.

86. See a study based on 1950s data by Warren E. Miller and Donald E. Stokes, "Constituency Influence in Congress," *American Political Science Review* 57 (March 1963):45–66.

87. Thomas E. Mann, "Making Foreign Policy: President and Congress," in ed. Thomas E. Mann, *A Question of Balance: The President, the Congress, and Foreign Policy* (Washington, D.C.: Brookings Institution, 1990), pp. 11, 14.

88. Sinclair, *The Transformation of the U.S. Senate,* p. 57.

89. See, for example, Bill Keller, "Soviet Loyalists in Charge after Attack in Lithuania; 13 Killed: Crowds Defiant," *New York Times* (January 14, 1991), p. 1.

90. See David C. Morrison, "The Big Thaw," *National Journal* (August 31, 1991), p. 2115; Carroll J. Doherty, "Soviet Republics in Spotlight as Hill Mulls U.S. Policy," *Congressional Quarterly Weekly Report* (August 24, 1991), pp. 2322–24; and Janet Hook, "Democrats Look to Defense for Funds to Aid Soviets," *Congressional Quarterly Weekly Report* (August 31, 1991), p. 2365.

91. Gallup polls of October 1954 and May 1990, cited in "Opinion Outlook," *National Journal* (June 9, 1990), p. 1428.

92. ABC News/*Washington Post* poll, cited in "Opinion Outlook," *National Journal* (June 9, 1990), p. 1428.

93. Charles William Maynes, "Coping with the '90s," *Foreign Policy* (Spring 1989), pp. 42–62.

94. Ibid.

95. Christopher J. Deering, "Congress, the President, and War Powers: The Perennial Debate," in Thurber, ed., *Divided Democracy,* p. 76. See also Harold M. Hyman, *Quiet Past and Stormy Present? War Powers in American History* (Washington, D.C.: American Historical Association, 1986).

96. Deering, "Congress, the President, and War Powers," pp. 175, 176.

97. Quoted in Adler, "The Constitution and Presidential Warmaking," pp. 20, 21.

98. Ibid., pp. 17–24.

99. Ibid., p. 24.

100. See "Declared Wars," *Congressional Quarterly Weekly Report* (November 17, 1990), p. 3882.

101. See Deering, "Congress, the President, and War Powers," pp. 178, 184, 185.

102. Ibid., p. 184.

103. Ibid., pp. 185, 186.

104. *Public Papers of the Presidents of the United States, 1955* (Washington, D.C.: U.S. Government Printing Office, 1959), pp. 209, 210.

105. *Public Papers of the Presidents of the United States, 1962* (Washington, D.C.: U.S. Government Printing Office, 1963), p. 674.

106. Ibid., pp. 50–52.

107. *Congressional Record* (1969), p. 17245. One of the two dissenters, Democratic Senator Wayne Morse of Oregon, made what turned out to be a highly accurate prediction. "Unpopular as it is," Morse told the Senate before the vote, "I am perfectly willing to make the statement for history that if we follow a course of action that bogs down thousands of American boys in Asia, the administration responsible for it will be rejected and repudiated by the American people."

108. Javits, "War Powers Reconsidered," p. 133.

109. See Daniel Paul Franklin, "War Powers in the Modern Context," *Congress and the Presidency* 14 (no. 1) (Spring 1987):82; and Robert A. Katzman, "War Powers: Toward a New Accommodation," in Mann, ed., *A Question of Balance,* pp. 52, 53.

110. *INS* v. *Chadha*, 462 U.S. 919 (1983).

111. Quoted in Madison, "No Blank Check," p. 2396.

112. *Crockett* v. *Reagan*, 558 F. Supp. 893 (1982).

113. Quoted in Javits, "War Powers Reconsidered," p. 135.

114. Ibid., p. 137.

115. *Lowry* v. *Reagan*, 676 F. Supp. 333 (D.D.C. 1987).

116. Katzman, "War Powers," p. 64.

117. In regard to the Panama invasion, see Pat Towell and John Felton, "Invasion, Noriega Ouster Win Support on Capitol Hill," *Congressional Quarterly Weekly Report* (December 23, 1989), pp. 3532–35.

118. See George C. Wilson, "Bush and the Gulf: Uncomfortable Parallels to Vietnam," *Washington Post National Weekly Edition* (December 10–16, 1990), p. 9.

119. See "An Interview with Foreign Policy Editor Charles William Maynes," in Fred R. Harris, *America's Democracy: The Ideal and the Reality* (Glenview, Ill.: Scott, Foresman, 3rd ed., 1986), pp. 570, 571.

120. Javits, "War Powers Reconsidered," p. 140.

121. I. M. Destler, Leslie Gelb, and Anthony Lake, *Our Own Worst Enemy: The Unmaking of American Foreign Policy* (New York: Simon and Schuster, 1984), p. 22.

122. McCormick and Wittkopf, "Bush and Bipartisanship," p. 15.

123. For the details of action in Congress to cut off funds for the Vietnam War see Graebner, "Foreign Policy Under the Constitution," pp. 50, 51.

124. Ibid.

125. *New York Times*/CBS News poll, cited in Michael R. Kagay, "Approval of Bush Soars," *New York Times* (January 19, 1991), p. 7.

126. *New York Times*/CBS News poll, taken January 17 to 21, reported in Adam Clymer, "Poll Finds Deep Backing While Optimism Fades," *New York Times* (January 22, 1991), p. A8.

127. See Richard J. Stoll, "The Sound of the Guns: Is There a Congressional Rally Effect?" *American Politics Quarterly* 15 (no. 2) (April 1987):223–37.

128. Gerald R. Ford, "Congress, the Presidency and National Security Policy," *Presidential Studies Quarterly* 16 (no. 2) (Spring 1986):203.

129. See Doherty, "Consultation on the Gulf Crisis Is Hit-or-Miss for Congress," pp. 3440, 3441; David C. Morrison, "New Phase for War Powers Debate," *National Journal* (June 25, 1988), p. 1690; and Katzman, "War Powers," pp. 67, 68.

130. See Javits, "War Powers Reconsidered," p. 139.

131. J. Brian Atwood, "Sharing War Powers," *New York Times* (October 14, 1987), p. 27.

132. Katzman, "War Powers," p. 68.

133. See Randall B. Ripley and Grace A. Franklin, *Congress, the Bureaucracy, and Public Policy* (Chicago: Dorsey Press, 1987).

134. On these definitions see Davidson and Oleszek, *Congress and Its Members,* pp. 396–99.

135. Ibid., p. 397, citing Kenneth H. Bacon, "The Congressional-Industrial Complex," *Wall Street Journal* (February 14, 1978), p. 22.

136. Davidson and Oleszek, *Congress and Its Members,* p. 397.

137. See Mike Mills, "Base Closings: The Political Pain Is Limited," *Congressional Quarterly Weekly Report* (December 31, 1988), pp. 3625, 3626; and Mike Mills, "A Dogged, if Futile Trench War Is Planned by Some on Hill," *Congressional Quarterly Weekly Report* (March 25, 1989), pp. 660–62.

138. See Sheila Maxwell, "The Base-Closing Czar's Fresh View," *National Journal* (April 13, 1991); and Elizabeth A. Palmer, "Commission Comes to Life, Vowing a 'Fresh Look,'" *Congressional Quarterly Weekly Report* (April 20, 1991), pp. 994–97.

139. Davidson and Oleszek, *Congress and Its Members,* p. 407.

140. See David C. Morrison, "Defense Contractors Trying to Hold On," *National Journal* (April 14, 1990), pp. 885–89.

141. Investigative story by UPI reporter Greg Gordon, cited in Jean Cobb, "Top Brass," *Common Cause Magazine* (May–June 1989), p. 23.

142. Ibid.

143. Robert B. Reich, "The Real Economy," *Atlantic Monthly* (February 1991), p. 46.

144. Ibid.

145. Barry M. Blechman, "The New Congressional Role in Arms Control," in ed. Mann, *A Question of Balance,* p. 109.

146. Except as otherwise indicated, the discussion in this section is based on Blechman, "The New Congressional Role in Arms Control," pp. 109–45.

147. Personal interview with Senator Pell, Washington, D.C., September 21, 1988.

148. Personal interview with Senator Packwood, Washington, D.C., September 26, 1988.

149. In regard to the proposed 1990 budget, see Christopher Madison, "Bush's Breaks," *National Journal* (March 10, 1990), pp. 564–67. In regard to the proposed 1993 budget, see George Hager, "The Budget: Bush Throws Down Gauntlet with Election-Year Plan," *Congressional Quarterly Weekly Report* (February 1, 1992), pp. 222–26; and Pamela Fessler, "Defense: Economic Reality May Limit Hill's Urge to Outcut Bush," *Congressional Quarterly Weekly Report* (February 1, 1992), pp. 253–56.

150. Reported in Pat Towell, "Nunn Weighs In with a Call for New Planning Cuts," *Congressional Quarterly Weekly Report* (April 21, 1990), p. 1206.

151. See Fessler, "Defense," pp. 253–56.

152. See Cecil V. Crabb, Jr., and Pat M. Holt, *Invitation to Struggle: Congress, the President, and Foreign Policy* (Washington, D.C.: Congressional Quarterly Press, 3rd ed., 1989), pp. 157, 158.

153. See Barbara Kellerman and Ryan J. Barilleaux, *The President as World Leader* (New York: St. Martin's Press, 1991), p. 39.

154. See Ralph G. Carter, "Congressional Foreign Policy Behavior: Persistent Patterns of the Postwar Period," *Presidential Studies Quarterly* 16 (Spring 1986):329–59.

155. McCormick and Wittkopf, "Bush and Bipartisanship," p. 7.

156. Except as otherwise indicated, the material in this section is based on Bruce W. Jentleson, "American Diplomacy: Around the World and Along Pennsylvania Avenue," in Mann, ed., *A Question of Balance,* pp. 146–200.

157. See "Opinion Roundup: Public Opinion on Nicaragua," *Public Opinion* (September–October 1987), pp. 21–24.

158. See *Report of the Congressional Committees Investigating the Iran-Contra Affair with Supplemental, Minority, and Additional Views,* H. Rept. 100–433, S. Rept. 100–216, 100th Congress, 1st Session (Washington, D.C.: U.S. Government Printing Office, 1987), p. 411.

159. See Rochelle L. Stanfield, "Floating Power Centers," *National Journal* (December 1, 1990), pp. 2915–19.

160. See Helen Dewar, "Senate Foreign Relations Panel Founders," *Washington Post* (October 10, 1989), p. A1.

161. Personal interview with Senator McConnell, Washington, D.C., September 27, 1988.

162. See John Felton, "Senators Load Up State Department Measure," *Congressional Quarterly Weekly Report* (October 17, 1987), pp. 2535–38; and Don Oberdorfer and Helen Dewar, "The Capitol Hill Broth Is Being Seasoned by a Lot of Cooks," *Washington Post National Weekly Edition* (October 26, 1987), p. 12.

163. Quoted in McCormick and Wittkopf, "Bush and Bipartisanship," p. 5.

164. Helen Dewar, "Bill Gives Senators a Chance to Play Secretary of State," *Washington Post* (July 20, 1989), p. A6.

165. Quoted in Rochelle L. Stanfield, "Fixing Foreign Aid," *National Journal* (May 19, 1990), p. 1223.

166. See Pamela Fessler, "Baker Lobbies for Quick Action on Assistance for Ex-Soviets," *Congressional Quarterly Weekly Report* (April 11, 1992), pp. 960–61.

167. Ibid.

168. Larry N. George, "Tocqueville's Caveat: Centralized Executive Foreign Policy and American Democracy," *Polity* 22 (no. 3) (Spring 1990):441.

169. Charles William Maynes, "America Without the Cold War," *Foreign Policy* (Spring 1990), pp. 5, 6.

170. Ibid., p. 6.

171. Personal interview with Senator Lugar, Washington, D.C., September 16, 1988.

Epilogue

1. Samuel C. Patterson, "Congress the Peculiar Institution," in David C. Kozak and John D. McCartney, *Congress and Public Policy: A Source Book of Documents and Readings* (Chicago: Dorsey Press, 2nd ed., 1987), pp. 474–76.

2. Richard Allan Baker, *The Senate of the United States: A Bicentennial History* (Malabar, Fla.: Robert E. Krieger Publishing Co., 1988), pp. 109, 110.

Index